CONCEPTUAL CHALLENGES IN EVOLUTIONARY PSYCHOLOGY

STUDIES IN COGNITIVE SYSTEMS

VOLUME 27

CONCEPTUAL CHALLENGES IN EVOLUTIONARY PSYCHOLOGY

Innovative Research Strategies

Edited by

Harmon R. Holcomb III

University of Kentucky, Lexington, U.S.A

SPRINGER-SCIENCE+BUSINESS MEDIA, B.V.

A C.I.P. Catalogue record for this book is available from the Library of Congress.

ISBN 978-94-010-3890-4 ISBN 978-94-010-0618-7 (eBook)
DOI 10.1007/978-94-010-0618-7

Printed on acid-free paper

Originally published by Kluwer Academic Publishers in 2001
Softcover reprint of the hardcover 1st edition 2001

SERIES PREFACE

This series will include monographs and collections of studies devoted to the investigation and exploration of knowledge, information, and data-processing systems of all kinds, no matter whether human, (other) animal, or machine. Its scope is intended to span the full range of interests from classical problems in the philosophy of mind and philosophical psychology through issues in cognitive psychology and sociobiology (concerning the mental capabilities of other species) to ideas related to artificial intelligence and to computer science. While primary emphasis will be placed upon theoretical, conceptual, and epistemological aspects of these problems and domains, empirical, experimental, and methodological studies will also appear from time to time.

Few areas of inquiry have generated as much interest and enthusiasm in recent times as has the discipline known as "evolutionary psychology", but its pretentions and its accomlishments have not always been properly understood. This collection brings together important work in psychology, anthropology, and the philosophy of science that contributes toward that goal, especially by emphasizing the role of natural selection and sexual selection as crucial factors in the evolution of cognitive mechanisms for information processing. The methodological studies that are presented here are bound to enhance appreciation for the scope and limits of this fascinating domain. The editor has produced a fascinating volume that should appeal to a broad and diverse audience.

J.H.F.

CONTENTS

III. Adaptationist Research Strategies and Evolutionary History

ABOUT THE CONTRIBUTORS

Denise Dellarosa Cummins, Department of Philosophy, University of California at Davis, Davis, California.

Steven W. Gangestad, Department of Psychology, University of New Mexico, Albuquerque, New Mexico.

Paul E. Griffiths, Department of History and Philosophy of Science, University of Pittsburgh, Pittsburgh, Pennsylvania.

Harmon R. Holcomb III, Department of Philosophy, University of Kentucky, Lexington, Kentucky.

William R. Jankowiak, Department of Anthropology, University of Nevada, Las Vegas, Nevada.

Timothy Ketalaar, Department of Communication Studies, University of California at Los Angeles, Los Angeles, California.

Linda Mealey, Department of Psychology, College of St. Benedict, St. Joseph, Minnesota.

Robert Richardson, Department of Philosophy, University of Cincinnati, Cincinnati, Ohio.

Frank J. Sulloway, Department of Psychology, University of California, Berkeley, California.

Peter M. Todd, Center for Adaptive Behavior and Cognition, Max Planck Institute for Human Development, Berlin, Germany.

Claudia Woodman, University of Las Vegas, Las Vegas, Nevada.

INTRODUCTION

HARMON R. HOLCOMB III

The theory of evolution by natural selection is bearing fruit in studies of the mind. A main line of inquiry has been a concerted effort to rethink cognitive psychology. So-called "*evolutionary psychology,*" like cognitive psychology, aims to discover the information-processing mechanisms that underwrite our psychological capacities. The concepts and methods, however, are distinct, appealing to our understanding of the natural processes of evolution by natural selection and sexual selection. The present volume refines those concepts and methods, offering new ways of conducting inquiry.

This anthology contains all original peer-reviewed contributions to improving Darwinian perspectives on human behavior. They offer balance in devising innovative research methods that respond to conceptual challenges in evolutionary psychology. Leading researchers report their scientific findings or reflect on the methods and claims in the literature, thereby providing a mix of empirical and theoretical analyses. I solicited contributions from individuals in various sub-disciplines of psychology, in anthropology, and in philosophy of science, insuring a healthy spread of disciplinary perspectives. With chapters from both practitioners and critics of the larger project, this volume is for readers who are skeptical about the field as well as those who like the paradigm and use it. Some people avoid the field because they view anything friendly to the field as dogma or else find flaws in its concepts or methods. They forget that we can simultaneously do evolutionary psychology and challenge it to be better. For instance, the volume addresses crucial topics that critics find omitted in the most well known literature.

Basic to all evolutionary thought is a distinction between the fact, processes, and the course of evolution. This distinction is important for understanding research based on the factual premise that our minds, not just our bodies, evolved. All contributors to this volume agree on the *fact* of psychological evolution: Our psychological capacities are the products of evolutionary processes. All share a consensus on the *processes* of psychological evolution: Modern theory enjoys a consensus on how natural selection and sexual selection operate, and they played a vital role in our evolutionary history, shaping specialized reproductive tasks for our minds just as they do for our bodies.

The standard rationale for doing evolutionary psychology derives from this consensus. Bodies that exhibit a general capacity for behavior but whose behavior did not lead to reproductive success could not have evolved by selection. Similarly, minds having general capacities to think, feel, learn, and fit into a culture, but whose capacities did not lead to reproductive success, could not have evolved by selection. Attaining reproductive success requires executing a host of specialized tasks that require specialized and not just general capacities (Barkow, Cosmides, & Tooby, 1992). Those who abandon evolutionary psychology because of their criticisms, whether or not their criticisms are well taken, must have some view of human evolution. Does that assumed view hold up any better than evolutionary psychology? Upon inspection, it usually, or always, fails as an alternative. In criticizing the new work, people rely on older views of human evolution, which typically depicted us as evolving only general capacities for cognition and culture. Such views can be proved obsolete on the basis of current evolutionary knowledge (Crawford & Krebs, 1998). Although we know the difficulties inherent in identifying the selection pressures that long ago shaped the specialized tasks of the human mind, our commitment to make evolutionary sense of ourselves cautions us not to abandon Darwinism in studying the mind. If a critic has a view of human psychological evolution that can be tailored to fit current evolutionary knowledge, then evolutionary psychologists are happy to pursue it. It is moot whether the critic's view is called "evolutionary psychology" or something else, for the field should be broadly conceived as guided by this commitment rather than narrowly conceived as a doctrine.

All contributors acknowledge that we cannot reconstruct a detailed *course* of evolution; we try to do the best science we can, given what we know and do not know about the evolutionary history of the human species involving mind and behavior. This raises a problem for explaining human mind and behavior using as explanatory factors the *processes* of evolution (natural and sexual selection) together with evolved psychological mechanisms (task-specialized information-processing devices which count as unobservable, hypothesized entities).

The *processes* of evolution (which we have a lot of theoretical knowledge about) do not apply directly to current behaviors, but govern the *course* of mental evolution (which we know little about, especially historical events in hominids, primates, and mammals that occurred long before humans existed). Given this blend of knowledge and ignorance, it is not easy to decide the standards for what counts as an explanatory achievement or a compelling conclusion. Contributors with different perspectives address the problem of how to develop and combine various types of new methods and evidence across many psychological and biological sub-disciplines in the face of limitations on the available evidence about our evolutionary past.

The result is a mix of advocacy and skepticism, of results and ruminations—a mix that should interest anyone trying to understand the mind using a naturalistic approach to inquiry. My aim here is to sketch briefly the problems and ideas that frame the chapters, highlighting what is distinctive and novel in the various views so as to underscore the volume's significance and unity. My introductions to each part of the volume provide fuller elaboration of individual chapters. Collectively, the introductions provide a guide to many different aspects of the readings and take a position on long-standing key issues in order to promote critical discussion that pushes debate in new directions suggested by the chapters. The real substance is contained in the chapters themselves.

Evolutionary psychology encompasses all areas of psychology that make use of the biological theory of evolution. The field is too young to have a single dominant paradigm, but has a variety of related standard approaches. My introductions highlight the import of the challenges and innovations in these scientific studies for how to conduct research on the evolution of mind, constructing a philosophy of evolutionary psychology.

NARROW AND BROAD VISIONS OF EVOLUTIONARY PSYCHOLOGY

To call something "innovative" is to describe it as new and different, showing a marked departure from previous practice. So, we need an idea of standard practice. People wrongly assume that there is one dominant approach in evolutionary psychology, the Darwinian perspective popularized by Robert Wright in *The Moral Animal* (1994), whose trajectory is exemplified by only a handful of leading scientists in the field: Martin Daly and Margo Wilson's *Homicide* (1988), Jerry

Barkow's, Leda Cosmides' and John Tooby's *The Adapted Mind* (1992), Steven Pinker's *The Language Instinct* (1994) and *How the Mind Works* (1997), and David Buss's *The Evolution of Desire: Strategies of Human Mating* (1994) and *Evolutionary Psychology: The New Science of the Mind* (1999). To elevate the shared overlap of doctrines or themes in these works to the status of "the dominant approach" does an injustice to a host of other researchers whose work is vital to advancing the field, such as the scientists contributing to this volume. Public perception is largely shaped by the fact that reporters seek out the "leaders of the field" and report on their work. These same figures are referred to again and again because they were the subjects of previous reports. This media pattern leads to a narrow vision of the field; the so-called "dominant approach" is constructed by over-emphasizing the work of some scientists and under-emphasizing the work of others.

The problem is exacerbated by the fact that any description of the field is biased toward the interests and work of the researcher describing it. D.S. Wilson highlights this point in his (1999) review of Buss's textbook, *Evolutionary Psychology* (1999), the first text purporting to cover the field. D.S. Wilson's title "Tasty Slice—But Where is the Rest of the Pie?" conveys his worry: If the public identifies "evolutionary psychology" with "Buss-style research" and Buss's empirical work is guided by research strategies that ignore topics central to psychology, even the true information he reports will not be accepted until it is seen in the context of a larger story. In fact, some evolutionists want to avoid the term "evolutionary psychology" because they think it refers exclusively to a very specific innatist-adaptationist interpretation of how evolution operates: The mind is a set of evolved adaptations, encoded in discrete modules under genetic control and inherent in the human species, and identified by evidence of adaptive design. The whole pie is evolutionary psychology defined as "the study of mind and behavior from an evolutionary perspective." How much of the pie did Buss omit? "Using his index as a guide, Buss devotes six pages to 'culture,' six pages to 'development,' two pages to 'norms,' two pages to 'individual differences,' one page to 'learning,' and no pages to 'morality,' 'religion,' 'behavior genetics,' or, for that matter, 'brain'" (Wilson, 1999). Given the typical reader's preconceptions about human mind and behavior—involving culture, individual development, norms, learning and so forth—that seem to conflict with assumptions behind Buss's interpretations of the empirical findings, people often find it is more reasonable to retain their assumptions and reject his conclusions than to throw out their entrenched preconceptions and accept Buss's findings.

The present volume does not seek to cover the whole pie, but does aim to redress this imbalance by including contributions that pay attention to the particular omissions D.S. Wilson points out. Let me summarize how each part of the volume does so.

Part One is entitled, "Naturally Selected Development of Behavior, Personality, and Cognition," in order to emphasize views in evolutionary psychology that reveal Buss's approach as only a part of a larger story about what makes us human. Buss's text is committed to a universal human nature of behavioral strategies so strong that there is little need to discuss existing heritable variation (behavior genetics) or individual differences apart from sex differences. By contrast, in the first chapter Linda Mealey devises a general methodology that unites the ethological study of universal human nature with the behavioral genetics of heritable variation and individual differences. To emphasize the importance of behavioral genetics' findings, in the second chapter Frank Sulloway shows how personality development yields individual differences based on the family environment niches determined by birth order and sibling competition. Buss's text is committed to Cosmides' and Tooby's view that all our cognitive evolution took place in the distant past, emphasizing their metaphor for the mind as a jukebox of specialized mechanisms that are played when pushed by environmental buttons. By contrast, in the third chapter Denise Cummins shows that the ever-changing social environment of dominance hierarchies plays a very large role in how and whether certain biological predispositions get expressed. Connecting the topics of development to norms, learning, and brain neurophysiology, she provides an alternative to the Cosmides-Tooby view that complex social animals inherit modules fully formed; instead, fast-track acquisition of social rules is part of our biological preparedness to develop social norms quickly for classes of problems critical to survival and reproductive success.

Part Two is entitled, "Sexually Selected Decision-Making in Mating and Parenting," in order to emphasize views in evolutionary psychology that reveal Buss's approach to mating as a part of a larger story about how these evolved psychological mechanisms actually operate to yield behavior. Buss's text expounds the theory that the mate preferences that drive selection of mates are mainly based on resources (material benefits such as food, direct parental care, physical protection from predators or other members of the same species, etc.). By contrast, in the fourth chapter Steve Gangestad focuses on attempts to document and test the effects of mate preferences based on genetic benefits passed on to the offspring of the chooser, connecting good genes selection to resource selection. In the fifth chapter Timothy Ketelaar and Peter Todd redress Buss's neglect of how evolved psychological mechanisms operate beyond what Cosmides and Tooby have said about optimizing decision rules and Buss's de-emphasis of emotions in mate attraction and retention. They argue for what are called "satisficing" rather than "optimizing" rules (i.e., rules for obtaining satisfactory behaviors rather than the best possible behaviors). They hold that emotional representations of value are more salient and more accurate than "objective, rational" calculations of costs and benefits. Buss also espouses Cosmides' and Tooby's emphasis on development as mainly the switching on of evolved psychological mechanisms that are domain-specialized, and he treats mechanisms as functions in a way that ignores the manifest causes of behavior when he tests predictions about their evolved functions. To redress Buss's neglect of domain-general reasoning and causality, in the sixth chapter this editor uses anorexia to address the coexistence of both types of reasoning and argues that the functionality of evolved psychological mechanisms cannot be established without first addressing proximate causes and developmental causes. Unlike Buss's total neglect of morality and religion and his treatment of culture as an expression of phenotypic plasticity ("evoked culture"), in the seventh chapter William Jankowiak and Claudia Woodman use the behavior of naming offspring to investigate how American Mormon culture, morality, and religion form a unified whole that affects parental investment in offspring via female desire for autonomy.

Part Three is entitled, "Adaptationist Research Strategies and Evolutionary History," in order to address the concerns over whether evolutionary psychology can construct a respectable scientific methodology—concerns that Buss hides under the rug in the first part of his text that treats "The Foundations of Evolutionary Psychology." In the eighth chapter, Paul Griffiths argues that even if we suspend some notorious worries of Gould and Lewontin about the strength of natural selection in producing adaptive design, there remains the need to combine adaptationist appeals to adaptive design with comparative and phylogenetic studies in much more specificity and detail than is normally done. If we continue on as Buss and others do, namely, by inferring adaptive problems from alleged adaptive solutions or inferring adaptive solutions from alleged adaptive problems, even if we can find a hypothesis that explains the trait and that has no strong rival hypotheses, other areas of evolutionary biology show that our favored hypothesis can be, and often is, contradicted by the comparative data. Many examples demonstrate that, in order to support evolutionary hypotheses sufficiently to generate compelling conclusions about how natural selection has designed human mental adaptations, we need to compare anatomically modern humans to other species (such as primates and ancestral hominids) in ways pertinent to identifying the relevant specific selection pressures, ancestral and derived traits, actual variant traits selected among, etc. In the ninth chapter, Robert Richardson agrees with Griffiths in principle but is much more skeptical of the possibility of rigorous functional adaptationist explanations. Griffiths and Richardson both argue that lack of testing using comparative and phylogenetic data in Buss-style research risks loss of an informative scientific account of our evolved psychological capacities. Griffiths makes his case by reference to human emotions, which can be studied across species. Richardson makes his case primarily by reference to human language as (explained by Pinker), which is *not* shared by other species. This difference may play a role in generating Richardson's conclusion that the requisite evidence to qualify evolutionary psychology as legitimate science is not available. Richardson's skeptical conclusion implies that even if some hypotheses in evolutionary psychology are true, thinking about the evolutionary historical record (the parts of the *course* of evolution pertinent to saying what task a trait has been selected for) reveals that we have no sufficiently persuasive reason to accept those hypotheses.

All this goes to show that it is premature to claim that evolutionary psychology has a single dominant paradigm that enjoys a consensus on fundamentals, one exemplified by leading works of the biggest names in the field. In the absence of something that deserves the status of a single dominant paradigm, the field is better understood in terms of a heterogeneous but partially overlapping array of multiple basic research strategies in various research areas that are sometimes allies and other times rivals. For this reason, I invited original papers by widely respected psychologists and philosophers of evolutionary theory to advance basic issues in their research areas. The process of scientific advancement typically involves both negative and positive components, namely, critiquing existing paths of research and creating new ones. I invited contributors either to critically assess some basic aspect of current work in evolutionary psychology or to offer constructive suggestions toward an alternative application of evolutionary theory to human psychology, or both. Their chapters scrutinize the influential work of such leading exponents of evolutionary psychology as Cosmides and Tooby, Buss, Pinker, and so forth, or else use their work as a springboard for further revisions of the way evolutionary psychology is done.

The collective result is a set of innovations that clarifies the way evolution-based research on human mind, behavior, personality, cognition, development, mating, parenting, pathology, culture, language, and so forth can progress. These innovations are far more developed than the suggestions Buss makes in the last chapter of his (1999) textbook on how evolutionary psychology approaches the major branches of psychology: cognitive, social, developmental, personality, clinical, and cultural. Because historical issues in the development of psychology and of evolutionary studies are at stake, the future of psychological science is being carved out. The immediate effect is to revise our basic understanding of research methods for using natural selection and sexual selection to explain the evolved design of the human mind.

THE EVOLUTIONARY DESIGN OF THE MIND: ABOUT THE INDIVIDUAL CHAPTERS

In addition to the external organization of this volume as a counterweight to the partiality of standard expositions of evolutionary psychology, the chapters have their own internal organization. After describing the flow of thought that links the chapters, I shall state the main ideas of each chapter in sequence to make that overview concrete.

The unifying force in evolutionary psychology is not so much a particular methodology as it is a reliance on evolutionary theory as an explanatory and predictive tool for understanding the diversity of life and the apparent design of organisms. Hence, the volume is divided into three sections on theoretical grounds. Part One focuses on natural selection. Part Two focuses on sexual selection. These are the two main components of evolutionary theorizing about adaptation, i.e., the natural processes most responsible for explaining the exquisite design of evolved organisms, including the design of the mind. Part Three focuses on the methods of inquiry and testing in these two types of theoretical explanations of human nature. These critiques of method apply with nearly equal force to all the preceding chapters as well as to the literature they evaluate, thereby providing a balance between practitioners and critics of the field that advances critical thinking.

All contributors to Parts One and Two focus upon aspects of the human life cycle that bear directly on family structure and family relations. The working assumption is that the crucial stages in our life cycle—those that tie us to our families—are best understood from an evolutionary or at least a natural selective or sexual selective point of view. In Part One, Mealey's methodology is sufficiently general to be used in studying traits at any stage of the life-cycle; Sulloway studies childhood; and Cummins investigates how our minds develop to enable us to live in social groups and to recognize the existence of other minds than one's own mind in group members. In Part Two, Gangestad focuses on traits we look for in selecting a mate; Ketelaar and Todd model the way we decide how valuable a potential mate is to us; this author clarifies the role of one's self-image as a potential mate in anorexia's effect of preventing reproduction; and Jankowiak and Woodman discuss the ways in which parents in a culture invest in their offspring using the symbolic behavior of naming them. We are not attempting to construct a unified view of families or the comparative evolution of families in nonhuman animals. Rather, these chapters focus on life stages central to family relations. Biology cannot be ignored in understanding families because the existence and nature of families reflects the ways we solve problems basic to being organisms; evolved life-history strategies govern the way we are born, develop personalities and cognitive powers throughout childhood and adolescence, reach sexual maturity and adulthood, find mates, have our own children and invest in them as parents, and finally die.

Part One gains historic significance in the way it advances the nature-nurture controversy, namely, by tackling the foundational research problem of how evolution relates to individual development (the subject of evolutionary life-history theory).

Linda Mealey, who contributes the first chapter, is the author of *Sex Differences: Developmental and Evolutionary Strategies* (2000), perhaps the clearest and most informed evolutionary textbook on sex and gender. Mealey begins with a review of the history leading to traditionally separate paradigms for the study of evolution of human nature and for the study of individual differences that arise from the particulars of individual development. She suggests that the methods of evolutionary psychology, which tend to focus on capacities thought to be species-universal, and those of behavior genetics, which focus upon capacities thought to be unique to the individual, can be usefully synthesized by focusing on kinship relations throughout the life cycle. Evolutionary psychologists assume that inclusive fitness, which extends Darwinian fitness to include both personal and kinship components, is an important factor of evolution. An individual's genetic endowment predisposes one to behave in ways that maximize personal fitness, but also predisposes one to behave in ways that maximize the fitness of one's kin. We should thus expect to discover in us an array of selected psychological mechanisms underwriting the kin relations in which we develop. Kinship is of concern to behavior geneticists less directly, but kin relations often provide the raw material with which to discern the relative efficacy of genetic and environmental factors involved in ontogenesis. Hence, appeal to family (kinship) relations promises to aid progress in our attempts to understand the human mind. Mealey explains how, by focusing upon kinship relations, we can better understand the evolution of species-universal capacities and individual capacities. She proposes a new methodology for unifying both sorts of research, constructing a sequence of research decisions for relating genetic and environmental factors of individual development to evolved psychological mechanisms. In a field impaled on a dichotomy of existing methods so limited that the conclusions they issue are subject to numerous objections and suggested alternative methods so ideal and demanding that they are rarely carried out in practice, her feasible and inclusive methodology is most welcome. Any such proposal raises two questions. In terms of development, what counts as "the environment"? In terms of evolution, what is an "evolved developmental predisposition"? The next two articles offer new ideas for answering, respectively, these foundational questions.

Frank Sulloway, who contributes the second chapter, is the author of *Born to Rebel: Birth Order, Family Dynamics, and Creative Lives* (1996), which has been extensively reviewed and discussed. Sulloway's chapter provides an introduction to the key themes of that book. He focuses on childhood, in particular, sibling rivalry. Birth order shapes the family environment within which a child develops and competes for parental resources. Firstborns develop in one sort of environment; laterborns develop in an importantly different environment. Environmental differences produced by the order of birth, according to Sulloway, have dramatic effects on personality development. So, family structure is important for at least this dimension of the human mind. He extends the evolutionary theory of parental investment to sibling rivalry, shows how personality is a function of birth order, and clarifies the concept of the environment in which individuals develop to distinguish the different environments ("niches") of offspring within a family, i.e., different within-family environments according to birth order.

Denise Cummins, who contributes the third chapter, is co-editor with Colin Allen of *The Evolution of Mind* (1998), a pioneering interdisciplinary anthology on domain-specialized functions of the mind. She critiques the main components of the Cosmides-Tooby approach that has dominated much work in cognitive psychology. She marshals comparative, developmental, and neurological evidence for the impact of primate dominance-hierarchies on human cognitive abilities, thereby clarifying the way the social environment is incorporated into biologically prepared human cognition. Her chapter provides a bridge between Part One and Part Two by addressing the nature of human reasoning, a core problem in Part Two, in terms of social reasoning, which clarifies the nature of the environment that affects human development, a core problem in Part One. Cummins focuses on capacities used in social reasoning. Her thesis is that human beings have inherited a set of psychological capacities that enable us to reason in and about social interactions. These capacities are relatively autonomous from those involved in other forms of reasoning, and they came to us via distant primate ancestors who lived together in social structures marked by dominance hierarchies. Such hierarchies imposed a selective regime favoring an array of psychological capacities, including the capacity to recognize one's own rank and the ranks of others, the capacity to understand the prohibitions and permissions associated with such ranks, and more. These and related capacities became entrenched in the information-processing architecture of our ancestors and have come to us in a similar form. We thus should expect to find—and, Cummins's claims, we do find—both neurological and psychological mechanisms underwriting similar capacities in us.

In sum, these three authors collectively represent the best studies that critique old ways and supply new ways to do evolutionary psychology at its most fundamental level: to understand (A) the problems humans encountered in ancestral environments, (B) the psychological mechanisms natural selection has designed via the basic evolutionary principles of kinship and its ramifications, within-family conflict, reciprocity between non-kin, and social structure in the forms of coalition-building and dominance-striving to deal with those problems, and (C) the way the evolved psychological mechanisms develop in our current environments that are species-typical or individually variable. This formulation refines the evolutionary paradigm of Crawford and Krebs (1998, p. ix).

Part Two addresses debate over how the mind works in the particular evolutionary psychological context of the nature of adaptations as decision-makers, specifically, the character of psychological function aimed directly at reproductive success. The main type of Darwinian logic besides natural selection is "sexual selection," that is, differences in the ability of individuals to acquire matings and to turn matings into lifetime reproductive success. Both evolutionary mechanisms work via differential reproductive success, which is why sexual selection is often considered a form of natural selection rather than a separate process. Indeed, because winners of dominance contests generally gain advantages in acquiring a higher quantity and quality of potential mates, Cummins' chapter implicitly links both natural selection and sexual selection in her discussion of social dominance hierarchies.

Steven Gangestad, author of Chapter Four, is known for a target article followed by over forty commentaries in the most cited psychology journal, *Behavioral and Brain Sciences*, (Gangestad and Simpson, 2000). His focus is on animal and human mate choice for "good genes." The received view, as exemplified in Buss's textbook, divides mate choice into males choosing females for childbearing aptitude (youth, health, attractiveness) and females choosing males for resources (material benefits), with selection for good genes in humans given lip service only. Gangestad extends his research, drawing on recent theoretical and empirical advances in evolutionary ecology and behavioral ecology. He stresses the individual development of physical asymmetries as a result of parasites and pathogens whose importance in evolution is well known from Ridley's *The Red Queen* (1993). His examination of developmental imprecision bridges the study of individual differences and human nature, as called for by Mealey. Developmental imprecision not only testifies to the importance of individual differences of physical form for mate value, but it also serves as data for generating a more complete picture of human mate preferences, one that includes the interaction of preferences for investing material resources in offspring and mating for good genes, which play a role in creating the typical human condition of marriage plus opportunistic adultery.

Gangestad responds to challenges presented by a range of methodological problems facing adaptationist explanations, as well as empirical work on the evolution of mate choice, and he makes a case for adaptation explanations that are persuasive even though incomplete in various ways. He provides a model for how studies in evolutionary psychology in general should make their case for explanatory success through his pattern of argument that, among competing theories of mate choice, a preference for good genes is a compelling adaptationist theoretical explanation.

Tim Ketelaar and Peter Todd, co-authors of Chapter Five, are members of Gerd Gigerenzer's internationally influential German research team that explores ecological rationality. Ketelaar is well known for reviewing the structure of research in evolutionary psychology as a whole in his co-authored target article in the top journal *Psychological Inquiry* (Ketelaar and Ellis, 2000), clarifying the sense in which evolutionary psychology is falsifiable, despite frequent claims to the contrary. Todd is also co-author of a book on heuristics that develops the idea of the heuristic character of psychological mechanisms (Gigerenzer, Todd, and the ABC Research Group, 2000). Here, Ketelaar and Todd clarify the concept of computation, an idea central to the way Pinker, in *The Language Instinct* (1994) and *How the Mind Works* (1997), and other cognitive psychologists conceptualize the design of evolved psychological mechanisms as information-processing systems. They use thought experiments to show how heuristic decision-making about whom to mate with can be satisfactorily effective in the absence of optimal algorithms.

The Ketelaar-Todd chapter gains wider significance throughout psychology as an illustration of the way evolutionary psychology solves the notorious "frame problem," the mechanism organisms use to narrow down the computationally enormous set of possible behavioral options to the few practical possibilities organisms actually use. We regularly reason in real-life situations in a manner in which the possible courses of action are numerous and the possible consequences of such courses are greater still. On a naïve rendering of the computational model, such possibilities would lead to computational overload and ultimately to system failure. The question, then, is, by virtue of what mechanisms do we, as a matter of fact, solve or bypass the frame problem? Ketelaar and Todd offer two suggestions to help diminish the force of the problem. On the assumption that our evolved mechanisms are ecologically rational, it is plausible, first, that we employ specific "satisficing" (satisfactory as short cuts, though not the best possible algorithmic) strategies that require limited information. Second, it is plausible that our computational efforts are focused upon information rendered relevant by the efficacy of certain emotions. The authors recommend that the strategies employed in their research are applicable generally and provide significant theoretical potential.

In Chapter Six, I, **Harmon R. Holcomb**, examine anorexia, a topic of practical importance today as a culturally-influenced public health epidemic. Evolutionary psychology, basically, is sociobiology applied to psychological matters, with key revisions (see my book, *Sociobiol-*

ogy, Sex, and Science, 1993; the philosophical issues I explore there in terms of sociobiology have counterparts raised by evolutionary psychology, e.g., adaptationism, genetic determinism or innatism, reductionism, and the scientific rationalization of the status quo). Whereas sociobiology was alleged to support the status quo, the topic of anorexia illustrates how evolutionary theorizing holds the potential for creating a better world through evolutionarily informed private decisions and public policies. Anorexia is a pathological condition useful for developing and testing evolutionary psychology's theory of mismatch between evolved adaptations designed for ancestral environments and modern environments where novel inputs into evolved functions can be harmful to survival or reproduction. Mismatch theory suggests that eating disorders are extreme forms of evolved functional adaptations for postponing reproduction to increase fitness, i.e., currently maladaptive adaptations. The usual nonevolutionary explanation is that eating disorders are symbols of underlying distressful mind-sets involving low-self esteem. I explore the relation between the two explanations, advancing the hypothesis that the concept of "self-esteem" found in the nonevolutionary explanation can be tied to the concept of one's "mate-value" found in the evolutionary explanation. This hypothesis is developed to suggest numerous methodological insights pertinent to hypothesis formation prior to hypothesis testing.

Jankowiak and Woodman, co-authors of Chapter Seven, provide an anthropological perspective on evolutionary theorizing. Using ethnographic methods, Jankowiak develops ideas found in his article on inclusive fitness in the journal *Evolution and Human Behavior* (Jankowiak and Diderich, 2000). With Woodman, he takes up the challenge culture poses for evolutionary psychology by means of ethnographic analysis of American Mormon polygamy, advocating the study of simple symbolic behaviors as a first approach for connecting evolutionary theory to symbolic behavior. They address the issue of linguistically-mediated decision making that results in behavior, while retaining the commitment that mind and behavior are aimed at maximizing inclusive fitness. They show how women partly obey and partly violate both purely evolutionary strategies of mating and parenting, and purely cultural emphases on cosmology, gender, and spirituality; instead, women procure resources in a range of strategies involving female dominance-hierarchies, especially when women's autonomy is undermined by cultural conditions (such as the extreme patriarchy of Mormons).

The Jankowiak-Woodman chapter thus extends Cummins' emphasis on dominance hierarchies in primate social environments, defined as the situation that certain individuals have priority of access to resources in competitive situations. Cummins points out that across primate societies, because dominance rank correlates with reproductive success, reproductive success drives evolution. We can see this principle play out clearly in decision making about mating and parenting. In polygamous situations, not all wives are expected to be equal on theoretical grounds, with women's dominance hierarchies organized so that the higher the rank the greater the access to resources, which are then turned into greater quantity and quality of offspring. Jankowiak and Woodman assume, on principled grounds, that the naming of offspring is an aspect of parental investment of resources in their offspring. Just as Cummins emphasizes that in primates a high-ranking position is based on creating and maintaining alliances, they analyze naming practices in an American Mormon polygamous community in terms of the alliances with kin and non-kin inherent in the decision of whom to name one's offspring after. Moreover, their chapter adopts the emphasis on kinship found in Mealey's chapter. They also extend the concept of the environment, which Sulloway had refined to include the developmental niches within the family of first born and later born children and their effects on personality types, to include mating niches within the polygamous family's female hierarchy of co-wives based on the personality types of "assertive wife," "favorite wife," and "low-profile wife." This illustrates the fact that evolutionary studies overlap in many ways.

Although Jankowiak and Woodman focus on a quite specific case (naming practices in a Mormon community), the lessons they draw are quite general. Does the existence of human culture simply add a nuance to the ways primates negotiate dominance hierarchies? Or does culture change things so dramatically that purely evolutionary explanations of parental investment are incorrect or incomplete or should be replaced by a purely cultural explanation? The important lesson is that, among the competing hypotheses concerning the role of naming, the strictly evolutionary hypothesis, taken alone, is not compelling. Equally important, however, is that the strictly cultural explanation is not compelling either. The view endorsed by Jankowiak and Woodman—a broad maternal-investment hypothesis—appeals to an unusually wide array of motives drawing on evolution, cultural beliefs, identity formation, and more.

In sum, the authors in Part Two collectively advance the topics of mating and parenting by extending previous research in innovative ways, confronting unsolved problems that demand a combination of old and new approaches. Moreover, subjects in Part One—such as parenting, individual development toward sexual maturity, reasoning and emotion, and social dominance hierarchies—reappear in Part Two, revealing the interconnectedness of evolutionary psychological research via topics that connect psychological sub-disciplines. This gives a good sense of the vertical (multi-level) integration of fields from biology to psychology to anthropology, often proclaimed as key to evolutionary psychology's aim to unify the social/behavioral/human sciences.

Part Three renews the critique of adaptationism, which has been the most devastating and widely discussed objection to evolutionary theorizing about human mind and behavior since it was introduced in the 1970s by Stephen Jay Gould and Richard Lewontin. In a variety of books, interviews, magazines, and other media, they inaugurated a continual attack on the conclusions of E.O. Wilson's sociobiology and then on evolutionary psychology by attempting to discredit its methodology.

By now, each practicing evolutionary psychologist has her or his own defense against the charge that their "adaptationist" methodology involves so much speculation and so little of the necessary sorts of testing that their explanations of the origins of the human mind may be plausible, but incorrect—like "just-so stories." In this situation, it is important to determine whether the critique of adaptationism should be brushed aside as destructive rather than constructive criticism or whether it still applies and must be confronted more seriously through a basic change in methodology, despite the protests of evolutionary psychologists. These last three chapters are potentially crucial for the direction in which evolutionary psychological research goes. None of the existing defenses by evolutionary psychologists against the critique of adaptationism successfully rebut the new form of the critique as levied by Griffiths and Richardson.

Thus, the core issue concerning the adequacy of adaptationist methods of inquiry employed in evolutionary psychology is perhaps the most contentious issue in evolutionary psychology. Enthusiasts insist that, if our psychological capacities are the products of selection, then the best way to discover the information-processing architecture of those capacities is by first investigating our selective history. This is intuitively plausible. After all, if we know the process by which a complex, adapted system was constructed, it is reasonable that we should study the process of construction in order to better understand the mechanisms within. But critics demur. One problem, made famous by Gould and Lewontin (1979), resides in assessing the breadth of natural selection in effecting evolution. A further problem, addressed here by Griffiths and Richardson, is the adequacy of the historical evidence. We can suspend the worries raised by Gould and Lewontin. That is, we can grant that selection is the only natural process capable of producing adaptive complexity, as Pinker (1997) argues, and that selection plays a central role even in the evolution of relatively simple traits. But we nevertheless can worry that the requisite sorts of evidence concerning the evolution of our psychological capacities are not enough to sustain an informative scientific account of those capacities. The focus of Part Three, then, shifts the debate away from the worries about the breadth of selection and focuses instead on questions of evidence, particularly the evidence available in phylogenetic studies of human language and reasoning. Of particular interest are the differences between Griffiths and Richardson over the utility of such phylogenetic studies.

Paul Griffiths, a philosopher of science known for his book *What Emotions Really Are* (1997), uses emotions in humans and other species as a case in point in Chapter Eight to argue that evolutionary psychology should broaden its existing methodology. It should not rest content to merely infer design solutions from ancestral adaptive design problems ("engineering") or infer design problems from manifest psychological design ("reverse engineering"), as Pinker, Dennett, and other leaders of the field advocate. They need comparative (across-species) data of the right sort to adequately test their selectionist hypotheses about human mental adaptations. Griffiths thinks that some comparative, phylogenetic information can be obtained for topics such as the emotions. Preceding articles in this volume do contain or refer to data that makes cross-species comparisons.

Robert Richardson, a philosopher of science known for many articles, such as his piece on the prospects of evolutionary psychology in the journal *Minds and Machines* (Richardson, 1996), supplies in Chapter Nine destructive criticisms along similar lines, including a review Griffiths' work. Richardson argues that the sort of historical, phylogenetic information needed is much more specific than we realized, but it is neither available nor forthcoming. This is especially true of uniquely human traits (e.g., Pinker's theory of language), but also holds for traits humans have that are found in other species. If the same standards of validation for theoretical explanation in evolutionary biology in general are applied to evolutionary psychology, evolutionary psychology will pale by comparison for relying on quite incomplete, and hence, weak sorts of evidence.

Part Three forces evolutionary psychologists to confront the unsolved problem whether: (A) information about ancestral conditions and about the dynamics of natural selection relevant to explanations of human mind and behavior is necessary but lacking for testability and for justified conclusions, as Griffiths and Richardson argue, or (B) lack of that information means that evolutionary psychological explanations are incomplete; such comparative, historical, and phylogenetic information is unnecessary when multiple sources of data from various psychological sub-disciplines (i.e., as found in this volume) converge on the likely design problems and solutions. Position (B) is adopted by Mealey (2000) in her book on sex differences and is implied by Gangestad's defense of adaptationism in this volume. Thus evolutionary psychology's success as science that can produce genuine knowledge or justified belief, not just opinion, hangs in the balance.

Chapter Ten, written by this editor, wraps up the volume. It functions as a coda that lays out the terms in which we debate the charge that adaptationism hinders rather than helps us arrive at compelling evolutionary conclusions (e.g., should we take the critique of adaptationism seriously or sweep it aside?) and draws attention to various faulty preconceptions many readers of evolutionary psychology bring with them when evaluating the field that give them misleading impressions of it. The individual chapters of the book are interesting for the different perspectives researchers have on the evolution of mind, the various introductions identify a host of philosophical issues raised by these evolutionary analyses, and the last chapter forces us to critically reflect on the way we think about the evolutionary approach. It is my hope that the volume as a whole will push the science and its debate into new and fruitful areas of discussion, not only about what kind of creatures we are, but also about evolutionary science itself.

REFERENCES

Barkow, J., Cosmides, L., & Tooby, J., eds., 1992. *The Adapted Mind: Evolutionary Psychology and the Generation of Culture*. New York: Oxford University Press.

Buss, D.M. 1994. *The Evolution of Desire: Strategies of Human Mating*. New York: Basic Books.

Buss, D.M. 1999. *Evolutionary Psychology: The New Science of the Mind*. Boston: Allyn & Bacon Press.

Crawford, C. & Krebs, D.L., eds., 1998. *Handbook of Evolutionary Psychol-*

ogy: Ideas, Issues, and Applications. Mahwah, New Jersey: Lawrence Erlbaum.

Cummins, D.D. & Allen, C., eds., 1998. *The Evolution of Mind*. New York: Oxford University Press.

Daly, M. & Wilson, M. 1988. *Homicide*. New York: Aldine de Gruyter.

Gangestad, S. & Simpson, J.A. 1999. "The Evolution of Human Mating: Trade-Offs and Strategic Pluralism", *Behavioral and Brain Sciences*, 23 (4), 573-644.

Gigerenzer, G, Todd, P.M., & The ABC Research Group. 1999. *Simple Heuristics That Make Us Smart*. New York: Oxford.

Gould, S.J. & Lewontin, R.C. 1979. "The Spandrels of San Marco and the Panglossian Paradigm: A Critique of the Adaptationist Programme", *Proceedings of the Royal Academy of London*, B, 581-598.

Griffiths, P. 1997. *What Emotions Really Are: The Problem of Psychological Categories*. Chicago: University of Chicago Press.

Holcomb, H. 1993. *Sociobiology, Sex, and Science*. Albany, NY: State University of New York Press.

Jankowiak, W. & Diderich, M. 2000, "Sibling Solidarity in a Polygamous Community in the USA: Unpacking Inclusive Fitness", *Evolution and Human Behavior*, 2, 125-140.

Ketelaar, T. & Ellis, B. 2000. "Are Evolutionary Explanations Unfalsifiable?" Evolutionary Psychology and the Lakatosian Philosophy of Science", Psychological Inquiry, 11 (1), 1-21.

Mealey, L. 2000. *Sex Differences: Developmental and Evolutionary Strategies*. San Diego, California: Academic Press.

Pinker, S. 1994. *The Language Instinct*. New York: Morrow.

Pinker, S. 1997. *How the Mind Works*. New York: Norton.

Richardson, R.C. 1996. "The Prospects for an Evolutionary Psychology: Human Language and Human Reasoning", *Minds and Machines*, 6, 541-57.

Ridley, M. 1993. *The Red Queen*. New York: Macmillan.

Schmitt, D. 2001. "Review of Alas, *Poor Darwin: Arguments Against Evolutionary Psychology*", Human Ethology Bulletin, 16, No.1, 11-13.

Sulloway, F. 1996. *Born to Rebel: Birth Order, Family Dynamics, and Creative Lives*. New York: Pantheon.

Wilson, D.S. 1999. "Tasty Slice—But Where is the Rest of the Pie?", *Evolution and Human Behavior*, 20, 279-287.

Wright, R.1994. *The Moral Animal: The New Science of Evolutionary Psychology*. New York: Pantheon.

PART ONE

NATURALLY SELECTED DEVELOPMENT OF BEHAVIOR, PERSONALITY, AND COGNITION

INTRODUCTION TO PART I

HARMON R. HOLCOMB III

Evolutionary psychology aims to provide a naturalistic account of the mind. In Part One, we begin our interdisciplinary attempt to understand how natural selection acts on the developing individual to create the evolved architecture of the human mind. I shall present several important issues central to discussions of evolutionary psychology. For each chapter, I shall provide a philosophical analysis of the author's reasoning that has four aims: to briefly state the questions they ask and the answers they give, to exhibit the logical structure of the case they make for their main theses, to note at least one way in which each chapter is significant in the present research situation, and to address an objection that readers are likely to make in order to guide the evaluation of the chapters away from pseudo-issues and toward genuine issues.

Mealey's chapter devises a new method for addressing perhaps the biggest controversy in the field at the moment: the role of human variation. To what extent is it genetic (fixed, varies with gene variations)? To what extent is it facultative (flexible, contingent on an individual's developmental or immediate environment)? To what extent is it adaptive (fitness-promoting, increasing the individual's probability of reproductive success)? She connects questions of heritability and adaptiveness, emphasizing that the field is broader than an innatist-adaptationist approach: it is both a genetics-oriented and an environment-oriented field. She calls for an integration of evolutionary psychology with behavior genetics. Her proposed research method applies to all traits of the developing individual, from behavior to personality to cognition. Despite the vast chasm between the two disciplines, Mealey argues that kinship supplies common ground for connecting them. Kinship is key to the method of behavior genetics

H.R. Holcomb III (ed.), Conceptual Challenges in Evolutionary Psychology, 3–17.

for disentangling the complex multiple interactions that constitute individual development (e.g., twin studies). Kinship is key to the concept of inclusive fitness in evolutionary psychology. Keys to scientific advance here are to understand the historical reasons for the separation between the two disciplines and to find in kinship a combination of theory and method that overcomes problems induced by that separation.

Mealey makes her case in four parts. First, what are the historical reasons for separation? On one hand, comparative psychologists took from Darwin the idea of common ancestry across species, developed experimental methods to study innate learning abilities, and used family methodologies to study individual differences that can be attributed to genes, environment or gene-environment interactions. On the other hand, ethologists took from Darwin the idea of adaptation to local environments, developed correlational methods to study behaviors as adaptations, and used natural selection methodologies to study our evolved human nature. Second, what are the obstacles to overcoming this historical separation? Preconceptions flowing from the behavior genetics school lead to criticizing evolutionary psychology for downplaying individual and group differences, for attributing differences to psychological rather than biological sources, and for abandoning the experimental method. Preconceptions guided by the evolutionary psychology school lead to criticizing behavior genetics for the opposite faults: for ignoring human nature, for examining heritability to the exclusion of genes all normally developing humans have as their ancestral repertoire that underlie human psychological strategies, and for abandoning naturalistic observation. In other words, each side expects the other side to be like itself, which begs the question of proper method. Third, what needs fixing in the present state of both disciplines? Despite their pretensions, both disciplines are descriptive more often than explanatory and neither gives us any genuine (causal) mechanisms for either phenotypic differences or similarities. Fourth, with what methodology can we combine the two disciplines so that they may answer the question, as Mealey puts it: "What are the internal (genetic, physiological, psychological) inputs and the external (natural environmental, social environmental) triggers of human social behavior?" No one knows how environmental triggers activate internal evolved predispositions to yield behavior, but Mealey has a flow chart of research goals and tactics useful for putting us into a position from which we can answer this vital question.

Now, let us turn to the evaluation of Mealey's chapter. Mealey's advice can be interpreted as a departure from the present practice of theorizing about ultimate functions, a practice that bypasses genuine causal mechanisms. As a move toward a joint study of function and causation, this gambit is very much in the spirit of Nino Tinbergen, the co-founder of ethology along with Konrad Lorenz. Following Tinbergen, our aim should be to devise well-articulated and well-tested explanations in psychology that integrate explanations concerning the function, causation, development, and evolution of mind and behavior. Her book on sex differences (2000) exemplifies some of the best work guided by research strategies for fulfilling this aim. There, her methodological framework for studying sex and gender differences is a ground-breaking combination of comparative, phylogenetic analyses of the evolutionary history of the human species (for traits from anatomy and physiology to genetics and behavior) with life-history theory of the development of the individual human being from conception to death as affected by age (ontogeny), genetics (phylogeny), and social status (culture).

Mealey suggests that behavioral genetics methods can be combined with selectionist hypotheses to study whether a trait under study is heritable or not and currently adaptive or not. Indeed, a unification of behavior genetics and selection theory has recently been achieved in quantitative genetics in biology as applied to nonhuman organisms but, unknown to evolutionary psychologists, it has not often been extended to humans. One upshot is that evolutionary psychology need not restrict itself, as it often has, to the one possible evolutionary scenario of the strict innatist-adaptationist approach: the extreme case of developmentally canalized species-typical adaptations that have a genetic basis but show no genetic variation (and, by definition, zero heritability). Traits that show individual differences in heritability and may or may not be currently adaptive should also be treated as within the domain of evolutionary psychology. Specifying a science's domain is an important feature of scientific activity, so that the domain of phenomena to be explained changes as the science develops (Holcomb, 1993). Evolutionary psychology's domain includes both "canalized traits" (formed or stabilized in response to environmental circumstances during a critical period of individual development) and "facultative traits" (flexible, often adaptive responses that are either "developmentally contingent," i.e., expressed differentially depending on which environmental experiences occur during individual development, or "environmentally contingent," i.e., expressed differentially depending on what the immediate environmental circumstances of the individual may be).

Mealey has a simple insight that grounds her call for an integration of behavior genetics and evolutionary psychology through their joint emphasis on kinship. Evolutionary psychologists utilize kinship as part of the revised theory of natural selection that emphasizes inclusive fitness (Hamilton, 1964a, 1964b): Natural selection favors the spread of traits that raise inclusive fitness, which includes personal reproduction and reproductive success via genetic relatives. Behavior geneticists utilize kinship as part of their methodology, i.e., in the analysis of data on the topic under study. Mealey envisages kinship, which ties families together, as the common ground by which evolutionary psychology's emphasis on human universals and behavioral genetics' emphasis on individual and group differences can be reconciled and unified.

We anticipate a myriad of objections based on the presumption that when people envisage both disciplines they think of undue emphasis on genes: that behavioral genetics focuses on heritable traits, which have a direct genetic basis that varies ontogenetically among individuals or groups, and that evolutionary psychology focuses on inherited traits that form the phylogenetic heritage of the human species. Small wonder that critics view both disciplines as biased toward an overly genetic and biologically based view of human behavior. They chide evolutionary psychology additionally for assuming that the traits it studies are adaptations to our hunter-gatherer environments of the Pleistocene, which makes evolutionary histories speculative.

A major effect of Mealey's contribution is to undermine these false preconceptions by advocating a methodology that makes them obsolete. Instead of parallel investigations at the ultimate and proximate levels, research questions and designs integrate both levels of explanation. The first goal is to identify traits for which heritable variations are adaptive rather than being "genetic junk;" if a goal is to distinguish adaptive from nonadaptive traits, then the well-known charge of "assuming adaptation a priori" is simply off the mark. The second goal is to study traits meeting criteria for expressing adaptive heritable variation for how those heritable differences map onto life history strategies; this goal clearly connects behavioral genetics' studies of heritability to evolutionary psychology's studies of life history strategies. Similarly, we need to determine how traits that do not meet criteria for adaptive heritable variation map onto life history strategies; if so, behavioral genetics and evolutionary psychology study traits that are "environmental" as opposed to "genetic" yet adaptive. Mealey gives several ways for studying environmentally produced traits, which may either be species universals or subject to individual or group differences.

Clearly, this integrated methodology is not subject to the recurring criticism that evolutionists posit adaptation a priori and subscribe to an overly genetic, biological view of human behavior—at least in the usual manner that charge is made. If it is replied that any view that posits non-zero heritability is too genetic, then the reply is factually in error. "Bailey's Law" affirms the point of departure for the next chapter by Sulloway, namely, that heritable differences account for .4 +/ - .2 of a psychological or behavioral trait's statistical variance (for reproductively significant traits). If it is replied that any view that posits adaptive environmentally-induced variance is too adaptationist, then the reply is theoretically in error. Evolutionary psychologists posit a variety of arguments that natural selection has often made environmentally-induced, learned behaviors adaptive, as biologists have done for a host of nonhuman species. Thus, evolutionary psychologists can claim the virtue of grounding psychology in Darwinian biology, while escaping the damaging associations of "biological" with necessarily "genetic," "adaptive," and "species-wide," which would severely limit the phenomena evolution could explain. This grounding is not reductionist, because it is not supposed to extend to the sorts of traits psychologists have traditionally studied, namely, traits chosen for the very reason that they are thought not to be biologically prepared.

Sulloway's chapter exemplifies one way evolutionary psychologists can make a strong case for their Darwinian perspective of the family. In view of the reluctance to see evolution as relevant to and consistent with human behavior, he argues that competition for parental investment is the main cause of sibling rivalry and that sibling rivalry provides an ultimate cause of some key aspects of personality development. Notably, birth order differences are proxies for different within-family microenvironments that make firstborns and laterborns compete differently for parental investment and thereby develop differences in the "Big Five" personality traits. A conceptual originality here is to take Darwin's principles of adaptation to local environment and divergent evolution and apply them to the family, so that we think of different siblings as growing up in different family niches, developing niche-specific ways of dealing with problems of survival and reproduction as they mature, and tapping into these personality strategies when appropriate environmental contexts arise as adults.

Sulloway makes his case in two parts. First, what are the reasons for believing that competition for parental investment causes sibling rivalry? Sulloway describes the biology of sibling competition, both theoretically and empirically. He refines the standard theory of parent-offspring conflict to include sibling competition, so as to apply to both humans and nonhumans. He documents empirically that sibling competition occurs across the gamut of nonhuman species, that sibling competition is for parental investment, that adaptations for sibling competition are undeniable, that siblicide is common, and that birth order is connected to sibling competition and to siblicide. These nonhuman precedents plus general theory make the claim about humans plausible, as there is no good reason why humans should be "an exception to the rule." In addition, recent findings from behavioral genetics motivate the application to humans: Shared family environment only accounts for 5 percent of the variance in individual personality traits in humans, as compared to 35 percent due to non-shared family environment. Because birth order and gender are two systematic sorts of non-shared family environments, birth order differences should be important for environmentally–caused strategic differences among siblings inherent in their personality differences.

Second, what are the reasons for believing that birth-order affects personality? Criticisms of the pre-existing 2,000 or so studies on that subject claiming that confounding variables were not properly controlled can be overcome by using meta-analysis, a tool for aggregating studies to increase statistical reliability and power. Theoretically, birth order differences are proxies for variables such as age, size, status, and power within the family—and these variables manifestly do affect personality. For instance, first-borns maximize biological fitness by acting as surrogate parents, and so are more conscientious, neurotic, and extroverted than laterborns, whereas laterborns are more agreeable and open to experience than firstborns. Quite compelling evidence for birth-order differences in behavior comes from Sulloway's examination of revolutions in history. For example, his study of 28 scientific innovations (revolutions), shows statistically that the more socially radical the innovation, the more it was supported by laterborns and opposed by firstborns. Sulloway answers a number of methodological criticisms of such conclusions, as laid out in his famous *Born to Rebel* (1996). He deftly makes some insightful points on the proper use of meta-analysis, e.g., problems of context, how to take account of modulating factors, self-report versus other-report, and how to explain away seemingly contradictory or counter evidence.

In evaluating Sulloway's chapter, let's take a step back toward our own expectations about counts as "evolutionary" in the first place. What can evolutionary psychology offer in the way of providing insight into families? Sulloway's contribution combines with Mealey's to provide a research avenue toward a multi-disciplinary but unified picture of families emerging from the field, one based on scientific progress in elucidating the evolved nature of the human life cycle and family relations as found in human family life today. Previous evolutionary work on humans has time and time again been thought of as too restrictive. That judgment arises from an uncritically assumed dichotomy between the sources of behavior in genes versus environment, innate versus learned, invariant versus variable, etc., together with the misconception that evolution involves only the first of each pair. It follows that most of human behavior escapes evolutionary analysis. Ironically, these research advances make previous evolutionary work appear too restrictive, not because the evolutionary approach itself is in principle restrictive (as its critics believe who necessarily tie "evolution" to its connotations in our culture that if something evolved it necessarily is a "genetic, innate, invariant trait universal for the whole human species"). Instead, previous evolutionary work was too restrictive because it did not take advantage of the new evolutionary theories that make the evolutionary approach richer, more varied, and more inclusive than we had imagined: evolution applies to both sides of these pairs and reveals these dichotomies as false dichotomies.

More specifically, how does Sulloway's work advance the research agenda of reintegrating evolutionary and developmental biology so as to yield novel insights into the human psyche? As children we grow up in families and have to deal with siblings and parents. Sulloway's chapter extends his theory of individual development in terms of the ability of each sibling to respond to its own specific within-family environment so as to outcompete other siblings for parental investment (attention, help, training, resources, etc.). The upshot is that evolutionary psychology need not restrict itself to viewing the family as a single environment; due to birth order differences the same family will be experienced as one environment for a firstborn child and quite a different environment for a laterborn child. Such fine-grained distinctions concerning individual development always raise the nature-nurture problem, i.e., how much of human behavior is due to inherited genes and how much to experienced environment.

Everyone agrees that each trait in an individual organism is due to both nature and nurture in a causal but unquantifiable sense. That consensus leaves open a different issue, namely, to what extent differences between individuals are mainly due to nature or to nurture, which lands us squarely in the province of behavior genetics, as discussed by Mealey. There are three theoretically possible mechanisms of phenotypic diversity (D.S. Wilson, 1994). The mechanism of "genetic polymorphism" operates when phenotypic differences are caused directly by underlying genetic differences. The mechanism of "phenotypic plasticity" occurs when a single genotype produces multiple phenotypes. The mechanism of "ontogenetic shifts" works when an organism is genetically programmed to change forms throughout its life cycle. D.S. Wilson (1994) marshals a number of arguments that criticize the research agenda of Cosmides and Tooby, which treats phenotypic plasticity as the main or sole source of adaptive individual differences in human psychological traits. Instead, Mealey's proposed methodology provides a flow chart for studying whether individual differences of any sort (from behavior to personality to cognition) are due to genetic polymorphism, phenotypic plasticity, or ontogenetic shifts. Sulloway's study of the effect of birth order as proxy for a host of ontogenetic shifts that control personality development has hidden implications for how behavior genetics experiments are conducted and how their findings should be interpreted. Specifically, by ignoring within-family environments created by birth order, behavioral genetics has lumped different environments together, so that the genetic component of heritability is systematically underestimated. When behavioral genetics creates experimental methods that distinguish within-family environments, we expect findings that will reveal human individual psychological differences to be much more "genetic" (i.e. due to ontogenetic shifts) than critics of evolutionary psychology have assumed.

Sulloway extends the natural selection theory of parental investment, from which sibling rivalry is a corollary, to postulate strategies of growing up in a family that vary with birth order. Birth order, he claims, is the engine of history. Now, when researchers announce some thesis to the effect that natural selection on genes explains history, a knee jerk reaction of doubt sets in. Such doubt is typically based on projecting a way of connecting genes to history that is prima facie absurd. For instance, it has been objected that the transition from the feudal system to an entrepreneurial system did not occur because natural selection increased the frequency of entrepreneurs. A similar objection is that we cannot understand the Japanese attack on Pearl Harbor during World War II in evolutionary terms because the genes of the Japanese in 1940 were the same genes as in 1941. These negative responses assume, incorrectly, that evolutionists are positing genes as "the cause of" (in other words, "directly for") every individual trait or group trait they seek to explain, so behavioral changes require genetic changes. The cited reason—that there is no natural selection without genetic differences to be selected among—is fallaciously used to infer that all traits evolutionary psychologists study must be genetic polymorphisms. Evolutionists who seek to illuminate human nature or individual differences typically invoke environmental causes of behavior, just as their audience does. (Both canalized and facultative traits have internal and environmental inputs, as previously defined.)

The value of introducing Sulloway's work in such conceptual detail is to illustrate the fact that a proper analysis of an evolutionary study has direct implications for whether an evaluation of that study yields a good or a bad objection. Reconsider the objection that evolutionary psychological studies are "too genetic" in terms of our previous analysis. Birth order differences are proxies for variables such as age, size, status, and power within the family, and these combinations of organismal and environmental conditions manifestly do affect personality. So, birth order is a proxy for the sorts of variables that nonevolutionary psychologists use routinely to explain human behavior. There are no genes for birth order, i.e., for being a firstborn or a laterborn. So, Sulloway's explanation does not fit the uninformed reader's assumption that for natural selection to explain history in this case it would have to be direct selection on the two birth order types, i.e., for natural selection to increase the frequency of firstborns over laterborns, which would make no sense.

Instead, natural selection operates to create the personality differences between firstborns and laterborns in a different way. Competition for parental investment causes sibling rivalry. Firstborns who are conscientious and extroverted succeed better than firstborns who are not conscientious and extroverted at getting a greater share of parental investment (e.g.. resources), increasing their reproductive success. Similarly, with a family niche already occupied by a firstborn, laterborns who are agreeable and open to experience succeed better than laterborns who are not agreeable and open to experience at getting what's left of parental investment, thereby enjoying greater reproductive success. Natural selection on genes has created a set of strategies for procuring resources that make use of personality traits. All humans have the personality strategies of firstborns and laterborns as genetic potentials (which contrasts with "genetic determinism"). The family-microenvironment triggers the development of one or the other set of strategies. We finally arrive at what is postulated to be genetic: Although these suites of personality traits are environmentally caused, the ability to develop one suite of personality traits rather than another is due to a genetic ability to respond differentially to different family niches. This type of explanation should not endanger environmentalist scruples, because they are "environmentally caused." What explains the fact that humans, but not other species, can learn enough about their environments to behave in such a variety of ways, and develop the sorts of personalities we do? The ability to learn from and respond to environments so as to develop the "Big Five" personality traits is part of our species' heritage. Sulloway's topic is the environment, not genes; remaining neutral the issue beyond this is not to espouse a particular position on it, e.g., innatism. Whereas Sulloway clarifies the family environment via genetically based predispositions to form developmental reproductive strategies, Cummins clarifies the social environment in terms of a biological preparedness to develop mental capacities that is not committed to the notorious module view of Cosmides and Tooby.

Cummins' chapter exemplifies one way evolutionary psychologists can make a strong case for their view that the mind contains domain-specific reasoning mechanisms specialized for solving various adaptive problems that have recurred over human evolutionary history. In view of the reluctance to admit the existence of domain-specific psychological mechanisms of behavior, she argues that we have a biological preparedness to develop capacities for constructing representations of other minds and of ourselves as agents who possess intentionality. We use these capacities to deal with social norms and social dominance hierarchies in a biologically adaptive manner. A conceptual originality here is to reinterpret the well-known data involving the Wason selection task, which Cosmides and Tooby (1992) had interpreted as experimental evidence for the theory that humans possess specialized cognitive modules for reasoning about social exchange (reciprocity). Cummins reinterprets the data as indicating our possession of specialized cognitive faculties for reasoning about dominance hierarchy interactions. She provides a variety of evidence for dominance theory.

Cummins makes her case in two parts. First, she connects the data in primatology on social dominance to the data on humans concerning the Wason card selection task. How does the social environment shape primate societies? There are strong selective pressures for adhering to social norms, which entail different behaviors for subordinates and dominants. For instance, dominance rank is directly related to the cognitive capacity to detect cheaters. High-ranking organisms need to detect cheaters in order to protect their priority of access to reproductive resources, whereas low ranking organisms may sometimes need to cheat in order to procure resources they are not permitted to have because of their low rank. Dominance rank is also related to the cognitive capacity to create and maintain alliances based on reciprocal obligation. What is the imprint of social dominance hierarchies on the cognitive functions of human primates? One striking effect is the deontic effect, i.e., when reasoning about social rules people spontaneously look for cheaters, as evident in the Wason selection task. Social rank affects the details of how and when cheater detection strategies are employed. Selection for cheating detection by dominants gives rise to counterselection for deception by subordinates. Deceptions are useful data in inferring the minimum level of intentionality needed to perpetuate the deception (and thus for studying intentionality), as has been done for nonhuman primates, human children, and human adults. By assuming that the cognitive capacities emerging earliest in life have the deepest evolutionary roots, we can reconstruct a picture of human mental development that reveals the chronology of development of reasoning about agents in humans and its phylogenetic background.

Second, what are the reasons for thinking that Cummins' dominance theory is superior to other theories that purport to explain the deontic effect? There are two main kinds of theories: empiricist and nativist (hers is nativist, but not modular). Empiricist theories explain it in terms of general learning mechanisms only. Typically, induction from experience is applied to the domain of social interaction as a content-free method of reasoning. In other words, social interaction, like any other substantive topic, is merely one among an indefinitely large set of possible domains of application of the same general learning mechanism. She argues that dominance theory is explanatorily superior on the grounds that empiricist theories concede the nativist view that social reasoning is dissociable from the other types of reasoning and that the skills involved in social reasoning are fitness enhancing. Further, she states that empiricist theories cannot explain (a) why performance on the non-deontic version of the Wason selection card task correlates with cognitive ability while performance on the deontic version does not, or (b) why effective deontic reasoning strategies emerge earlier than effective truth-testing and false belief testing strategies, or (c) why the optimal strategy for the deontic version is obvious to three year olds but even philosophers of science disagree on what counts as an optimal truth-testing strategy, or (d) why social reasoning is selectively impaired in various neurological syndromes but truth-testing is not.

The other main nativist theory is the social contract theory, also termed "social exchange theory" of Cosmides and Tooby, e.g., you may obtain a benefit from me if you pay the cost to me or else if you do not pay the cost to me then you may not obtain the benefit from me. They explain the deontic effect in terms of cost/benefit analysis and cheater detection, which are posited to be innate strategies selected during human evolution (specifically, the Pleistocene conditions of our hunter-gatherer ancestors). Whereas social contract theory identifies the ultimate function of such reasoning to be effective for generating cooperative action for mutual benefit, dominance theory identifies the ultimate function with reasoning about hierarchies, which includes cooperative action for mutual benefit under the rubric of alliance building in social dominance hierarchies. Cummins argues that dominance theory is explanatorily superior on several grounds in addition to inclusiveness. Social contract theory cannot explain the continuity of primate and human adaptation in social reasoning. Dominance theory gives a better explanation of the selective pressure by showing that dominance (and thwarting it by subordinates) has a measurable impact on reproductive success. Dominance depends in higher primates more on social acumen than on physical size. Dominance theory explains the effects of natural selection on neurological architecture in terms of biological preparedness and canalization, which fit what we know about human developmental plasticity better than the innate special-purpose cognitive modules posited by social contract theory. Her closing line states a general view of human development that runs through all the chapters in this volume: "Our biological predispositions impose the framework that is necessary to learn the things most vital for survival in a complex social environment, while neurological plasticity allows our actual environmental experiences the final say in whether and how those predispositions are expressed."

The evaluation of Cummins's chapter not only raises substantive issues, but also issues in scientific method. Evolutionary research on the nature of the human mind promises to deliver insights into the purposes for which various mental faculties operate. Is she right that an evolutionary approach is promising in psychology for the way it illuminates how and why humans use mental representations of objects and agents to engage in adaptive or maladaptive behavior? To decide, reconsider on what basis she argues that this is a promising research avenue.

Heavily rooted in primatology, Cummins's chapter tells us something about how mental representations came to be in nonhuman and human primates, as specialized tools for social life. She draws on the theory and data concerning dominance hierarchies in primates to construct a picture of the human mind. She argues that social dominance hierarchies constitute the key selective pressure on the formation of human cognitive and emotional functions. Across primates and traditional human societies, dominance rank correlates with relative reproductive success. This fact should not be surprising because the hierarchies themselves are formed on the basis of priority of access to resources in competitive situations, and more resources make for more reproductive success, either through differential survival or mate selection for mates with the most resources. How have the exigencies of social living made a mark on the human mind? We should expect our mental and emotional capacities to guide us in thinking about social norms—about what is permitted, obligated, and forbidden. Human reasoning is oriented to reason about oneself and others as agents subject to social norms, rather than as mere objects. Such thought is functional—it is used so that we can follow social norms by representing to ourselves what is permitted, so that lower-ranking individuals can hide their forbidden activities of mating, alliance-formation, and resource-taking, and so that higher-ranking individuals can expose the deceptions of lower-ranking individuals and punish them.

Now, it would be a "just so story" to say without further ado that because such behaviors enhance reproductive success, these functions have been favored by natural selection, and exist in us as adaptations. Instead, it is the combination of the preceding considerations with the body of theory and data on primate hierarchies and the data on the deontic effect in humans that leads Cummins to conclude that natural selection established reasoning in humans that is specialized for negotiating life in dominance hierarchies. Cummins' work is best interpreted, I submit, according to the method of "inference to the best explanation" rather than justificationism or falsificationism (a method explained in detail as applied to sociobiology in Holcomb, 1993;

with the new literature in evolutionary psychology providing variation on the same methodological themes, Holcomb, 1998). She certainly hasn't tried to show dominance theory to be true according to our best standards of evidence. Nor has she tried and repeatedly failed to falsify it. Instead, she has argued along a variety of grounds that dominance theory is superior in its ability to explain the totality of the evidence—primate behavior, connections between primate and human behavior, the Wason selection task, facts about human development—better than the two main leading theories, namely, empiricist theory and the social contract theory championed by Cosmides and Tooby. Their theory seeks to explain the human data by reference to what would have been adaptive during the Pleistocene environments of our hunter-gatherer ancestors (here speculation enters because we don't know as much about the Pleistocene, a period of large climatic variation rather than one environment, as we need to know in order to identify which traits were selected for and which adaptations existed). Social contract theory, which had been the leading evolutionary theory, accounts for the Wason selection card task as revealing the existence of specialized learning mechanisms specific for ongoing reciprocal interaction.

Social contract theory had been devised as the only evolutionary alternative to various nonevolutionary theories in terms of general learning mechanisms and empiricist strictures, i.e., against innate content-specific learning mechanisms and for general associational learning that in principle applies across domains regardless of biological fitness considerations. Of course, advocates of social contract theory or associational theory have made replies that Cummins does not include in her chapter. In evaluating her theory it is important to notice that she gives her best arguments, but leaves it up to the readers to decide for themselves which arguments are persuasive. In the absence of proof and disproof, she realizes that rebuttals are possible.

Social contract theory can be defended, for instance, by maintaining that social contract is the driving force behind the deontic effect revealed in the Wason selection card task and that effects of living in dominance hierarchies are not easily empirically separated from effects of social exchange. Associationism can be defended, e.g., by acknowledging that social reasoning is dissociable from other types of reasoning, as types of reasoning often are, but that does not mean it does not rely on some sort of general learning mechanism. Not only can general learning mechanisms (e.g., operant conditioning) be fitness enhancing, but the only thing that needs to be domain-specific is the input device that takes information from the environment and puts it into the general processor. However, Cummins has performed experiments whose results are best explained by her hypothesis that cheater detection is modified by social rank via "dominance detection" and "cheater detection" mechanisms that are not "innate modules" (not inherited intact), but are heavily canalized (the end-states of normal genetic and environmental factors of developmental trajectories) so that "the environment plays a very large role in how and whether biological predispositions get expressed" (Cummins, 1999, pp. 241-42). In the ensuing battle of objection and rebuttal, each side hopes to find decisive evidence, to show its theory to be the best explanation of the total available evidence, and to find evidence that only the favored theory can explain.

REFERENCES

Cosmides, L. & Tooby, J. 1992. "Cognitive Adaptations for Social Exchange", in Barkow, J., Cosmides, L., & Tooby, J., eds., *The Adapted Mind: Evolutionary Psychology and the Generation of Culture*. New York: Oxford University Press, pp.19-136.

Cummins, D.D. 1999. "Cheater Detection is Modified by Social Rank: The Impact of Dominance on the Evolution of Cognitive Functions", *Evolution and Human Behavior*, 20, 229-48.

Hamilton, W.D. 1964a. "The Genetical Theory of Social Behavior. I", *Journal of Theoretical Biology*, 7, 1-16.

Hamilton, W.D. 1964b. "The Genetical Theory of Social Behavior. II", *Journal of Theoretical Biology*, 7, 17-52.

Holcomb, H. 1993. *Sociobiology, Sex, and Science*. Albany, NY: SUNY Press.

Holcomb, H. 1998. "Testing Evolutionary Hypotheses", in Crawford, C. & Krebs, D., eds., *Handbook of Evolutionary Psychology: Ideas, Issues, and Applications*. Mahwah, New Jersey: Lawrence Erlbaum.

Mealey, L. 2000. *Sex Differences: Developmental and Evolutionary Strategies*. San Diego, California: Academic Press.

Sulloway, F. 1996. *Born to Rebel: Birth Order, Family Dynamics, and Creative Lives*. New York: Pantheon Books.

Wilson, D.S. 1994. "Adaptive Genetic Variation and Human Evolutionary Psychology", *Ethology and Sociobiology*, 15, 219-235.

CHAPTER 1

KINSHIP: THE TIE THAT BINDS (DISCIPLINES)

LINDA MEALEY
College of St. Benedict

Abstract. As disciplines, evolutionary psychology and behavior genetics have independent—and even antagonistic—histories, assumptions, and methodologies. One point of intersection, however, is their mutual investment in, and reliance on, the concept of kinship. I argue that this mutual concern might serve as common ground supporting an interdisciplinary investigation into the developmental and evolutionary causes of individual differences. In particular, I argue that the statistical main effects reported by each discipline are not particularly illuminating, and that what is needed is a combined effort to unravel the nature of nurture—the rules of epigenesis. To this end I outline a research program which would: (a) identify traits for which heritable variations are adaptive, versus traits for which heritable variations are simply "genetic junk;" then (b) determine how both heritable and non-heritable differences map onto life history strategies.

(1) INTRODUCTION: A TALE OF TWO DISCIPLINES

"Evolutionary Psychology is Not Behavior Genetics". So pronounces a section subheading of a recent discussion of evolutionary psychology (Cosmides & Tooby, 1997, p. 89). True, indeed. Evolutionary psychologists seek to explain behavior from an ultimate, phylogenetic perspective while behavior geneticists seek to explain behavior from a proximal, ontogenetic perspective. Evolutionary psychologists seek to identify and delineate the mechanisms underlying human universals (Buss, 1995; Cosmides, Tooby & Barkow, 1992; Pinker, 1994) while behavior geneticists seek to quantify human diversity and to identify the causes of that diversity (Gottlieb, 1995; Loehlin, 1989; Thiessen,

H.R. Holcomb III (ed.), Conceptual Challenges in Evolutionary Psychology, 19–38.

1972). As a result, evolutionary psychologists study cross-cultural similarities and focus on statistical means and main effects, while behavior geneticists study individual and group differences and focus on statistical variances and covariances (Bailey, 1998; Miller 2000a). Over the years, what has provided "wheat" for one discipline has generally been treated as "chaff" (or worse) by the other (Barash, 1997; Tooby & Cosmides 1990; Wilson 1994). In fact, it could easily be argued that these two disciplines represent paradigmatic extremes of the different ways of addressing the fundamental questions of psychology. Bailey (1995/1997) reported being able to find only six published studies that included both the word "evolution" and the phrase "behavior genetics" among their suggested index terms. Yet, despite a history of acrimony and debate between practitioners of these two disciplines, it is my belief that the most exciting vista in psychology today involves their interface and, ultimately, their integration.

In 1984, personality theorist David Buss published an essay in North America's most widely-read psychology journal, *American Psychologist*, outlining the theoretical and methodological differences between research programs investigating the two fundamental and essential questions of psychology: "What is human nature?" and "What makes each individual unique?" In that essay, Buss made a plea for rapprochement between what could, both then and now, be considered to be two completely independent research programs.

Today the two paradigms can be typified by the labels "evolutionary psychology" and "behavior genetics," respectively, but in fact, the schism between those studying species-typical behavior and those studying individual differences dates back to the beginnings of scientific psychology in the late nineteenth century (Jaynes, 1969). Even at that time, psychologists could be classified as belonging to one of two already-antagonistic groups: comparative psychologists and ethologists. Ironically, each had its roots in the Darwinian revolution, but each took something different from Darwin's insights—with the result that we, their intellectual descendents, find ourselves at diverging points on two fundamentally different theoretical trajectories.

Comparative psychologists took from Darwin the idea that humans share a common ancestry with other animals, particularly mammals. This insight led to the notion that we could learn something about the human psyche by studying the behavior of other animals, and it paved the way for experimentation that could not be ethically or practically pursued using human subjects. Which species were studied didn't matter to comparative psychologists, as long as we and they shared a significant common heritage; the result is that, because of its popularity among animal fanciers in England at the time, Rattus norwegicus, the common "lab rat," became the animal of choice. Similarly, the setting in which an animal was housed, reared and studied was not considered to be of particular importance because both instinctive behavior and the "laws of learning" were thought to be fixed by heredity; thus appeared the experimental research lab with its breeding colonies, mazes and, eventually, Skinner boxes and t-tests.

Ethologists took from Darwin not the idea of a common heritage, but the idea that each species (or type) was adapted to its unique niche, and that variation in form (phenotype) covaried in a functional way with variation in the environment. Ethologists claimed that in order to study this adaptive variation, behavior had to be studied in a natural environment and in a developmental context; thus appeared the field study, naturalistic observations, longitudinal studies and, from Darwin's cousin, Francis Galton, the correlation statistic.

Even though both groups of investigators broadened their methods and scope to include more and more studies of our own species, throughout the twentieth century the philosophical distance between experimental and correlational psychology was maintained (Cronbach, 1975). When Buss published his essay in 1984, he concluded that while "the relations between important species-typical characteristics and important individual differences should be identified and their implications understood," the gulf between the two approaches was so large that rapprochement could be envisioned only in terms of increased cross-disciplinary communication and parallel, but not integrated, study.

Five years after Buss' essay (but notably, without citing it), Charles Crawford and Judith Anderson (1989) proffered a specific methodological approach for the study of adaptive "differences and similarities." They suggested using a Darwinian life-history framework and a quasi-experimental design that had previously been adopted only in studies of non-human animals. Although they, like Buss, published in *American Psychologist*, their essay included little in the way of examples relevant to human behavior, and the only version of their design which they suggested as being applicable to humans involved the difficult study of monozygotic (identical) twins reared apart (MZAs); as a consequence, their important message, like that of Buss, went largely unappreciated.

After almost five more years, in 1993, developmental psychologist Nancy Segal published an essay (again in *American Psychologist*) exhorting psychologists to take advantage of not just the "natural experiment" of MZAs, but the wide variety of designs utilizing twins and other family constellations that had, to that point, been used fairly exclusively in the context of behavior genetics research. Segal provided as examples, studies that utilized family designs to test questions derived from what was, by then, labeled "evolutionary psychology" (Tooby & Cosmides, 1989).

Now, again with the passage of more than five years, the purpose of this essay is to show that the rapprochement between those who study species-typical behavior and those who study individual differences can entail more than just the parallel study envisioned by Buss; we have the means to achieve a total integration of these two, previously independent, research paradigms. The impending integration of the Darwinian life history approach (suggested by Crawford and Anderson) with the behavior genetic methodology (suggested by Segal) is poised to open a vast new space of possibilities for the study of both human nature and individual differences (see also Bailey, 1998; Buss & Greiling, 1999; Segal & MacDonald, 1998).

(2) COMMON GROUND: THE TIE THAT BINDS

Despite their real differences in philosophy and method, evolutionary psychologists and behavior geneticists are often perceived by others to be a single group with a unitary perspective. (It is this misperception that provided the motivation for the cautionary section subheading quoted at the beginning of this essay). The common perception of evolutionary psychology, like sociobiology before it, is that its emphasis on human universals invariably and inexorably leads to 1) a glossing over of important individual and group differences, and 2) attribution of any acknowledged differences to biological rather than to environmental sources (Barash, 1997; Dawkins, 1982; Plomin, DeFries, & McClearn, 1990; Thiessen, 1998). Likewise, the common perception of behavior genetics is that, while it focuses on differences rather than similarities, it, too, emphasizes biological as opposed to environmental "causes" of behavior (Bronfenbrenner & Ceci, 1994; Loehlin, 1989). Like all stereotypes, these perceptions have led to an oversimplified view of both disciplines and, because of their (presumed) shared emphasis on biological explanations, to the inaccurate categorizing of the two disciplines as one and the same.

One outcome of this stereotyping is the belief that both evolutionary psychology and behavior genetics are necessarily biologically determinist. In fact, as Crawford and Anderson (1989) noted, the evolutionary perspective cries out for systematic study of environmental effects and, as Segal (1993) pointed out, behavior genetic methods need not remain restricted to the single task of partitioning and quantifying genetic "versus" environmental sources of variance. The potential of both disciplines to shed explanatory light on the role of environmental factors in cognitive and behavioral development is prodigious, yet generally ignored (Loehlin, 1989).

The route to realizing this potential lies in the one acknowledged bit of common ground shared by evolutionary psychologists and behavior geneticists: Albeit for different reasons, both acknowledge the critical role of genetic kinship. For evolutionary psychologists, kinship, via the effect of inclusive fitness (Hamilton, 1964), constitutes a core construct of relevance to all social interaction. A quick browse through any animal behavior, behavioral ecology or evolutionary psychology journal will yield a plethora of papers on familial interactions from pair bonding to divorce, parental investment to parent-offspring conflict, sibling cooperation to sibling rivalry. For behavior geneticists, analysis of kinship is the key to a set of methods used to disentangle the complex multiple interactions that constitute development. Kin interactions, while not, per se, the topic of behavior genetic investigation, are studied as a matter of course in the analysis of data from across the gamut of topics of interest to students of human behavior (e.g., Reiss, Plomin, Hetherington, Howe, Rovine, Tryon & Hagen, 1994).

Furthermore, from the perspective of both disciplines—and, for that matter, almost every discipline in the social sciences—kinship can be seen as the central feature in the unfolding of each individual's life story. For the great majority of people, the family, in its various forms, is the most salient feature of the day-to-day environment (Salmon & Daly, 1996; Wilson & Daly, 1997; see also Emlen 1994, 1995). Kinship, the tie that binds, represents the common ground between social science disciplines (Daly, Salmon & Wilson, 1997). As such, it can provide the basis to forge new collaborative links and develop a fresh approach to the study of the role of the environment in the production of both human universals and individual differences (Fuller, 1983).

(3) CURRENT GROUND: THE STATE OF THE UNION

In a 1995 address (see 1995/1997), Bailey presented a fairly noncontroversial answer to the once-controversial question: "To what extent are individual differences attributable to heritable genetic differences?" The answer, "Bailey's Law," is: .4 +/- .2. That is, for almost any individual difference one wants to assess in the psychological realm, the percentage of within-population variance that can be attributed to underlying, within-population, genetic variance is about 40 percent, plus or minus 20 percent. (The mass of studies which has accrued over a long period suggests that heritability for cognitive abilities is probably on the high end of that range, heritability for attitudes is probably on the low end, and heritability for personality is about in the middle; see, e.g., Plomin, Owen & McGuffin, 1994.) Retrospectively, the fact that these "middling" numbers keep appearing as results of behavior genetic studies should not be surprising, as we have all long acknowledged that both nature and nurture contribute significantly to the processes of individual development and change (Turkheimer, 1998).

Recent advances in behavioral (and especially medical) genetics have facilitated the ability to identify particular genes that have major effects on particular phenotypes (Barinaga, 1994; Martin, Boomsma & Machin, 1997). Unfortunately, no matter how complex or elegant the mathematics, such designs are still not able to isolate the contributions of specific gene-gene interactions (dominance, heterosis and epistasis), gene-environment interactions, or gene-environment correlations (different from gene-environment interactions; see, e.g., Neale & Cardon, 1992; Plomin, DeFries & Loehlin, 1977). Neither can such designs identify "regulator genes," critical life events, threshold phenomena, feedback loops or other non-linear or dynamic interactions (Gottlieb, 1995; Mealey, 1995a, 1998), let alone identify particular proximal mechanisms that produce, or result from, such interactions (Bronfenbrenner & Ceci, 1994; Carey, 1991). Furthermore, by definition, heritability itself will differ in differing environments (Bailey, 1998; Hunt, McNeill & Rowe, 1997; Jacobs, 1981; Nicholson, 1990; Wahlsten, 1990). As David Wasserman, convener of the widely-publicized conference "Research on Genetics and Criminal Behavior" (September 1995, Queenstown, Maryland) Stated, "... the sun is setting on traditional heritability research on social behavior ... it has served its purpose" (1996, p. 108; but see Martin et al. 1997 for a qualified counter-argument).

Compared to behavior genetics, evolutionary psychology is a relative newcomer and, as a stand-alone discipline, still has much to offer. Yet despite its fresh face, the shortcoming that plagues traditional behavior genetics plagues evolutionary psychology as well: Too often, it tells us something we already know. Worse, it often tells us what we already think we know. Evolutionary psychology, like human sociobiology before it, has a bad habit of "documenting" folk knowledge and providing post-hoc "Darwinian" interpretations of the "findings." While one of the major motivations for coinage of the phrase "evolutionary psychology" was to distinguish a new paradigm, different from the "Just so" models of human sociobiology (Tooby & Cosmides, 1992), not many have taken up the challenge for a truly new approach and, so far, (with apologies to those whose names are too many to list, but still, in quantity, somewhat underwhelming), little has been documented that is truly new. Despite a few novel insights (see examples in Barkow, Cosmides & Tooby, 1992; Bock & Cardew, 1997; Crawford & Krebs, 1998; and Simpson & Kenrick, 1996), evolutionary psychology has not lived up to its promise (Mealey, 1994; Miller, 2000a; Thiessen, 1998).

In short, the literatures from both behavior genetics and evolutionary psychology are more often descriptive than explanatory. Neither gets down to the nitty gritty level of mechanism, and neither (yet) tells us anything about the intervening variables, structures, processes, or pathways that mediate phenotypic differences *or* similarities.

(4) GROUND PLAN: STRATEGIES AND TACTICS FOR STUDYING STRATEGIES AND TACTICS

The big question we all are waiting to hear answered is: "How do cognitive dynamics map onto physiological systems laid down by interacting genes and, in the other direction, onto inputs from the larger social environment?" (Kenrick & Brown, 1995, p. 561). To answer this question requires the rapprochement of philosophy and paradigm hoped for by Buss (1984) and the integration of methods outlined by Crawford and Anderson (1989) and Segal (1993). Rather than pursuing parallel investigations at the ultimate and proximate levels, research questions and designs need to integrate both paradigms by setting the following goals.

Research Goal #1

The first goal is to identify traits for which heritable variations are adaptive versus traits for which heritable variations are simply "genetic junk." Genetic variation may reflect: 1) adaptive polymorphisms at evolutionary equilibrium; 2) non-equilibrium variation resulting from temporally- or geographically-varying directional selection, migration, or outbreeding; 3) selection-irrelevant variation that has accumulated through random mutation; or 4) deleterious mutational load (Bailey, 1998; Gangestad, 1997; Miller, 2000b; Tooby & Cosmides 1990; Wilson, 1994). Which traits have variability maintained by which process can be determined by discovering a) for which traits variations actually map onto reproductive success (or proxies thereof) across varied environmental conditions and b) which traits exhibit what kinds of genetic interactions.

Specifically, in traits for which heritable differences reflect adaptive polymorphism, genotypic differences should produce mean differences in reproductive success (RS) when all individuals are constrained to a single environment or strategy, but similar RS when genetic variation is allowed to co-vary with salient environmental features (Crawford & Anderson, 1989; Mealey & Segal, 1993; Perusse, 1993). Furthermore, adaptive heritable differences are more likely to be mediated by dominance, epistatic effects, correlated gene complexes, and gene-environment interactions, than by purely additive genetic effects (Bailey, 1998; Bruell, 1964, 1967; Plomin, DeFries & McClearn, 1990, p. 284). New techniques have recently emerged that can be used to quantify some of these effects (Barinaga, 1994; Fenster, Galloway & Chao, 1997; Gottesman, 1997), but even when they cannot be quantified, simple knowledge of their existence provides powerful evidence for inferring adaptive value (Stearns, de Jong & Newman, 1991). Adaptive trait variation might, thus, be identified by systematic discrepancies between the "broad heritability" estimates provided by twin studies and the "narrow heritability" estimates provided by adoption studies (see Bailey, 1998; Mealey, 1995b, footnote 7; and Segal & MacDonald, 1998).

Research Goal #2a

If a trait is identified that meets the criteria for expressing adaptive heritable variation, the next goal is to determine how those heritable differences map onto life history strategies. Various psychological attributes, styles, and behaviors that correlate with underlying genetic variation might be considered to be tactics within the framework of an overall reproductive strategy- for example: personality (Buss, 1991; MacDonald, 1995; Wilson, Clark, Coleman & Dearstyne, 1994); sociosexuality (Buss & Schmitt, 1993; Gangestad & Simpson, 1990; Simpson & Gangestad, 1992); attachment style (Belsky, 1997, 2000;

Goldsmith & Harman, 1994; MacDonald, 1997a&b; Mealey, 1995c); and even psychopathology (Allen, 1997; Allen & Sarich, 1988; Kofoed & MacMillan, 1986; Mealey, 1995a&b; 2000; Jensen, Mrazek, Knapp, Steinberg, Pfeffer, Schowalter & Shapiro, 1997). Mapping of heritable differences onto life history strategy will require the traditional, but under-utilized, adoption and blended-family studies from behavior genetics (e.g., O'Connor, Hetherington, Reiss, & Plomin, 1995; Segal 1993), as well as longitudinal study (e.g., Di Lalla & Bishop, 1995) of the previously acknowledged but little-studied processes of self-selection into varied environments (Caspi, Elder & Bem, 1987; Cleveland, Wiebe, van den Oord, & Rowe, 2000; Ickes, Snyder & Garcia, 1997; Rowe, 1990; Scarr, 1992; Scarr & McCartney, 1983) and differential parental or other social treatment of individuals of different genotypes—what Scarr (1992) calls "evocative gene-environment correlations" and Tooby and Cosmides (1990) call "reactive heritability."[1]

Research Goal #2b

For traits that do not meet the criteria for expressing adaptive heritable variation, the goal is to determine how non-heritable differences map onto life history strategies. Keeping in mind Bailey's law, if heritable differences account for .4 +/- .2 of a trait's statistical variance, then non-heritable differences must account for most (specifically, .6 +/- .2) of that variance. Evolutionary psychologists postulate that a significant portion of this environmentally-induced variance is likely to be adaptive (e.g., Belsky, Steinberg & Draper, 1991; Buss & Greiling, 1999; Chisholm, 1996; Draper & Belsky, 1990; Hill, Young & Nord, 1994; Lamb,Thompson, Gardner, Charnov & Estes, 1984; MacDonald, 1995; Sulloway, 1995, 1996; Wilson & Daly, 1997). As Wilson et al. (1994) note: "It seems that the natural environment acts to create and maintain individual differences in phenotypically plastic individuals..." (p. 443).

Adaptive, environmentally-induced individual differences might be developmentally-contingent, such that each individual has the genetic potential to realize every phenotype, but the trait in question becomes fixed after the individual is exposed to certain environmental stimuli in the course of development. Alternatively, such differences might be environmentally-contingent, such that every individual has the genetic capacity to exhibit every phenotype, but does so in direct relation to immediate circumstances. In the former case we say that the phenotype is "canalized;" in the latter we say that the phenotype is "facultative." Both canalized and facultative attributes will appear to be heritable when certain genotypes are regularly restricted to certain environments (Jacobs, 1981). This set of circumstances includes situations in which offspring acquire parental attributes through non-genetic forms of transmission (e.g., Freedman & Gorman, 1993; Kochanska, Clark, & Goldman, 1997; van Ijzendoorn, 1995; van IJzendoorn & Kroonenberg, 1988), meaning that studies of parent-offspring interactions can be unenlightening if authors don't consider all three alternatives (Buss & Greiling, 1999; Halverson & Wampler, 1997; Rowe, 1994, 2000).

It is unfortunate from a methodological perspective, but behavior genetics research has discovered that it is not the easily measured, shared, within-family environmental features (such as parental education, religion and socioeconomic status) that are most critical in terms of producing individual differences. Rather, it is each individual's unique experiences or unique perceptions of shared experiences that are key (Dunn, 1992; Dunn & Plomin, 1990; LaLumiere, Quinsey & Craig, 1995; Marjoribanks, 1985; Plomin & Daniels, 1987; Rowe 1983, 1990, 1994; Rowe & Plomin, 1981). This means that non-traditional approaches to personality and developmental psychology must be used to study the effects of environment on individual development (Halverson & Wampler, 1997; Phillips & Matheny, 1997; Segal & MacDonald, 1998). Given these new facts, what methodology can we turn to?

(5) FUTURE GROUND: TO GO WHERE NO ONE HAS GONE BEFORE

Crawford and Anderson (1989) pointed out that "the operation of contingent environmental strategies and the mechanisms mediating them can be seen most clearly when there is no genetic variation, and hence, zero heritability for the behavior in question" (p. 1454). Indeed this is true. Yet we know that most traits have significant non-zero heritability. For this reason, Crawford and Anderson suggested studying identical twins reared apart: By studying differences between MZ twins reared apart, one can control for those ubiquitous and troublesome genetic differences, isolate the effect of different environments on developmental outcomes, and identify adaptive, environmentally-induced strategies.

While one has to admit the beauty of the design, practically speaking, it is not of much use for those of us who want to study the human animal; identical twins reared apart are rare, and the environments to which they have been exposed constitute a restricted, small, and not well-documented sample of all possible, or even all common, environments. Indeed, only one study, so far, has attempted to use this design. In 1993, Mealey and Segal reported that in MZA co-twin pairs where one twin had been raised by kin and the other by non-kin, the twin reared by kin was significantly more likely than his or her co-twin to report feeling wanted, less likely to report family conflict, and more likely to report parental encouragement of dating. These family dynamics were, in turn, significantly related to age at first marriage and birth of first child (for men) and number of marriages, desire for children, and number of children (for women)—findings which reflect life history patterns reported by others using non-genetic designs (e.g., Rholes, Simpson, Blakely, Lanigan, & Allen, 1997). Though based on small samples with restricted environmental variability, the Mealey and Segal results at least suggest that environmental variables relating to family dynamics have the potential to trigger different psychological mechanisms and life-history strategies in genetically similar individuals.

The bad news is that MZAs will continue to be rare, the range of their rearing environments will continue to be restricted, and the documentation of their respective environments will continue to be based on post-hoc reconstructions with built-in (even genetically-contaminated) biases (Hur & Bouchard, 1995). The good news is that because we now know that non-shared environmental factors contribute more to the development of individual differences than do shared environmental factors, and since siblings—even identical twins—do have different experiences and do differ as a result, the same information can be obtained from the study of twins from intact families (MZTs) as from MZAs! Indeed, the study of MZTs can provide even better information than the study of MZAs, since longitudinal studies can begin at very early ages and provide better, more detailed, and more objective information on the differential experiences and differential perceptions of young twins. Only this kind of early and extensive data collection will allow researchers to map early experiences onto subsequently-expressed individual differences in personality and life history strategy (Mealey, Bridgstock & Townsend, 1999; Pelps, Davis & Schartz, 1997).

Little data of this sort currently exists for very young twins, but it can be collected much more easily than data from MZAs. Furthermore, as a place to start, plenty of twin registries already have data on adolescent or adult twins that can be re-analyzed with a focus on the correlations between environmental differences and phenotypic differences in monozygotic co-twins. For example, two studies that have analyzed data in this way reported significant correlations between co-twin differences in early environment[2] and co-twin differences in adult attributes. Rose, Reed and Bogle (1987) reported correlations between co-twin differences in symmetry and co-twin differences in personality, and Mealey and Townsend (1999) and Mealey et al. (1999) reported significant correlations between co-twin differences in symmetry and co-twin differences in physical attractiveness.

Investigators can use several tactics to study the environmental events that influence life history strategies. Some might take a "shotgun" approach, beginning with a large, unselected sample of twins, and searching for statistical patterns in reports of differential early experience as related to later-occurring phenotypic differences. Others might take a "bullet" approach: (1) investigators interested in a certain developmental outcome can use a retrospective strategy, starting out with twins known to differ on the trait of interest, and then searching backward for expected differences in early experience (or, more specifically, and not at all less meaningfully, different perceptions of early experience), while (2) investigators interested in the influence of particularly salient life events can use a prospective strategy, starting out with twins known to differ with regard to the relevant experience (or perceived experience), and then searching forward for the predicted, differential outcomes. The retrospective approach could be applied to the study of individual differences such as personality, sexual orientation, and mental illness, while the prospective approach could be applied to the study of developmental variables such as environmental predictability/unpredictability, childhood illness, and differential parental solicitude/favoritism (Daly & Wilson, 1980, 1982; Mann, 1992; Wilson, Daly, & Weghorst, 1981; Zervas & Sherman, 1994).

Coming full circle, we see that despite their differing philosophical questions, historically separate paradigms, and methodological apartheid, behavior geneticists and evolutionary psychologists can find common ground by focusing on that most evident and powerful construct of the evolutionary and social sciences: kinship. The family methodologies that have, so far, been the hallmark of behavior genetics, can become powerful tools for evolutionary psychologists in the study of the role of the environment in inducing and maintaining adaptive variation. The rapprochement of these one-time foes is poised to generate the next scientific frontier: the elucidation of the internal and external environmental triggers that activate the genetic, physiological, and psychological mechanisms contributing to social behavior.

NOTES

[1] Mazur & Booth's 1998 review of the reciprocal interrelations between genes, testosterone, environment and behavior provides an exemplar of the kind of multilevel research that is needed to integrate ultimate and proximate explanations of adaptive variataion: it addresses four kinds of gene-environment interaction, three kinds of feedback loop, and three kinds of evolutionarily stable strategy for maintaining population variability (Mealey, 1998).

[2] Both of these studies used fluctuating symmetry (FA) as a proxy messure of early environmental influences. FA is defined as deviations from perfect symmetry for which population differences between the right and left sides have a mean of zero and a statistically normal distribution. When both sides of a bilaterally symmetric feature are encoded by the same genetic instructions, deviations from symmetry reflect the inability of the individual to maintain developmental homeostasis. This means that individual differences in symmetry can be used as a measure of both susceptibility to, and exposure to, environmental challenges such as toxins and infection. When the individuals being compared are genetically identical, any differences between them must, therefore, reflect differences in exposure to relevant features of the environment. Differential FA of monozygotic twins thus reflects differential exposure to environmental stressors, with greater asymmetry reflecting greater exposure.

Parts of this chapter are based on a talk given at the 7th annual meeting of the Human Behavior and Evolution Society, Evanston, Illinois, June 1996.

This chapter was written while the author was at the School of Psychology of the University of Queensland, Brisbane, Australia 4072.

REFERENCES

Allen, J.S. 1997. "At Issue: Are Traditional Societies Schizophrenogenic?", Schizophrenia Bulletin, 23, 357-364.

Allen, J.S. & Sarich, V.M. 1988. "Schizophrenia in Evolutionary Perspective", Perspectives in Biology and Medicine, 32, 132-151.

Bailey, J.M. 1995/1997. "Are Genetically-Based Individual Differences Compatible with Species-Wide Adaptations?", keynote address at the APA-Sponsored Festschrift for Daniel G. Freedman: Uniting Psychology and Biology: Genetic, Ethological, Cultural and Evolutionary Perspectives on Human Development, University of Chicago, October 1995. In Segal, N.L., Weisfeld, G.E. & Weisfeld, C.C., eds., Uniting Psychology and Biology:

Integrated Perspectives on Human Development. Washington, DC: APA Press, 1997.
Bailey, J.M. 1998. "Can Behavior Genetics Contribute to Evolutionary Behavioral Science?", in Crawford, C. & Krebs, D.L., eds., Handbook of Evolutionary Psychology: Ideas, Issues, and Applications. Mahwah, NJ: Lawrence Erlbaum.
Barash, D.P. 1997. "In Search of Behavioral Individuality", Human Nature, 8, 153-169.
Barinaga, M. 1994. "A New Tool for Examining Multigenic Traits", Science, 264, 1691.
Barkow, J., Cosmides, L. & Tooby, J., eds., 1992. The Adapted Mind. New York, NY: Oxford University.
Belsky, J. 1997. "Attachment, Mating and Parenting: An Evolutionary Perspective", Human Nature, 8, 361-381.
Belsky, J. 2000. "Conditional and Alternative Reproductive Strategies: Individual Differences in Susceptibility to Rearing Experience", in Rodgers, J.L., Rowe, D.C. & Miller, W., eds., Genetic Influences on Human Fertility and Sexuality. London, England: Kluwer Academic Publishers.
Belsky, J., Steinberg, L., & Draper, P. 1991. "Childhood Experience, Interpersonal Development and Reproductive Strategy: An Evolutionary Theory of Socialization", Child Development, 62, 647-670.
Bock, G.R. & Cardew, G., eds., 1997. Characterizing Human Psychological Adaptations. Chichester, England: John Wiley & Sons.
Bronfenbrenner, U. & Ceci, S.J. 1994. "Nature-Nurture Reconceptualized in Developmental Perspective: A Bio-Ecological Model", Psychological Review, 101, 568-586.
Bruell, J.H. 1964. "Inheritance of Behavioral and Physiological Characters of Mice and the Problem of Heterosis", American Zoologist, 4, 125-138.
Bruell, J.H. 1967. "Behavioral Heterosis", in Hirsch, J., ed., Behavior-Genetic Analysis. New York, NY: McGraw-Hill.
Buss, D.M. 1984. "Evolutionary Biology and Personality Psychology", American Psychologist, 39, 1135-1147.
Buss, D.M. 1991. "Evolutionary Personality Psychology", Annual Review of Psychology, 42, 459-491.
Buss, D.M. 1995. "Evolutionary Psychology: A New Paradigm for Psychological Science", Psychological Inquiry, 6, 1-30.
Buss, D.M. & Greiling, H. 1999. "Adaptive individual differences", Journal of Personality, 67, 209-244.
Buss, D.M. & Schmitt, D.P. 1993. "Sexual Strategies Theory: An Evolutionary Perspective on Human Mating", Psychological Review, 100, 204-232.
Carey, G. 1991. "Evolution and Path Models in Human Behavioral Genetics", Behavior Genetics, 21, 433-444.
Caspi, A., Elder, G.H., & Bem, D.J. 1987. "Moving Against the World: Life-Course Patterns of Explosive Children", Developmental Psychology, 23, 308-313.

Chisholm, J.S. 1996. "The Evolutionary Ecology of Attachment Organization", Human Nature, 7, 1-37.

Cleveland, H.H., Wiebe, R.P., van den Oord, E.J.C.G., & Rowe, D.C. 2000. "Behavior Problems among Children from Different Family Structures: The Influence of Genetic Self-Selection. Child Development", 71, 733-751.

Cosmides, L. & Tooby, J. 1997. "The Modular Nature of Human Intelligence", in Scheibel, A.B. & Schopf, J.W., eds., The Origin and Evolution of Intelligence. Boston, MA: Jones & Bartlett.

Cosmides, L., Tooby, J., & Barkow, J. 1992. "Evolutionary Psychology and Conceptual Integration", in Barkow, J., Cosmides, L., & Tooby, J., eds., The Adapted Mind. New York, NY: Oxford University.

Crawford, C.B. & Anderson, J.L. 1989. "Sociobiology: An Environmentalist Discipline?", American Psychologist, 44, 1449-1459.

Crawford, C. & Krebs, D.L. 1998. Evolutionary Psychology: Ideas, Issues and Applications. Mahwah, NJ: Lawrence Erlbaum.

Cronbach, L.J. 1975. "The Two Disciplines of Scientific Psychology", American Psychologist, 12, 671-684.

Daly, M. & Wilson, M. 1980. "Discriminative Parental Solicitude: A Biological Perspective", Journal of Marriage and the Family, 42, 277-288.

Daly, M. & Wilson, M. 1982. "Who Are Newborn Babies Said to Resemble?", Ethology & Sociobiology, 3, 69-78.

Dawkins, R. 1982. The Extended Phenotype: The Gene as the Unit of Selection. Oxford: Oxford University.

DiLalla, L.F. & Bishop, E.G. 1996. "Differential Maternal Treatment of Infant Twins: Effects on Infant Behaviors", Behavior Genetics, 26, 535-542.

Draper, P. & Belsky, J. 1990. "Personality Development in Evolutionary Perspective", Journal of Personality, 58, 141-161.

Dunn, J. 1992. "Siblings and Development", Current Directions in Psychological Science, 1, 6-9.

Dunn, J. & Plomin, R. 1990. Separate Lives: Why Siblings Are So Different? New York, NY: Basic Books.

Emlen, S.T. 1994. "Benefits, Constraints and the Evolution of the Family", Trends in Ecology and Evolution, 9, 281-285.

Emlen, S.T. 1995. "An Evolutionary Theory of the Family", Proceedings of the National Academy of Science, 92, 8092-8099.

Fenster, C.B., Galloway, L.F., & Chao, L. 1997. "Epistasis and its Consequences for the Evolution of Natural Populations", Trends in Ecology and Evolution, 12, 282-286.

Freedman, D.G. & Gorman, J. 1993. "Attachment and the Transmission of Culture-An Evolutionary Perspective", Journal of Social and Evolutionary Systems, 16, 297-329.

Fuller, J.L. 1983. "Sociobiology and Behavior Genetics", in Fuller, J.L. & Simmel, E.C., eds., Behavior Genetics: Principles and Applications. Hillsdale, NJ: Lawrence Erlbaum.

Gangestad, S.W. 1997. "Evolutionary Psychology and Genetic Variation: Non-

Adaptive, Fitness-Related and Adaptive", in Bock, G.R. & Cardew, G., eds., Characterizing Human Psychological Adaptations. Chichester, England: John Wiley & Sons.

Gangestad, S.W. & Simpson, J.A. 1990. "Toward an Evolutionary History of Female Sociosexual Variation", Journal of Personality, 58, 69-95.

Goldsmith, H.H. & Harman, C. 1994. "Temperament and Attachment: Individuals and Relationships", Current Directions in Psychological Science, 3, 53-57.

Gottesman, I.I. 1997. "Twins: En Route to QTLs for Cognition", Science, 276, 1522-1523.

Gottlieb, G. 1995. "Some Conceptual Deficiencies in "Developmental" Behavior Genetics", Human Development, 38, 131-169.

Halverson, C.F., Jr. & Wampler, K.S. 1997. "Family Influences on Personality Development", in Hogan, R., Johnson, J., & Briggs, S, eds., Handbook of Personality Psychology, New York, NY: Academic Press.

Hamilton, W.D. 1964. "The Genetical Evolution of Social Behavior: I & II", Journal of Theoretical Biology, 7, 1-52.

Hill, E.M., Ross, L.T., & Low, B.S. 1997. "The Role of Future Unpredictability in Human Risk-Taking", Human Nature, 8, 287-325.

Hill, E.M., Young, J.P., & Nord, J.L. 1994. "Childhood Adversity, Attachment Security, and Adult Relationships: A Preliminary Study", Ethology & Sociobiology, 15, 323-338.

Hrdy, S. 1990. "Sex Bias in Nature and in History: A Late 1980's Reexamination of the "Biological Origins" Argument", Yearbook of Physical Anthropology, 33, 25-37.

Hunt, C., McNeill, P., & Rowe, D. 1997, June. "Does Environment Condition the Heritability of Menarchy?", paper presented at the Human Behavior and Evolution Society, Tucson, AZ.

Hur, Y.-M. & Bouchard, T.J., Jr. 1995. "Genetic Influences on Perceptions of Childhood Family Environment: A Reared Apart Twin Study", Child Development, 66, 330-345.

Ickes, W., Snyder, M., & Garcia, S. 1997. "Personality Influences on the Choice of Situations", in Hogan, R., Johnson, J., & Briggs, S., eds., Handbook of Personality Psychology. New York, NY: Academic Press.

Jacobs, J. 1981. "How Heritable Is Innate Behaviour?", Zeitschrift fur Tierpsychologie, 55, 1-18.

Jaynes, J. 1969. "The Historical Origins of 'Ethology' and 'Comparative Psychology'", Animal Behaviour, 17, 601-606.

Jensen, P.S., Mrazek, D., Knapp, P.K., Steinberg, L., Pfeffer, C., Schowalter, J., & Shapiro, T. 1997. "Evolution and Revolution in Child Psychiatry: ADHD as a Disorder of Adaptation", Journal of the American Academy of Child and Adolescent Psychiatry, 36, 1672-1681.

Kenrick, D.T. & Brown, S. 1995. "Al Capone, Discrete Morphs, and Complex Dynamic Systems", Behavioral and Brain Sciences, 18, 560-561.

Kochanska, G., Clark, L.A., & Goldman, M.S. 1997. "Implications of Mothers'

Personality for Their Parenting and Their Children's Developmental Outcomes", Journal of Personality, 65, 387-420.
Kofoed, L. & MacMillan, J. 1986. "Alcoholism and Antisocial Personality: The Sociobiology of an Addiction", The Journal of Nervous and Ment al Disease, 174, 332-335.
LaLumiere, M.L., Quinsey, V.L., & Craig, W.M. 1995. "Why Children from the Same Family Are So Different from One Another", Human Nature, 7, 281-290.
Lamb, M.E., Thompson, R.A., Gardner, W.P., Charnov, E.L., & Estes, D. 1984. "Security of Infantile Attachment as Assessed in the "Strange Situation": Its Study and Biological Interpretation", Behavioral and Brain Sciences, 7, 127-171.
Loehlin, J.C. 1989. "Partitioning Environmental and Genetic Contributions to Behavioral Development", American Psychologist, 44, 1285-1292.
MacDonald, K. 1995. "Evolution, the Five-Factor Model, and Levels of Personality", Journal of Personality, 63, 525-567.
MacDonald, K. 1997a. "The Coherence of Individual Development: An Evolutionary Perspective on Children's Internalization of Parental Values", in Grusec, J.E. & Kuczynski, L., eds., Parenting Strategies and Children's Internalization of Values: A Handbook of Theoretical and Research Perspectives. New York, NY: Wiley.
MacDonald, K. 1997b. "Life History and Human Reproductive Behavior: Environmental/Contextual Influences and Heritable Variation", Human Nature, 8, 327-359.
Mann, J. 1992. "Nurturance or Negligence: Maternal Psychology and Behavioral Preference among Preterm Twins", in Barkow, J.H., Cosmides, L., & Tooby, J., eds., The Adapted Mind. New York, NY: Oxford University Press.
Martin, N., Boomsma, D., & Machin, G. 1997. "A Twin-Pronged Attack on Complex Traits", Nature Genetics, 17, 387-392.
Mazur, A. & Booth, A. 1998. "Testosterone and Dominance in Men" (with commentary and rejoinder), Behavioral and Brain Sciences, 21, 353-397.
Mealey, L. 1994. "Review of The Adapted Mind: Evolutionary Psychology and the Generation of Culture", Politics and the Life Sciences, 13, 294-295.
Mealey, L. 1995a. "Primary Sociopathy is a Type, Secondary Is Not", Behavioral and Brain Sciences, 18, 579-599.
Mealey, L. 1995b. "The Sociobiology of Sociopathy: An Integrated Evolutionary Model", Behavioral and Brain Sciences, 18, 523-541. (Reprinted in Baron-Cohen, S., ed., The Maladapted Mind. Erlbaum, Taylor & Francis, 1997).
Mealey, L. 1995c. "Internal Working Models: Cause, Effect or Co-Variate?", ASCAP: Newsletter of the Society for Sociophysiological Integration, 8, 5,22.
Mealey, L. 1998. "Testosterone and Aggression: An Exemplar of Interactionism", Behavioral and Brain Sciences, 21, 380-381.

Mealey, L. 2000. "Anorexia: A "Losing" Strategy?", Human Nature, 11, 105-116.

Mealey, L., Bridgstock, R., & Townsend, G.C. 2000. "Symmetry and Perceived Facial Attractiveness: A Monozygotic Co-Twin Comparison", Journal of Personality and Social Psychology, 76, 157-165.

Mealey, L. & Segal, N.L. 1993. "Heritable and Environmental Variables Affect Reproduction-Related Behaviors, but Not Ultimate Reproductive Success", Personality and Individual Differences, 14, 783-794.

Mealey, L. & Townsend, G.C. 1999. "The Role of Fluctuating Asymmetry on Judgments of Physical Attractiveness: A Monozygotic Co-Twin Comparison", Perspectives in Human Biology, 4, 219-224.

Miller, G. 2000a. "How to Keep Our Metatheories Adaptive: Beyond Cosmides, Tooby and Lakatos", Psychological Inquiry, 11, 42-46.

Miller, G.F. 2000b. The Mating Mind: How Sexual Choice Shaped the Evolution of Human Nature. N.Y.: NY, Random House.

Neale, M.C. & Cardon, L.R. 1992. Methodology for Genetic Studies of Twins and Families. London, England: Kluwer.

O'Connor, T.G., Hetherington, M., Reiss, D., & Plomin, R. 1995. "A Twin-Sibling Study of Observed Parent-Adolescent Interactions", Child Development, 66, 812-829.

Perusse, D. 1993. "Cultural and Reproductive Success in Industrial Societies: Testing the Relationship at the Ultimate and Proximate Levels", Behavioral and Brain Sciences, 16, 267-282.

Phelps, J.A., Davis, J.O., & Schartz, K.M. 1997. "Nature, Nurture and Twin Research Strategies", Current Directions in Psychological Science, 6, 117-121.

Phillips, K. & Matheny, A.P., Jr. 1997. "Evidence for Genetic Influence on Both Cross-Situation and Situation-Specific Components of Behavior", Journal of Personality and Social Psychology, 73, 129-138.

Pinker, S. 1994. The Language Instinct. Cambridge, MA: MIT Press.

Plomin, R. & Daniels, D. 1987. "Why Are Children in the Same Family So Different from Each Other?", Behavioral and Brain Sciences, 10, 1-16.

Plomin, R., DeFries, J.C., & Loehlin, J.C. 1977. "Genotype-Environment Interaction and Correlation in the Analysis of Human Behavior", Psychological Bulletin, 84, 309-322.

Plomin, R., DeFries, J.C., & McClearn, G.E. 1990. Behavioral Genetics: A Primer, 2nd edition. New York, NY: W.H. Freeman.

Plomin, R., Owen, M.J., & McGuffin, P. 1994. "The Genetic Basis of Complex Human Behaviors", Science, 264, 1733-1739.

Reiss, D., Plomin, R., Hetherington, E.M., Howe, G.W., Rovine, M., Tryon, M.A., & Hagan, M.S. 1994. "The Separate World of Teenage Twins: An Introduction to the Study of Non-Shared Environment and Adolescent Development", in Hetherington, E.M., Reiss, D., & Plomin, R., eds., Separate Worlds of Siblings: The Impact of Nonshared Environment on Development. Hillsdale, NJ: Lawrence Erlbaum.

Rholes, W.S., Simpson, J.A., Blakely, B.S., Lanigan, L., & Allen, E.A. 1997. "Adult Attachment Styles, the Desire to Have Children, and Working Models of Parenthood", Journal of Personality, 65, 357-385.

Rose, R.J., Reed, T., & Bogle, A. 1987. "Asymmetry of A-B Ridge Count and Behavioural Discordance of Monozygotic Twins", Behavior Genetics, 17, 125-140.

Rowe, D.C. 1983. "A Biometrical Analysis of Perceptions of Family Environment: A Study of Twin and Singleton Sibling Kinships", Child Development, 54, 416-423.

Rowe, D.C. 1990. "As the Twig Is Bent? The Myth of Child-Rearing Influences on Personality Development", Journal of Counseling and Development, 68, 606-611.

Rowe, D.C. 1994. The Limits of Family Influence: Genes, Experience, and Behaviour. New York, NY: Guilford.

Rowe, D.C. 2000. "Environmental and Genetic Influences on Pubertal Development: Evolutionary Life History Traits?", in Rodgers, J.L., Rowe, D.C., & Miller, W., eds., Genetic Influences on Human Fertility and Sexuality. London, England: Kluwer Academic Publishers.

Rowe, D.C. & Plomin, R. 1981. "The Importance of Nonshared Environmental Influences on Behavioral Development", Developmental Psychology, 17, 517-531.

Salmon, C.A. & Daly, M. 1996. "On the Importance of Kin Relations to Canadian Men and Women", Ethology and Sociobiology, 17, 289-298.

Scarr, S. 1992. "Developmental Theories for the 1990s: Development and Individual Differences", Child Development, 63, 1-19.

Scarr, S. & McCartney, K. 1983. "How People Make Their Own Environments: A Theory of Genotype-Environment Effects", Child Development, 54, 424-435.

Segal, N.L. 1993. "Twin, Sibling and Adoption Methods: Tests of Evolutionary Hypotheses", American Psychologist, 48, 943-956.

Segal, N.L. & MacDonald, K.B. 1998. "Behavior Genetics and Evolutionary Psychology: A Unified Perspective on Personality Research", Human Biology, 70, 157-182.

Simpson, J.A. & Gangestad, S.W. 1992. "Sociosexuality and Romantic Partner Choice", Journal of Personality, 60, 31-51.

Simpson, J.A. & Kenrick, D.T. 1996. Evolutionary Social Psychology. Hillsdale, NJ: Erlbaum.

Stearns, S., de Jong, G., & Newman, B. 1991. "The Effects of Phenotypic Plasticity on Genetic Correlations", Trends in Ecology and Evolution, 6, 122-126.

Sulloway, F.J. 1995. "Birth Order and Evolutionary Psychology: A Meta-Analytic Overview", Psychological Inquiry, 6, 75-80.

Sulloway, F.J. 1996. Born to Rebel: Birth Order, Family Dynamics, and Creative Lives. New York, NY: Random House.

Thiessen, D.D. 1972. "A Move Toward Species-Specific Analyses in Behav-

ior Genetics", Behavior Genetics, 2, 115-126.
Thiessen, D.T. 1998. "Expanding the Boundaries of Evolutionary Psychology: The Context of Domain-Specific Adaptations", Mankind Quarterly, 38, 337-362.
Tooby, J. & Cosmides, L. 1989. "Evolutionary Psychology and the Generation of Culture I: Theoretical Considerations", Ethology & Sociobiology, 10, 29-49.
Tooby, J. & Cosmides, L. 1990. "On the Universality of Human Nature and the Uniqueness of the Individual: The Role of Genetics and Adaptation", Journal of Personality, 58, 17-67.
Tooby, J. & Cosmides, L. 1992. "The Psychological Foundations of Culture", in Barkow, J., Cosmides, L., & Tooby, J., eds., The Adapted Mind. New York, NY: Oxford University.
Turkheimer, E. 1998. "Heritability and Biological Explanation", Psychological Review, 105, 782-791.
van IJzendoorn, M.H. 1995. "Adult Attachment Representations, Parental Responsiveness, and Infant Attachment: A Meta-Analysis on the Predictive Validity of the Adult Attachment Interview", Psychological Bulletin, 117, 387-403.
van IJzendoorn, M.H. & Kroonenberg, P.M. 1988. "Cross-Cultural Patterns of Attachment: A Meta-Analysis of the Strange Situation", Child Development, 59, 147-156.
Wahlsten, D. 1990. "Insensitivity of the Analysis of Variance to Heredity-Environment Interaction", Behavioral and Brian Sciences, 13, 109-161.
Wasserman, D. 1996. "Research into Genetics and Crime: Consensus and Controversy", Politics and the Life Sciences, 15, 107-109.
Wilson, D.S. 1994. "Adaptive Genetic Variation and Human Evolutionary Psychology", Ethology and Sociobiology, 15, 219-235.
Wilson, D.S., Clark, A.B., Coleman, K., & Dearstyne, T. 1994. "Shyness and Boldness in Humans and Other Animals", Trends in Ecology and Evolution, 9, 442-446.
Wilson, M. & Daly, M. 1997. "Life Expectancy, Economic Inequality, Homicide, and Reproductive Timing in Chicago Neighbourhoods", British Medical Journal, 314, 1271-1274.
Wilson, M. & Daly, M. 1997. "Relationship-Specific Psychological Mechanisms", in Bock, G.R. & Cardew, G., eds., Characterizing Human Psychological Adaptations. Chichester, England: John Wiley & Sons.
Wilson, M., Daly, M., & Weghorst, S.J. 1981. "Differential Maltreatment of Girls and Boys", Victimology, 6, 249-261.
Zervas, L.J. & Sherman, M.F. 1994. "The Relationship Between Perceived Parental Favoritism and Self-Esteem", Journal of Genetic Psychology, 155, 25-33.

CHAPTER 2

BIRTH ORDER, SIBLING COMPETITION, AND HUMAN BEHAVIOR

FRANK J. SULLOWAY
University of California, Berkeley

Abstract: Sibling competition is widespread among bird and animal species and sometimes leads to siblicide. By influencing the strategies that siblings employ in their struggles for dominance, birth order affects the outcomes of such contests. In our own species, birth order is a proxy for disparities in age, physical size, and status, all of which contribute to personality. In addition, birth order is related to the roles and niches available to offspring within the family system. On average, firstborns—who tend to act as surrogate parents—are more conscientious than laterborns, whereas laterborns are more agreeable, extraverted, and nonconforming. As strategies for dealing with rivals in a dominance hierarchy, as well as for optimizing parental investment, these sibling differences are consistent with a Darwinian perspective on family life. So are other links between personality and family dynamics, particularly those associated with parental investment and parent-offspring conflict. In adulthood, human behavior continues to reflect these formative influences, although such behavioral dispositions generally need to be catalyzed by appropriate situations in order to be fully expressed.

1. THE BIOLOGY OF SIBLING COMPETITION

A wide variety of animal species exhibit birth-order differences in behavior, usually in competition for parental investment. These behavioral effects are influenced by two distinct kinds of biological causes: *ultimate* and *proximate*. Ultimate causes include adaptive tendencies that have evolved by natural selection. Proximate causes comprise influences operating during the lifetime of the organism and encom-

H.R. Holcomb III (ed.), Conceptual Challenges in Evolutionary Psychology, 39–83.

pass biological as well as environmental factors, which almost always interact with one another. For example, some avian species possess an instinct to migrate during the autumn and spring, an adaptation that has its ultimate cause in natural selection. Temperature and day length, along with the various neuropsychological mechanisms they trigger, supply the proximate causes of bird migration (Mayr, 1961).

Viewed in these conceptual terms, a biological propensity to engage in sibling rivalry is one of the ultimate causes of personality development. Darwin's theory of natural selection explains this part of the story, which focuses on the biological dispositions that most offspring have to compete for parental favor. As William Hamilton (1964) recognized, natural selection maximizes inclusive fitness. This form of Darwinian fitness can be defined as an organism's own reproductive success, together with its contribution to the reproductive success of close relatives, discounted according to their coefficient of relatedness. On average, siblings share half of their genes. Hamilton's theory asserts that siblings will compete for scarce resources whenever the benefits of doing so are greater than twice the costs. Competition for parental investment is the main cause of sibling rivalry.

From a Darwinian point of view, sibling competition and parent-offspring conflict are flip sides of the same coin. Parents are equally related to all of their offspring and generally favor equal sharing among them, whereas siblings usually prefer a bias in their own favor. Among animals, weaning conflicts are an example of such disagreements. At the time of weaning, offspring want parents to continue investing in them, and them alone, whereas parents are inclined to reserve additional investment for future offspring (Trivers, 1974).

Darwinian and Freudian theory supply contrasting explanations for parent-offspring conflict, as Daly and Wilson (1990) have pointed out. In Freudian theory, such conflicts have their origins in the child's desire for sexual access to the opposite-sex parent—an urge that constitutes the Oedipus complex. In Darwinian theory, sexual desires have nothing to do with these conflicts. Rather, siblings compete to optimize parental investment and hence to get out of childhood alive. Siblings may be inclined to harm their rivals for parental investment, but they have no Darwinian incentive to harm a parent—at least not for the reasons Freud himself envisioned.[1] As Daly and Wilson have argued, Freud systematically misinterpreted evidence for parent-offspring conflict—which is generally nonsexual in nature—to fit his theoretical expectations.[2]

Siblicide

Fatal sibling competition has been documented in insects, fish, and mammals. Even plants exhibit competition that sometimes ends in siblicide. The Indian black plum tree (*Syzygium cuminii*) develops seeds having 25 to 30 ovules—all siblings. Only one ovule survives to become the pit of the mature fruit—usually the ovule that is fertilized first. This first-fertilized seed secretes a "death chemical" that prevents its siblings from being able to metabolize sucrose, causing them to starve to death (Krishnamurthy, Shaanker, & Ganeshaiah, 1997).

Among animals, sibling competition usually centers around parental investment. Among sea birds and predatory birds, siblicide is particularly common (Mock, Drummond, & Stinson, 1990; Mock & Parker, 1998). In some species, such as African black eagles (*Aquila verreauxi*), siblicide is "obligate"—occurring in almost every instance. The female of this species lays two eggs, and the first-hatched chick pecks its younger sibling to death during the first days of life. In one documented case, an elder chick delivered more than 1,500 pecks to its younger sibling during the latter's three-day lifespan. "In all siblicidal species studied to date," report Mock et al., "there is a striking tendency for the victim to be the youngest member of the brood" (1990, p. 445). It is noteworthy that avian parents never intervene when chicks are engaged in siblicidal aggression. The parents' own genetic interests are generally best served by raising one healthy chick rather than two undernourished ones.

Among blue-footed boobies (*Sula nebouxii*), females lay two or even three eggs. Siblicide is conditional in this species, depending on the food supply, which is therefore a proximate cause of siblicide. Aggressive pecking of a younger chick by an elder begins when the elder's body weight drops below 80 percent of normal. In experimental studies in which the necks of booby chicks have been taped to prevent them from ingesting food, aggression increases sharply and is especially pronounced in the elder chick (Drummond & García-Chavelas, 1989).

Hatching order in boobies is associated with learned behaviors. In one experimental study, junior chicks that had developed subordinate behaviors were removed from their nests and paired with smaller dominants from another nest. Size normally decides dominance in boobies. In this experiment, however, smaller chicks that had previously been seniors were successful in achieving dominance over larger but previously junior chicks. The superiority of the smaller senior chicks owed itself to their refusal to submit when attacked, and by their generally being more capable fighters than chicks that had previously been subordinate (Drummond & Osorno, 1992).

Whether siblicide is conditional or unconditional is determined by various ecological considerations that have shaped the genetic predispositions of each species. Blue-footed boobies do not have fixed territories, and, when the food supply is plentiful, they can generally rear more than one chick in a year. By contrast, African black eagles occupy fixed territories and their young require unusually large amounts of food, circumstances that limit the parents' ability to raise more than one chick, even in a very good year.

Why do avian parents regularly lay more eggs than are needed in any given breeding season? Two adaptive benefits are associated with this practice. First, an additional chick represents an insurance policy, in case an older chick dies of disease or predation. In species in which siblicide is conditional, the parents' reproductive success is enhanced whenever the food supply allows them to raise more than one offspring. Blue-footed boobies, for example, sometimes successfully rear three chicks, and many birds of prey are able to fledge two offspring.

Even in species in which siblicide is absent, proximate-causal mechanisms often regulate sibling competition. Female canaries (*Serinus canaria*) lay four or five eggs, which hatch on different days. Compared with their earlier-hatched siblings, the later-hatched chicks are physically underdeveloped. By lacing each successive egg with greater amounts of testosterone, female canaries even the competition. The fifth egg, for example, may receive twice as much testosterone as the first. Testosterone accelerates neural development and also makes chicks more pugnacious, allowing the later-hatched chicks to compete more effectively for food (Schwabl, 1996; Schwabl, Mock, & Gieg, 1997).

Specialized Adaptations for Sibling Competition

Among species that exhibit intense sibling competition, specialized adaptations have sometimes evolved to enhance the individual's chances of survival. Such adaptations for sibling competition often include weaponry in the form of teeth and other sharp structures. Tadpoles of the spadefoot toad develop teeth with which they cannibalize their broodmates (Bragg, 1954). Piglets are born with eye teeth that they shed after having competed for the sow's most nourishing teats. The earliest-born piglets head directly for the anterior teats, which are richest in milk supply, and they fiercely defend these teats against encroachment by laterborn piglets. Compared with firstborn piglets, piglets born in the latter half of the litter are twice as likely to die before the third week (Trivers, 1985, p. 23).

Another striking case of adaptation for sibling competition involves spotted hyenas (*Crocuta crocuta*). Sibling competition begins as soon as the second pup is born, and fighting has been observed while the younger pup is still encased in its amniotic sac. Unlike other carnivores, hyena pups are born with fully erupted teeth, which assist them in these brutal struggles. Among spotted hyenas, 25 percent of offspring succumb to sibling aggression. In same-sex litters, the mortality rate is 50 percent. One explanation for this difference in mortality rates draws on theories about adult competition over reproduction (Hamilton, 1967). Female hyenas compete for the right to reproduce, and offspring generally acquire the rank of their mothers. Killing a sister eliminates a close-ranking competitor (Frank, Glickman, & Licht, 1991). This hypothesis works less well for siblicide among males, who disperse in early adulthood, so one must be cautious about endorsing an adaptationist interpretation for both sexes. Still, juvenile males face the task of integrating themselves into a new clan, and physical size is positively correlated with rank and reproductive success. Singleton pups experience greater rapid weight gain during the first year of life. As adults, they may also enjoy a reproductive advantage in competition with other males.

Evolution sometimes leads to specialized adaptations promoting sibling cooperation. In the Taiwanese aphid (*Pseudoregma alexanderi*), offspring exist in two forms, one of which is a soldier caste that defends the other caste from attack. Because members of this soldier caste remain in the first larval stage, they do not reproduce (Trivers, 1985, p. 42). Such morphological adaptations for altruistic behavior are explained by Hamilton's theory of kin selection. Because Taiwanese aphids reproduce parthenogenetically, offspring carry the same genes. The soldier caste's genes are therefore passed on by their reproductively active twins.

Hamilton's theory was prompted by his own study of another particularly altruistic insect group, namely, the social insects. Owing to the unusual genetic system of these species—called haplodiploidy—females are more closely related to their sisters (by 3/4) than they are either to their brothers (by 1/4) or to their own offspring (by 1/2). The unusually cooperative nature of social insect societies revolves around the fact that sisters, who do most of the work, suppress their reproductive potential in favor of a queen, who rewards their self-sacrifice by producing more sisters (Hamilton, 1964).[3]

2. BIRTH ORDER AND SIBLING COMPETITION AMONG HUMANS

Just as with many species of lower animals, human offspring compete for parental favor. Birth order is just one of many factors that influence the ways in which this competition is expressed. By itself, competition among siblings does not cause birth-order differences in personality. But birth order is a powerful proximate (environmental) source of sibling strategies. These strategic variations arise because birth order is correlated with differences in age, physical size, power, and status within the family. These disparities cause siblings to experience family relationships in dissimilar ways and to pursue differing ways of maximizing their parents' investments in their welfare.

Competition for parental love has been an important driving force in human evolution, just as have been parental decisions about how to invest in offspring. Before 1800, half of all children did not survive childhood, and differences in parental favor, mediated through nutrition and health care, influenced which children reached adulthood (Voland, 1988, 1990). Children living long enough to become the eldest in a family were often a better Darwinian bet for their parents. Having survived the most perilous years of life, these children were more likely than their younger brothers and sisters to reach the age of reproduction and to pass on their parents' genes. In every society surveyed by anthropologists, eldest children are accorded higher status (Rosenblatt & Skoogberg, 1974). For example, many traditional societies permit infanticide, especially when a child is deformed or when a slightly older infant is still breast-feeding, but no society allows the killing of the older of two siblings (Daly & Wilson, 1988, pp. 41-46).

Parental investment strategies tend to be variable because parents themselves do not always share the same interests and because birth order is only one of many relevant factors in these decisions. In addition to taking into account the relative quality of their offspring, parents may invest differentially in children based on such considerations as the parents' age and available resources. Leaving property exclusively to the eldest child or son (primogeniture) is a policy that has been practiced by affluent parents in agrarian societies, where wealth is tied to land and where talent does not matter much. This inheritance system is much less common in mercantile societies where wealth can be acquired rapidly through entrepreneurship. Under these conditions, parents tend to invest equally in all of their offspring (Hrdy & Judge, 1992; Sulloway, in press-b).

Even when parents do not favor one child over another, sibling competition influences the dynamics of family life because it promotes diversity. Such competition generally involves the cultivation and exploitation of family niches that correspond to differences in birth order. That families provide offspring with a series of niches is a conclusion that is also suggested by research in behavioral genetics (Plomin & Daniels, 1987). During the last two decades psychologists have discovered that brothers and sisters raised together are almost as different in their personalities as people who grow up in separate families. From studies of twins raised together and apart, behavioral geneticists have concluded that only about 5 percent of the variance in individual personality traits is attributable to the shared environment—that is, growing up in the same family—whereas 35 percent is associated with the nonshared environment. About 40 percent of the overall variance is believed to be genetic, and the remaining 20 percent is attributable to errors of measurement (Loehlin, 1992).

By suggesting that the family is not a single environment, but rather a collection of microenvironments or "niches," these research findings have begun to reshape the understanding of personality development. The main reason why the shared family environment does not have a substantial impact on personality is that very little of the family experience is actually shared. For example, siblings often interpret shared experiences differently, something that is reinforced by the circumstance that brothers and sisters are at different ages when they experience the same events within the family. One particularly important and systematic source of nonshared experiences is birth order.

3. BIRTH ORDER AND PERSONALITY

Ever since Charles Darwin's cousin Francis Galton (1874) reported that eldest sons were overrepresented as members of the Royal Society, psychologists have been investigating the consequences of birth order. Alfred Adler (1927, 1928) highlighted social influences on personality, including birth order, as part of his challenge to Sigmund Freud's biologically based theory of psychosexual development. Adler regarded firstborns as "power-hungry conservatives," middleborns as competitive, and youngest children as spoiled and lazy.

Psychologists have conducted more than two thousand studies on the subject of birth order since Adler set forth his own theories on the subject. Critics of this literature have rightly argued that the findings conflict and that most studies are inadequately controlled for social class, sibship size, and other background influences that, because they correlate with birth order, can lead to false conclusions. Nevertheless, meta-analysis—a technique for aggregating findings from different studies in order to increase statistical power and reliability—suggests that these differences are robust. If we consider those well-designed studies that control for sibship size and social class, meta-analysis points to consistent birth-order differences for many personality traits. These conclusions may be summarized in terms of the Five Factor Model of personality (Sulloway 1995, 1996, in press-a).[4]

Studies generally show that firstborns are more *conscientious* than laterborns, a difference that is exemplified by their being more responsible, ambitious, organized, and academically successful. Laterborns emerge as being more *agreeable* than firstborns, in the sense of being more tender-minded, accommodating, and altruistic. Differences by birth order are more limited and mixed for the three remaining dimensions of the Five Factor Model. Laterborns appear to be more *open to experience*, as expressed by their being more nonconforming and unconventional; by contrast, firstborns appear to be more open to experience in ways that reflect intellectuality. Compared with laterborns, firstborns also appear to be more *neurotic* in the sense of being temperamental and anxious about their status. Lastly, firstborns are more *extraverted* than laterborns, in the sense of being assertive and dominant; whereas laterborns are more extraverted in the sense of being fun-loving and sociable. Sociability and assertiveness are substantially different personality traits, even though they are classified together within the Five Factor Model.[5]

Firstborns tend to have higher IQs than laterborns, but this difference is small, especially after one controls for differences in sibship size. On average, IQ declines one point with each increase in birth rank. Proponents of the Five Factor Model consider IQ to be a sixth factor, largely independent of personality. The causes of these reported IQ differences are controversial and have given rise to several competing theories. According to the confluence model (also known as the resource dilution hypothesis), firstborns experience an environment that is intellectually richer than the one experienced by laterborns, who progressively dilute this environment with their own relative lack of intellectual ability (Zajonc & Mullally, 1997; Zajonc, 2001; and

Downey, 2001). Considerable evidence—both developmental and cross-cultural—appears to support the validity of this hypothesis in samples fully controlled for social class and sibship size, although some critics remain unconvinced (Retherford & Sewell, 1991; Rodgers, Cleveland, van den Oord, & Rowe, 2000; Rodgers, 2001).

Psychological Mechanisms: A Family Dynamics Model

The personality differences I have just reviewed are consistent with a Darwinian framework, albeit one that gives preeminence to adaptation through learning. Unlike the biologically driven propensity to compete with one's siblings, which is an ultimate cause of sibling conflicts, personality is shaped by various proximate causes that spur individuals to adapt themselves to the surrounding world. Firstborns often seek the favor of their parents by serving as a surrogate parent for their younger siblings. As a result, firstborns tend to be conscientious, parent-identified, and respectful of authority. Laterborns cannot babysit themselves, so they seek out an unoccupied family niche, in part by cultivating latent talents that can be discovered only through experimentation. For these reasons, laterborns are generally more exploratory, unconventional, and tolerant of risk.

Another reason for the disparate personalities of siblings is the different strategies they employ in their relations with one another. These strategies derive from behaviors that are typical of mammalian dominance hierarchies. Because firstborns are bigger than their siblings, they are more likely to employ intimidation and physical aggression; and in general they are more inclined to boss and dominate their brothers and sisters. Laterborns tend to use low-power strategies, such as whining, pleading, humor, social intelligence, offers of reciprocal altruism, and, whenever expedient, appealing to parents for help. Two or more laterborns may also join forces against the firstborn, or laterborns may team up with their elder siblings in an effort to dominate their juniors.

A Darwinian approach to personality leads to specific predictions about middle children, who lack the advantages of being either first or last. Whenever resources are scarce and children are still largely dependent on parental care, parents are expected to invest preferentially in eldest surviving children because they are the first to reproduce. Parents are also expected to invest preferentially in youngest children because these offspring are the most needy and vulnerable to disease and, after parents have ceased reproducing, are the last children they will ever have. Even when parents treat all of their offspring equally, middle children still receive fewer resources than firstborns and lastborns. This counterintuitive conclusion follows from the fact that firstborns and lastborns generally experience some period as only children, whereas middle children always share parental resources with another sibling (Hertwig, Davis, & Sulloway, 2001).

Middle children often respond to their Darwinian handicap by becoming peer oriented and independent of the family. Compared with firstborns and lastborns, middle children are less closely attached to the family, less likely to turn to their parents for help in an emergency, and less likely to report having been loved during childhood (Salmon, 1999; Salmon & Daly, 1998). Middle children are also likely to live farther away from their parents. In addition, they are less likely than their siblings to visit close kin.

Because only children experience no sibling rivalry, they are not driven to occupy a specific family niche and effectively represent a controlled experiment in birth-order research. Like other firstborns, they are generally achievement oriented and conform to parental authority, because these attributes are esteemed by parents. Contrary to psychological folklore, only children do not appear to be less sociable or more neurotic than other children (Ernst & Angst, 1983, p. 259).

Schachter, Gilutz, Shore, and Adler (1978) found that there is a greater difference in personality and interests between a firstborn and a secondborn child, or between a secondborn and a thirdborn, than there is between the firstborn and thirdborn. The reason is that sibling competition promotes mutual differentiation in order to avoid direct conflicts, and children who are farther apart in age have less need to compete. This process of sibling differentiation (or "deidentification") extends to relationships with parents. When a firstborn identifies more strongly with one parent, the next younger sibling is likely to identify more strongly with the other parent (Schachter, 1982).

Some of these contrasts are remarkable. Although the following evidence is purely anecdotal, it illustrates trends that have been documented in formal scientific studies. The youngest of three children, Voltaire (François-Marie Arouet) had an acrimonious relationship with his elder brother Armand. The elder brother became a follower of the Jansenists, a fanatical Catholic sect. Voltaire was particularly repelled by Armand's attempt to forgo life's pleasures in order to win God's grace. As a leader of the French Enlightenment, Voltaire became noted for his attacks on the Catholic Church. He also chose literature as a profession in part to spite his brother, whom he had repeatedly defeated in impromptu poetry contests devised by his family.

Another example of sibling contrasts involves the consumer rights advocate Ralph Nader and his three older siblings. In early adolescence, the Nader children divided a globe of the world into four equal portions and assigned one to each sibling. Thereafter, each specialized in the culture, history, and languages of his or her own quarter of the globe. By cooperatively pooling their resources as a family unit, the Naders were able to learn far more about the world than if they had all chosen to compete directly with one another in the same domain. As Darwin recognized in the *Origin of Species* (1859), diversification is an effective way to reduce competition while also realizing the benefits arising from the division of labor.

In addition to affecting personality and interests, sibling deidentification has been shown to influence social attitudes. Among Chinese families living in Indonesia, Skinner (1992) found that traditionalism and filial loyalty were both related to firstborn status. He also found that social attitudes followed a zigzag pattern among siblings, with subjects generally being higher or lower on these attributes in direct contrast to the social attitudes of their nearest siblings in age.

Owing to this general process of sibling deidentification, laterborns may become more socially conservative than firstborns if parents themselves are unusually liberal. This is because the eldest child is likely to adopt the parental perspective on social issues (which, in these families, will be the "conservative" thing to do). If younger siblings wish to be different, they must adopt a more conservative social position. Much of the tendency toward rebellion among laterborns is probably attributable to sibling deidentification rather than to rejection of parental authority. In rebelling, laterborns are often repudiating the "surrogate parental" authority of their elder siblings rather than the authority of parents per se. For this reason, the nature of "rebellion" needs to be considered within the general context of family values (Sulloway, 1996, p. 507).

In sum, at least four causal mechanisms are candidates for explaining the environmental sources of sibling differences in personality and social behavior (Table 1). Some sibling differences are attributable to disparities in parental investment. Other sibling differences are associated with the occupation of disparate family niches. Still other behavioral differences have their origins in sibling-sibling interactions. Lastly, some differences arise because of deidentification among adjacent siblings. Each of these four behavioral mechanisms

Table 1. A family dynamics model of sibling differences in personality and social behavior

Causal Mechanism	Expected Sibling Differences	Direction of Expected Trends
1. Parental Investment	Compared with siblings receiving less parental investment, siblings receiving more investment are expected to be more conscientious, agreeable, and extraverted (in the sense of having positive emotions), less neurotic, and less open to experience (especially in the sense of being unconventional and rebellious). Also, offspring receiving greater parental investment should tend to identify more closely with the family and to be more strongly attached to it.	**Predominantly quadratic trends**, reflecting greater parental investment in firstborns and lastborns; but trends are expected to vary with the age of parents and also with the developmental timing of investments (Sulloway, 1996; Salmon & Daly, 1998; Hertwig, Davis, & Sulloway, 2001).
2. Niche Partitioning (firstborns as surrogate parents, laterborns as family newcomers seeking an open niche)	Siblings who act as surrogate parents should be more conscientious than those who do not. They should also be less extraverted (in the sense of being less fun-loving and excitement-seeking), less open to experience (in the sense of being less unconventional and rebellious), but more open to experience (in the sense of being more intellectually oriented—reflecting Zajonc's [1976] "teaching function"). Surrogate parents should also tend to be more bossy and hence higher in extraversion (in the sense of being more assertive) but lower in agreeableness.	**Predominantly linear trends**, based on differences in opportunities for surrogate parenting. Quadratic trends will be associated with some family niches, such as the middleborn role as "peace-maker" (agreeableness) (Sulloway, 1996, pp. 298-305, 322). Niche partitioning is also expected to occur based on other contingent circum-stances, including genetic dispositions.
3. Dominance Hierarchy Strategies	Compared with nondominant siblings, dominant ones (typically firstborns) should be less agreeable, less open to experience (in the sense of being less rebellious against authority), and less neurotic, but more conscientious and extraverted (in the sense of being more assertive).	**Predominantly linear trends**, based on sibling differences in age and size, but also reflecting individual differences in temperament (Sulloway, 1996, pp. 69, 430-31).
4. Deidentification	Among adjacent siblings, patterns of deidentification should foster small differences on most personality dimen-sions, in social attitudes, and in family sentiments.	**Zigzag trends**, as siblings seek to maximize differences between themselves (Schachter et al., 1978; Skinner, 1992; Sulloway, 1996, pp. 483, 506)

is expected to produce somewhat different trends by birth order. For example, trends associated with parental investment will tend to exhibit quadratic (U-shaped) forms, whereas trends associated with surrogate parenting and dominance hierarchies among siblings will tend to be linear. Finally, behavioral differences that owe themselves to deidentification are expected to follow zigzag patterns. Because the four behavioral mechanisms I have outlined are predicted to engender different birth-order trends, tests for the specific nature of these trends provide a potentially useful way of estimating the relative contribution of each psychological mechanism. For example, a predominantly quadratic trend in birth-order effects is not likely to be the result of differences in surrogate parenting, just as a predominantly linear trend is not likely to be the result of differences in parental investment (Hertwig, Davis, & Sulloway, 2001).

4. DIRECT SIBLING COMPARISONS

Birth-order differences in personality vary in size, and sometimes even in direction, depending on how they are measured. When assessed by self-report questionnaires, birth-order effects are generally modest and nonsignificant. Yet significant differences are typically found when parents rate their own offspring or when siblings compare themselves with one another. A comparative method of assessment has several advantages over customary methods of self-report. Direct comparison serves to anchor the scales. In addition, comparative judgments among siblings obviate any confounding effects associated with differences between families.

In one study, I employed unanchored as well as anchored scales in a survey involving 660 business leaders (Sulloway, 1999). In self-report personality ratings, firstborn CEOs did not differ from laterborns on 10 of the 11 personality traits included in my survey. After providing these self-assessments, respondents were asked to compare themselves with their siblings, using the same scales. In these direct comparisons, 8 of the 11 scales included in the survey elicited significant birth-order differences. Relative to their younger siblings, firstborn business leaders were more dominant, tough-minded, uncooperative, inflexible, conservative, conventional, temperamental, and lacking in empathy. These comparative ratings produced birth-order differences that were five times larger than those previously obtained, using unanchored scales.[6]

A much larger follow-up study involving 6,053 individuals aged 8 to

95 (M=36.8, SD=17.1) has produced similar results for a broad array of personality traits.[7] Firstborns were asked to rate themselves and an immediately younger sibling, whereas laterborns were asked to rate themselves and an immediately older sibling. Subjects made their assessments on 9-step scales using bipolar adjective pairs. Thirty adjective pairs were selected to represent the 30 facets of the NEO PI-R, a comprehensive personality inventory based on the Five Factor Model (Costa & McCrae, 1992). In direct sibling comparisons, 23 of these 30 bipolar adjective pairs yielded significant differences—all in the expected direction. As anticipated, firstborns were judged to be more conscientious than their younger siblings, whereas laterborns were judged as being more agreeable, extraverted, and open to experience.[8] For Neuroticism, a dimension for which birth-order differences were expected to be minimal and mixed, firstborns were found to be more anxious and quicker to anger, whereas laterborns emerged as more self-conscious (Table 2).[9]

After controlling for age, sex, sibship size, and social class, the partial correlation between birth order and a scale score of predicted differences was .20, with birth order accounting for 4.1 percent of the variance. Two other family background variables—sibship size and social class—each accounted for less than 0.1 percent of the variance in this same scale score of predicted differences, as did age. By comparison, sex accounted for 2.1 percent of the variance.

Controlled for the linear effect in the scale score, there was also a significant quadratic trend in all scale scores except those for Neuroticism: Middleborns scored higher than firstborns or lastborns, particularly on Agreeableness, where the quadratic trend was even larger than the linear trend. As expected, the 548 only children in my study were intermediate between firstborns and laterborns on most personality traits, although they were generally more similar to firstborns on traits associated with Conscientiousness.[10]

When correlated with the overall scale score for expected personality differences, which was designed to test specific predictions about birth order, age and sex both accounted for substantially less variance in personality scores than did birth order. It is important to note, however, that age and sex explain considerably more about personality in general than they do about trait predictions specifically related to birth order. Sex differences in my study accounted for 8.3 percent of the variance in overall dimension scores for the Big Five personality dimensions, and age explained another 2.6 percent of the variance. In accounting for 4.1 percent of the variance in these same dimension

Table 2. Birth-order effects in scale scores for the Big Five personality dimensions, based on direct sibling comparisons (Sulloway, 1999)

Personality dimension[a]	Partial correlation with birth order[b]	N	p<
CONSCIENTIOUSNESS Firstborns are more *deliberate*, *dutiful*, *effective*, *energetic*,[a] *hardworking*, *organized*, *self-disciplined*, and *under control*[a]	**-.18**	**4,507**	**.0001**
AGREEABLENESS Laterborns are more *acquiescent*, cooperative, *easygoing*,[a] *modest*, straightforward,[c] *unassertive/submissive*,[a] *tender-minded*, and *trusting*	**.10**	**4,510**	**.0001**
OPENNESS TO EXPERIENCE Laterborns are more aesthetically inclined, *prone to fantasy*, *attentive to inner feelings*, *untraditional*, *attracted by novelty*, and drawn to ideas[c]	**.08**	**4,484**	**.0001**
EXTRAVERSION[a] Laterborns are more *affectionate*, *excitement-seeking*, *fun-loving*, and gregarious	**.14**	**4,404**	**.0001**
NEUROTICISM	**-.04**	**4,278**	**.001**
Firstborns are more *anxious*, as well as more prone to depression and feelings of vulnerability.[c]	-.04	4,278	.001
Laterborns are more *self-conscious*	.05	3,548	.005
SCALE SCORE FOR PERSONALITY DIFFERENCES, AS PREDICTED[d]	**.20**	**4,177**	**.0001**

a. Each of the 30 bipolar adjective pairs, representing the 30 facets of the NEO PI-R, is classified under the Big Five personality dimension on which it has its highest factor loading. Four of the 30 adjective pairs have their highest loading on a personality dimension other than the one for which they were originally selected. "Assertive (dominant)/unassertive (submissive)" has its highest loading on Agreeableness (-.54) rather than Extraversion (.32). Similarly, "quick to anger/easygoing" has its highest loading on Agreeableness (-.55) rather than Neuroticism (.47). "Impulsive/under control" has its highest loading on Conscientiousness (-.59) rather than Neuroticism (.24). "Energetic/leisurely" has its highest loading on Conscientiousness (.47) rather than Extraversion (.28).

b. A positive partial correlation denotes a higher score for laterborns. For each bipolar trait, I have calculated a sibling difference score using z-scores and have then used this value to compute scale scores for the predicted differences. *Italicized traits* represent significant birth-order differences. (All statistical tests are two tailed.) All partial correlations with birth order (coded dichotomously as firstborn/laterborn) are controlled for age, sex, sibship size, and social class.

c. Four traits, scored as predicted, exhibit nonsignificant partial correlations in a direction opposite to the anticipated one: being straightforward (-.004), prone to depression (-.03), and inclined toward feelings of vulnerability (-.03)—all expected to be laterborn attributes; and being drawn to ideas (.004)—expected to be a firstborn attribute.

d. Some respondents made ratings on fewer than 30 bipolar adjective pairs. In these cases, scale scores have been computed from the observed data.

scores, birth order appears to be substantially more influential than age, but less so than sex. Nevertheless, on two of the Big Five dimensions (Conscientiousness and Extraversion), birth order exerted greater influence than did either age or sex.

Family Niches and Other Moderating Influences

Birth order is just a proxy for the real causes that lie behind sibling differences in personality, namely, disparities in age, physical size, status, and power within the family system. Not all firstborns choose to assume the role of a surrogate parent toward their younger siblings, and some firstborns are less bossy than others. Individual disparities in genetic endowment also help to explain why some people deviate from patterns of personality expected by birth order. In order to test the hypothesis that family niches are a principal source of sibling differences in personality, I asked the subjects in my study to what extent they acted as a surrogate parent toward their siblings during childhood, and also to what extent they bossed their siblings around. From these two variables I constructed a composite indicator, which correlates strongly with birth order (r=-.56). This composite indicator accounts for 10.5 percent of the variance in personality scores, making it a substantially better predictor of personality than any other variable in my study. The greater predictive success of this variable lies in significant part in its ability to account for exceptions based on birth order alone. For example, some laterborns (typically eldest daughters) report having engaged in substantial surrogate parenting of their younger siblings. Such people also tend to describe themselves as having firstborn personality traits.

Additional variables that affect personality and that sometimes modify the effects of birth order include sex, sex of siblings, age spacing between siblings, parental favoritism, conflict with parents, and patterns of deidentification among siblings. For example, the influence of birth order is muted when the age gap between siblings is so large that siblings do not interact much with one another and do not compete directly for the attention of parents (Sulloway, 1996, pp. 119-47).

Parent-offspring conflict and patterns of favoritism are moderately associated with individual differences in personality, in confirmation of the predictions based on a family dynamics model (see Table 1). In the study whose results I have already summarized in Table 2, parent-offspring conflict was negatively correlated with Conscientiousness,

Agreeableness, and Extraversion, and positively correlated with Openness to Experience and Neuroticism (Table 3). A related question in my study asked whether respondents were favored by their parents. As expected, respondents who reported that they were favored attained higher scores on Conscientiousness, and lower scores on Openness to Experience and Neuroticism. Also as expected, middle children were less likely than either firstborns or lastborns to report that they had been favored by their parents.[11]

Genetic factors play a substantial role in personality development (Loehlin, 1992). At least one prenatal factor that is under genetic control is linked with birth order. Among males (but not among females), laterborns are more likely to become homosexuals (Blanchard, 1997; Jones & Blanchard, 1998; and Williams et al., 2000). Unlike other behavioral attributes associated with birth order, tendencies toward homosexuality are influenced by the number of older brothers, not by relative birth rank from eldest to youngest child. In other words, lastborn males who are the eldest of their sex are no more likely to become homosexuals than are firstborns. These findings are consistent with the hypothesis that some mothers develop antibodies either to the hormones responsible for masculinizing the fetus or to one of the male-specific minor histocompatibility antigens. Such immunological responses, Blanchard and his colleagues have argued, prevent male fetuses from being fully transformed from female to male.[12]

Because personality is shaped by so many different influences—

Table 3. Personality, as it relates to parent-offspring conflict and favoritism

Personality Dimension[a]	Partial Correlation with Parent-Offspring Conflict	Partial Correlation with Parental Favoritism	*N*s	*p*s<
Conscientiousness	-.13	.08	4,839/2,112	.0001/.0003
Agreeableness	-.11	-.01	4,783/2,111	.0001/.7722
Openness	.16	-.09	5,001/2,111	.0001/.0001
Extraversion	-.05	.02	5,155/2,111	.0006/.4909
Neuroticism	.20	-.10	5,010/2,108	.0001/.0001

a. The dependent variables represent scale scores of self-ratings on each personality dimension (see Table 2). All correlations are controlled for age, sex, sibship size, and social class.

genetic and environmental—multivariate models are substantially better at predicting personality than are models that rely on single predictors such as birth order. In particular, variables that reflect within-family differences in family environments appear to explain a substantial portion of the variance in personality.[13] We currently know almost nothing about how such within-family differences in home environments interact with genetic factors, a process that is likely to moderate the influence of variables such as birth order. For example, firstborns who are physically big for their age are more likely to be effective in pursuing strategies associated with domination than are firstborns who are physically small for their age.

5. BIRTH-ORDER EFFECTS OUTSIDE OF THE FAMILY MILIEU

The method of direct sibling comparisons does not prove that reported birth-order differences in personality are real, as opposed to being the product of stereotypes.[14] Nevertheless, if the documented birth-order effects within the family are based solely on stereotypes, such stereotypes appear to be surprisingly powerful. Even if they do exist, however, birth-order stereotypes may also contribute to observed birth-order differences in behavior. Comparative assessments of personality by family members may also be susceptible to "contrast effects," whereby small differences are magnified into larger differences (Saudino, 1997). In my own study, contrast effects appear to be small, inasmuch as results are nearly the same even when variance on the scales has been reduced to a bare minimum.[15]

Other evidence, especially from studies that have documented birth-order effects in behavior outside the family, supports the conclusion that birth order is associated with real differences in personality and behavior, not just with stereotypes. In an effort to address this issue, I asked nearly two thousand subjects in my own study the following question: "What would your friends consider to be the two or three most unconventional or rebellious things, if any, that you have done during your life?" Compared with firstborns, laterborns tended to produce significantly longer written responses. They also tended to list more examples of genuinely unconventional behaviors, as assessed by two independent judges.[16] Because these findings are not based on direct sibling comparisons, and because they draw on real-life experiences, they are less likely to be affected either by contrast effects or by birth-order stereotypes. Also, most of the behaviors that subjects reported involved activities occurring outside the family of origin.

Other recent studies using the NEO Personality Inventory and comparable instruments have generally yielded null results for birth order. Most of these studies, however, have possessed only moderate statistical power and only one of them has used the method of direct sibling comparisons.[17] Basing their conclusions on similar discrepancies between self-report ratings, which often yield meager results, and more substantial differences as judged by family members, some researchers have argued that birth-order effects are *parent specific* and do not hold up outside the family (Ernst & Angst 1983, p. 171).

There is considerable plausibility to this last viewpoint, which has also been argued by Harris (1995, 1998) in connection with her theory that personality is mostly shaped by peer groups. In support of this position, studies in which family members rate one another exhibit more confirmatory findings, and larger effect sizes, than do other kinds of studies. But studies involving assessments by nonfamily members also exhibit more confirmatory findings than would be expected by chance. In addition, significant findings are especially likely to occur whenever studies involve real-life behavior rather than self-report (Sulloway, in press-a). Also, if birth-order effects are specific to childhood and the family milieu, one might well expect these differences to diminish with age, which is not the case (Sulloway, 1999).

Spousal Ratings

In order to ascertain whether birth-order differences in behavior are recognized by people other than siblings, I asked respondents to my survey to rate a variety of other people, including spouses, roommates, and friends. Among spouses, birth-order differences emerged in the expected direction, although the magnitude of these effects is smaller than it is for direct sibling comparisons by the same individuals. After one controls for age, sex, sibship size, and social class, the partial correlation between birth order and a scale score for predicted personality differences among spouses is .12 (N=822, p<.001). Significant differences in the expected direction were found for Conscientiousness, Agreeableness, and Extraversion. These findings suggest that the behavioral differences that are associated with birth order within the family are also being expressed, at about half the magnitude, in marital relationships.

Analysis of scores for individual traits reveals that subjects are detecting the same birth-order differences in their spouses that they do when they compare themselves with a sibling. An effective way of demonstrating this point is to compare the within-family effect sizes for birth order, trait by trait, with the effect sizes for the same traits as judged by spouses. These two sets of effect sizes are strongly correlated (r=.63, N=30 traits, p<.001). In other words, those traits that are the most strongly associated with birth order in sibling relationships are the same traits that are the most strongly associated with birth order in married couples (Figure 1).[18]

Roommates

Spousal ratings of personality might be affected by birth-order stereotypes being communicated to subjects by the spouse's family. Alternatively, spouses may have the opportunity to observe real birth-order differences as they watch one another interacting with their siblings. These two possibilities are less likely, however, among the college-age roommates in my study (mean age 20.5 years). After one controls for age, sex, sibship size, and social class, the partial correlation between birth order and a scale score for predicted personality traits among roommates is .16 (N=165, p<.05). As with ratings of spouses, there is substantial consistency between the birth-order effects observed among roommates and those observed among siblings (r=.76, N=30 traits, p<.0001—see Figure 1).

In short, birth-order differences do not appear to be restricted to family members or confined to childhood, as Ernst and Angst (1983) and Harris (1998) have asserted. Rather, these differences seem to manifest themselves in a variety of intimate living situations. It might still be argued that birth-order stereotypes are somehow exerting an influence on ratings by roommates and especially by spouses, and that the observed birth-order effects are therefore artifacts of these stereotypes. If this is the case, however, the same stereotypes ought to be operative in my study among those people who rated a close friend. For these peer ratings (N=1,002), birth-order effects are negligible.[19] For this reason, it seems likely that most of the birth-order effects documented in this study owe themselves to the interpersonal context in which they are being measured. In particular, birth-order effects seem to emerge most clearly in relationships where people are living together. Unlike friendships, for example, intimate living relationships share important commonalties with the behavioral circumstances under which siblings strategies were originally learned. These

Effect Sizes for Birth Order and Personality Traits: Direct Comparisons by Siblings versus Ratings by Spouses and Roommates

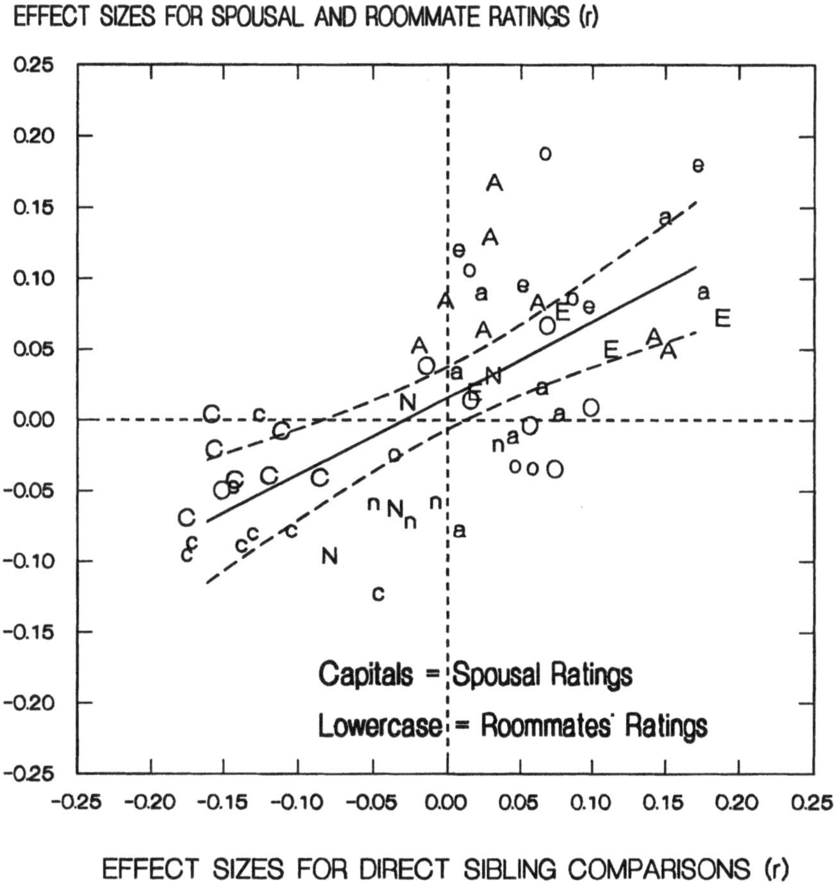

Figure 1. Birth-order effect sizes (partial correlations) for personality traits, as manifested in self-ratings based on direct sibling comparisons (N=4,177) and in ratings of spouses (N=822) and roommates (N=165). Results for 30 bipolar adjective pairs are indicated by the letter of the personality dimension on which each trait has its highest factor loading in the Five Factor Model: **C**=Conscientiousness, **A**=Agreeableness, **E**=Extraversion, **O**=Openness to Experience, and **N**=Neuroticism. Findings for spouses are indicated by capital letters; findings for roommates are indicated by lowercase letters. Within each sample, the two sets of effect sizes are strongly correlated (r=.63, p<.001, for spouses; and r=.76, p<.0001, for roommates). Traits that produce the largest birth-order effects in ratings by siblings also produce the largest birth-order effects when people rate a spouse or a roommate. Congruent clustering is particularly evident for traits related to Conscientiousness (*displayed on the left*). All correlations are controlled for age, sex, sibship size, and social class.

common elements include a shared living space, competition over shared resources, and features of a dominance hierarchy.

Perhaps the most compelling evidence supporting the existence of birth-order effects outside the family of origin comes from a meta-analytic review of the birth-order literature. More than two thousand published studies on birth order, which have most of which have been conducted outside the family of origin, have consistently shown small but significant effects (Sulloway, 1995, 1996, in press-a). What is most noteworthy about these collective findings is how closely they resemble the basic pattern of birth-order effects that I have documented in Table 2, using within-family data. For example, the correlation between the birth-order effect sizes in Table 2 and the proportion of significant results found in the overall birth-order literature for each dimension of the Five Factor Model of personality is .92. This statistic reflects more than two hundred and fifty meta-analytic outcomes (and more than nine hundred individual findings).[20]

Some critics of birth-order research, such as Harris (1998), may argue that such a high level of concordance is mostly or entirely attributable to results from *within-family* studies. For the more than two hundred meta-analytic outcomes that do not involve within-family comparisons, the correlation between the proportion of confirming outcomes for each of the Big Five personality dimensions and the results set forth in Table 1 is .95. Hence the conclusions drawn from extrafamilial studies agree closely with those based on within-family designs.

Understanding the Importance of "Small" Effect Sizes

Another common objection to arguments about the influence of birth order centers around the interpretation of "small" effect sizes. Harris (1998, p. 19) exemplifies this line of argument with her claim that "a correlation of .19, even if it is significant in the statistical sense, is all but useless." Since most birth-order effects for personality are about half the size of the particular correlation that Harris dismisses as useless, it is important to address this issue. Just how useless is a correlation of .10? Squared, a correlation of this magnitude may be said to explain one percent of the variance in the attribute being studied. Considering that about 40 percent of the variance in most personality traits appears to be genetic in origin and that another 20 percent of the variance is generally associated with errors of measurement, any nongenetic finding that explains 1 percent of the variance in a particular personality trait leaves only 39 additional influences of this magnitude before we know everything there is to know about the environmental sources of that trait. Given what is already known about the influence of age, sex, social class, and other determinants of behavior, the number of additional influences that actually explain this much environmental variance in personality has to be substantially fewer than 39.

Some behavioral scientists have effectively argued that "variance explained" grossly underestimates the importance of most behavioral phenomena (Rosenthal & Rubin 1982; Ozer, 1985). A much better way of conveying the importance of small correlations is to convert them to odds ratios, as is generally done in medicine to illustrate the effectiveness of drugs. A correlation of .10, for example, is equivalent to a medicine that increases survival among treated individuals by 49 percent. By this general metric, even very small correlations can be said to represent impressive effects. For instance, a correlation as small as .05 is equivalent to a 26 percent increase in survivorship among individuals receiving medical treatment. Most behavioral scientists would consider it noteworthy if birth order, which regularly produces correlations of this magnitude and larger, increases the likelihood of a person behaving in a certain manner by 26 percent over the base rate for this behavior. In short, most critics of birth-order research fail to understand just how substantial supposedly "small" effect sizes really are. When we are able to document such seemingly modest effects in large controlled studies, we generally ought to point to them with pride and to affirm that they are not so small after all.

It is largely because of Harris's (1995, 1998) own failure to understand the meaning of "small" effect sizes that she has put forward her dramatic thesis that the family exerts little influence on personality. Properly understood, the accumulated evidence belies her claims. Behavioral genetic studies indicate that about 5 percent of the variance in personality is attributable to the shared family environment (Loehlin, 1992). Interpreted by means of odds ratios, this particular influence is equivalent to a medicine that would increase survival among treated individuals by a factor of 2.5. In other words, siblings who grow up with warm and affectionate parents, for example, are 2.5 times more likely to partake of these traits than are children who have not done so, based on environmental influences alone. This particular measure of *shared* family influences does not even begin to take into account the even greater influence that can be ascribed to the *nonshared* family environment. (As a whole, the nonshared environment appears to exert about seven times the influence on personality that the shared environment does.) Such nonshared experiences are equivalent to a medicine that would increase survival in a medically treated population by a factor of 15. Although it has not been established exactly what proportion of these nonshared environmental influences stem from experiences within the family, this proportion is probably much greater than the 5 percent that can generally be ascribed to the shared environment. In short, the family clearly matters in personality development, in part by causing siblings to share the personality characteristics of their parents, but even more so by causing siblings to become different from one another. The influence of the family on social attitudes and values is another matter entirely, and here the effects are even more substantial, regularly explaining 30 percent or more of the variance (Eaves, Eysenck, & Martin, 1989, p. 363). Hence, based on common environmental influences, siblings from the same family are at least 12 times more likely than nonfamily members to share the same social attitudes.

Radical Revolutions in History

Other compelling evidence for birth-order differences in behavior comes from intellectual and social history. Considerable research shows that laterborns are more inclined than firstborns to change their views during times of radical political, social, or intellectual change, a tendency that reflects birth-order differences in the nonconformist component of openness (Sulloway, 1996). During the Protestant Reformation, for example, laterborns gave their lives to bring about theological change and firstborns did so to preserve orthodoxy. Compared with firstborns, laterborns were nine times more likely than firstborns to suffer martyrdom in support of the Reformed faith. In countries that turned Protestant, such as Henry VIII's England, firstborns were five times more likely than laterborns to suffer martyrdom by refusing to abandon Catholicism. (These statistics are corrected for the greater number of laterborns in the population.)

Responses of scientists to radical conceptual changes show similar differences by birth order. The Copernican revolution overthrew church doctrine by asserting that the earth rotates around the sun. During the first half-century of this controversy, laterborns were five times more likely than firstborns to endorse this heretical view. Nicholas Copernicus himself was the youngest of four children. In the period before Darwin himself became an evolutionist, younger siblings were nearly ten times more likely than eldest siblings to endorse that theory. Darwin himself was the fifth of six children, as was Alfred Russel Wallace, who codiscovered the theory of natural selection (Figure 2).

During other notable revolutions in science, including those led by Bacon, Descartes, Newton, Einstein, and Heisenberg, laterborns have generally been twice as likely as firstborns to initiate or to endorse the new point of view. This trend holds true even when the initiators of revolutions, such as Newton and Einstein, happen to be firstborns. Laterborns are more likely to endorse radical revolutions even when scientific stance has been controlled for social attitudes (which are themselves a good predictor of the acceptance of such events). Accordingly, such birth-order effects cannot be reduced to attitudinal differences, although birth order does appear to influence social attitudes, which in turn influence openness to radical innovations.[21]

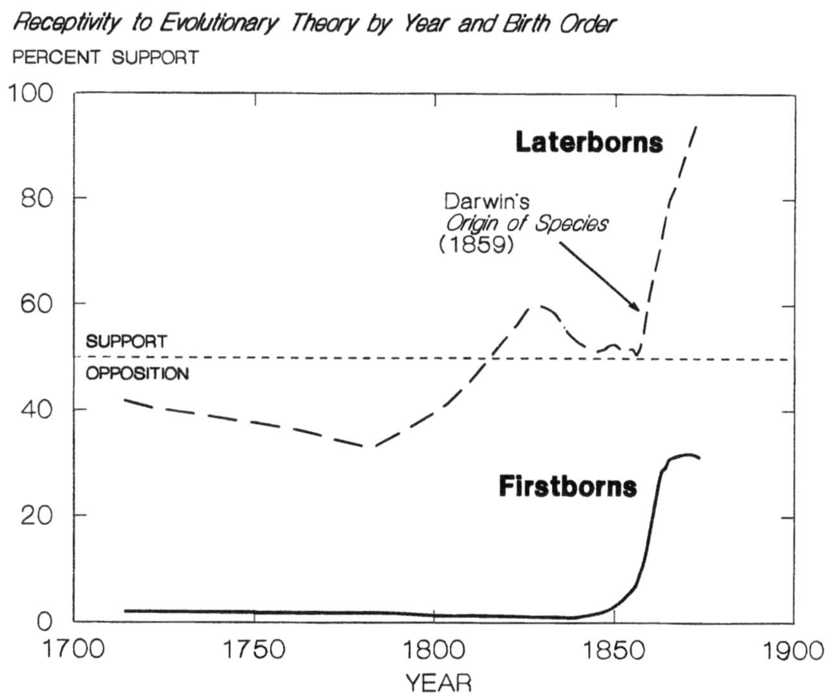

Figure 2. The reception of evolutionary theory from 1700 to 1875 by birth order (r=.35, N=263, p<.0001; controlled for sibship size and social class). During the long period of debate preceding publication of Darwin's *Origin of Species* (1859), individual laterborns were 9.7 times more likely than individual firstborns to endorse evolution. These group differences are corrected for the greater frequency of laterborns in the population. (From Sulloway, 1996, p. 33.)

Consistent with their tendency to ally themselves with their peers rather than with the family, middle children in history appear to have been particularly inclined to engage in diplomacy, cooperation, and nonviolent kinds of reforms—strategies that seem to reflect previously learned proclivities for mediating disputes between their siblings. Martin Luther King, Jr., the middle of three children, first developed his penchant for nonviolent reform by trying to prevent his younger brother from teasing their older sister (Lewis, 1970). During the French Revolution middleborn deputies within the National Convention were more likely than either firstborns or lastborns to resist the extreme measures imposed during the Reign of Terror, and I have documented a similar trend in the rejection of violent methods of reform during the American Abolitionist cause (Sulloway, 1996).

Several recent studies have confirmed these historical trends. My own findings on the Darwinian revolution have been replicated by Ronald Numbers (1998), who systematically examined those members of the National Academy of Sciences who spoke out on Darwin's theories, and by Michael Ruse, who rated 84 prominent participants in this scientific revolution drawn from his 1979 book on the subject (Sulloway, in press-a).[22] Using a contemporary sample, Salmon and Daly (1998) asked 100 middle-aged Canadian subjects, "Do you think that you are open to new and radical ideas (such as cold fusion)?" Of the firstborn respondents, 47 percent answered "yes" to this question, whereas 86 percent of the middle children answered in the affirmative, and 89 percent of the lastborns did so (partial r=.38, p<.001, controlled for age, sex, and sibship size). In another recent study Zweigenhaft and Von Ammon (2000) found that laterborns were more likely than firstborns to engage in civil disobedience during a labor dispute taking place in Greensboro, North Carolina. These two investigators also found that laterborns were more likely than firstborns to undergo multiple arrests as a result of their protests (pr=.24, N=73, p<.05; controlled for sibship size).

The tendency for middleborns to behave differently from firstborns and lastborns, seen in history during radical reforms, has also been documented in contemporary samples. Using an experimental format, Salmon (1998) found that middle children were more likely than firstborns or lastborns to identify with political speeches that employed peer-group terms such as "friend." By contrast, firstborns and lastborns preferred political speeches that appealed to them as "brothers" and "sisters." Salmon and Daly (1998) have also shown that middleborns are less invested in the family than their siblings generally are, as indicated by a variety of different measures.

It is useful to sum up the results of all known studies that have dealt with the relationship between birth order and social attitudes, or between birth order and responses to social or intellectual change. Based on 27 controlled studies involving these topics, I have found a mean-weighted correlation of r=.09 (N=11,240) for the 20 studies that reported an effect size (Sulloway, in press-a).[23] These collective findings indicate that laterborns are 43 percent more likely than firstborns to hold liberal social attitudes or to campaign for a liberal or unconventional social change. It is noteworthy that reported effect sizes appear to be substantially larger for real-life studies than they are when ascertained from self-report surveys. For the eight real-life studies, laterborns are 2.3 times more likely than firstborns to support the radical social alternative (r=.20, N=1,952).

Birth order is only one of many predictors of individual responses to social, political, or intellectual change. Being young and socially liberal are two other good predictors. So is parent-offspring conflict, which, by causing people to reject parental authority, makes some firstborns into "honorary laterborns" in their tendency toward nonconformity (Sulloway, 1996, pp. 120-33). Compared with laterborns who join radical revolutions, firstborns who do so are roughly twice as likely to have experienced high conflict with a parent.[24] History abounds with biographical examples illustrating the liberalizing consequences of parent-offspring conflict. Mao Zedong, the eldest of four children, was radicalized by strife with his father, a cruel and tyrannical man who mistreated his wife, his children, and the workers on his farm. There were two parties in his family, Mao once said, and he was the leader of the opposition (Rejai & Phillips, 1979, p. 177). In contrast to firstborns, laterborns tend to rebel even if they do not have Attila the Hun for a father or the Wicked Witch of the West for a mother. They have their older brothers and sisters to induce them to identify with the underdog.

As with multivariate models of personality, models that employ many predictors of openness to radical innovation are substantially more powerful than models using single predictors. In my own study of 28 scientific innovations during the last four centuries, the following six variables were the best predictors of scientific stance: social attitudes (pr=.24), birth order (pr=.22), age (pr=-.19), personal friendship with the leader of a revolution (pr=-.19), national differences (pr=.14), and parent-offspring conflict (pr=.07).[25]

These findings do not mean that young people, laterborns, social liberals, and people who have experienced serious conflicts with their parents have a monopoly on scientific creativity or truth. For example, laterborns and social liberals run the risk of accepting new and radical ideas too quickly, just as firstborns and social conservatives run the risk of resisting certain kinds of necessary changes until the mounting evidence in their favor can no longer be ignored. Laterborns were nine times more likely than firstborns to support Franz Joseph Gall's false theory of phrenology—the notion that character can be read by examining bumps on the head. Firstborns correctly opposed this theory as pseudoscientific. (They also rejected phrenology because of its materialistic implications.)

In the case of everyday "normal" science, firstborns possess a small but consistent advantage over laterborns. They tend to be more successful academically and are more likely to become scientists in the first place. In addition, firstborns win more Nobel prizes, which are generally awarded for creative puzzle-solving (openness in the sense of "intellect") rather than for radical innovations (openness in the sense of "nonconformity"). With their discovery of the double helix structure of DNA, Watson and Crick (both firstborns) revolutionized the field of molecular biology, but the particular form of this "revolution" is very different from most radical revolutions in science (Sulloway, 1996, pp. 330-36). A good indicator of a truly radical scientific revolution is the widespread opprobrium, not the accolades, that initially greets the instigators. Owing to publication of the *Origin of Species*, Charles Darwin lost a knighthood that he had previously been slated to receive, and he was never subsequently knighted in his lifetime. Although Darwin did receive other notable recognition (culminating with his burial in Westminster Abbey), his ultimate scientific acclaim should not obscure the essential point. Darwin's success would not have been possible without the many anxious years that he spent developing a theory whose public revelation he once compared to "confessing a murder" (Darwin, 1887, 2:23).

6. SITUATIONAL INFLUENCES

Historical Context

Whenever scientific innovations have involved ideologically conservative implications—as occurred, for example, with various vitalistic doctrines during the seventeenth and eighteenth centuries—firstborns have led the way. Firstborns readily adopted eugenics and spiritualism, theories that likewise appealed strongly to religious and political conservatives. Similarly, firstborns have repeatedly championed new theories, such as idealist theories of biological classification, that bolstered God's role in the Creation. By contrast, laterborns have gener-

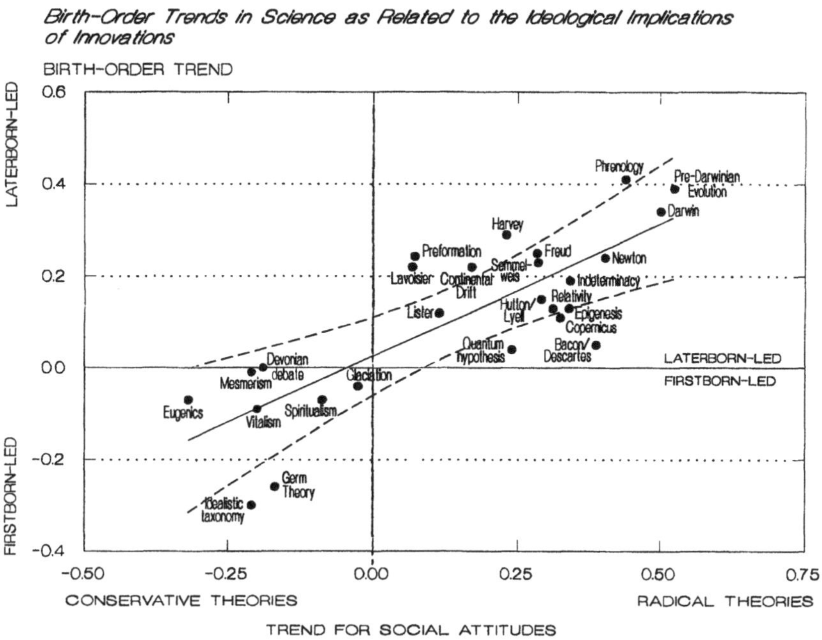

Figure 3. Birth-order effects in science, as they relate to the religious and political ramifications of 26 different innovations. The vertical axis depicts the correlation of *birth order* with support for scientific innovation (N=2,232). All events above the horizontal line (0.0) were generally endorsed by laterborns and generally opposed by firstborns, whereas all events below that line drew greater support from firstborns. The horizontal axis depicts the correlation of *social attitudes* with support for each innovation (N=1,881). Events to the left of the vertical line (0.00) were generally endorsed by social conservatives and were generally rejected by social liberals. Events to the right of this vertical line reflect support by social liberals and opposition by conservatives. Assessments of social attitudes involve more than 19,000 ratings made by expert historians, who judged the religious and political attitudes of participants in these 26 debates. The dashed lines indicate the 99-percent confidence limits for the regression line.

This analysis illustrates a general pattern: *The more socially radical the innovation, the more it is likely to be supported by laterborns and opposed by firstborns* (r=.80, N=26, p<.0001). Generally missing from the history of science are two classes of potential events. Within expected margins of statistical error, there appear to be no radical revolutions that are backed primarily by firstborns, just as there are no conservative theories that are backed primarily by laterborns.

ally led radical revolutions, such as Copernicanism and Darwinism, that strongly challenged social and religious authority. Thus the relevance of birth order to scientific innovation is highly dependent on the nature of the innovation (Figure 3).[26]

Considerable evidence reinforces the importance of situational factors in the emergence of birth-order effects. During radical scientific revolutions, birth-order effects tend to fade, as new and initially controversial ideas become more widely accepted. In addition, some new ideas are more controversial than others, and they tend to evoke correspondingly larger birth-order effects. National differences sometimes mediate these controversiality effects. Given their support of Descartes's rival theory of celestial mechanics, French physicists—especially firstborns—manifested strong opposition to Isaac Newton's theory of universal gravitation. By contrast, British scientists, including firstborns, welcomed Newton's ideas. Ultimately, the relationship between birth order and openness to experience appears to be strongly mediated by the nature of the innovation, as well as by the social and intellectual contexts in which such innovations arise.

Other Context-Sensitive Influences

Evidence from history reaffirms a particularly useful lesson about birth-order effects. As a rule, the consequences of birth order almost always depend on the behavioral context, which is one of the most important moderators of human behavior. Because of this circumstance, overt behavior—although generally a better indicator than self-report measures—can still be a problematic guide to birth-order effects. For example, a firstborn may take greater risks than a laterborn in an effort to impress an authority figure, whereas a laterborn may take greater risks in order to help a friend. Similarly, firstborns might be expected to dominate their inferiors, act graciously toward peers, and behave in a subordinate manner toward their superiors—a behavioral style known as a "pecking order personality" (Block, 1995). Personality tests are not particularly effective at capturing such context-sensitive dispositions, which means that the influence of variables such as birth order will tend to be underestimated.

Social categories such as gender, social class, marital status, and job status entail standards of behavior that often affect the expression of personality in context-sensitive ways. In connection with the extensive questionnaire study I have discussed in this chapter, more than a thousand respondents assessed the personality of a close friend who was neither a spouse nor a roommate. Overall, there were no significant differences by birth order—a confirmation of the null hypothesis. Significant birth-order differences did emerge, however, in interaction with other variables. Respondents saw their peers as exhibiting the kinds of differences expected by birth order if these peers were from middle- and especially upper-class families. In addition, birth-order effects emerged in older subjects more strongly than they did in younger (college-age) subjects.[27]

These kinds of contradictory results are perhaps best understood by considering the attributes that are important for success within each socioeconomic class, and during different stages of life. College students, especially from lower-class backgrounds, are likely to prefer friends who are sociable and who know how to have a good time. These criteria for popularity will tend to encourage the projection of an appropriate "persona"—one emphasizing agreeableness, extraversion, and openness to experience rather than conscientious dedication to work or career. When people graduate from college, take a job, and finally marry, they assume new life roles and greater responsibility. Within these maturing populations, birth-order effects in personality increasingly appear to conform to the expected pattern as the behavioral context becomes more consonant with their expression.

Firstborns appear to be especially affected by such life transitions. Compared with other subjects in my study, firstborns were significantly more variable in how their personalities were perceived across the social categories of age, class, and marital status. The more responsibility and status firstborns accrued—for example, by getting married—the more they were seen by their friends as manifesting a typical firstborn personality.[28] Firstborns therefore appear to conform more strongly than laterborns to age-related social expectations.

Another good example of such context-sensitive results is provided by data on dominance in interpersonal relationships. In relationships with peers, firstborns do not appear to be more dominant than laterborns. In marriages, however, spouses appear to reveal a different side of themselves because they report this expected birth-order difference. Similarly, college-age roommates report that firstborns are more dominant than laterborns.[29] In short, birth-order differences do not appear to be parent specific; but they are often situation specific, and they also seem to be catalyzed in varying ways by different stages of the life course.

These findings make sense from the perspective of evolutionary psychology (which, at the proximate-causal level, is a form of social psychology). Personality traits associated with birth order develop in the service of competition for scarce resources, principally parental investment. Adolescents do not usually command scarce resources, and adult friends do not generally compete over them. For these and other reasons, the behavioral priorities of adolescents and adult friends are substantially different from those of siblings. Similarly, in college-age populations dominant behaviors are not particularly effective routes to popularity, even if these same traits once brought success in sibling competition and may do so again in corporate board rooms, politics, and military life.

7. CONCLUSION

In species that reproduce sexually, sibling competition is widespread. Paradoxically, so is a disposition for siblings to behave altruistically toward one another. As William Hamilton (1964) has argued, these alternating behavioral dispositions are to be expected in organisms that are genetically related, but not genetically identical. Sibling competition is particularly prevalent in species that care for their young, and these conflicts sometimes lead to siblicide. By serving as an effective proxy for differences in age, physical size, and power, birth order among animal species often determines the outcomes of these contests.

Human beings are no exception to these behavioral trends. By influencing the choice of adaptive strategies within the family system, birth order and sibling competition contribute to personality in modest but apparently enduring ways. This is a Darwinian story, albeit one with a predominantly environmental twist. Siblings may be predisposed to compete for parental favor, but the particular strategies they adopt within their own family are shaped by the specific niche in which they have grown up. In general, firstborns tend to be more conscientious than their younger siblings—mainly because they serve as surrogate parents—whereas laterborns tend to be more agreeable, extraverted, and open to experience (in the sense of being unconventional). Measured as direct sibling comparisons, birth-order differences in personality are somewhat larger than those associated with age but somewhat smaller than those associated with gender. The extent to which these birth-order effects may be inflated by stereotypes is still an open question. So too is the question of whether such stereotypes reflect real behavioral propensities, and whether these stereotypes may also augment these propensities.

An even better predictor of personality than either birth order or sex is the kind of family niche that a person has occupied during childhood. This finding follows logically from the fact that birth order is an imperfect proxy for the many contingent experiences associated with family life. These experiences, not birth order, are the key to how personality develops in individual cases. Hence birth order and family niches more generally are part of a larger and still inadequately understood story about the nature of sibling interactions and their effects on personality development.

Birth-order studies that employ direct sibling comparisons produce significantly larger effects than do self-report measures using unanchored scales. This circumstance is sometimes taken as evidence that birth-order effects are confined to behavior within the family. Nevertheless, ratings by roommates and spouses reveal similar results by birth order, although effect sizes for these relationships are generally smaller than are found using direct sibling comparisons. In ratings of peers, where birth-order effects are smaller still, the effects nevertheless appear to be moderated by variables such as age and social class, emerging most strongly among upper-class subjects and among older subjects in all social classes. Such disparities in the findings for various subpopulations suggest that traits related to birth order are closely associated with the behavioral context, especially contexts involving intimate living situations and dominance hierarchies.

Given this kind of evidence, it would seem that a particular behavioral situation is likely to elicit birth-order effects in adulthood if it resembles situations previously experienced within the family. For example, firstborns do not usually dominate their adult friends and peers; but when placed in a position that calls for the exercise of authority, these same firstborns might be expected to behave more assertively than younger siblings generally would. It also seems likely that birth-order differences will manifest themselves in adult behavior whenever patterns of identification dating from childhood—either with parents and authority or, conversely, with the underdog—are tapped by real-life situations. Radical revolutions in history, which have regularly involved challenges to family authority, provide repeated demonstrations of this assertion.

In adult behavior, a particularly important key to documenting the influence of birth order and family niches is to examine people in multiple contexts. This conclusion is consistent with a Darwinian perspective on human behavior, which sees personality as a collection of strategies for dealing with the diverse problems associated with survival and reproduction. As adults, we are far more flexible in our strategies than we once were as children, but we still seem to take advantage of childhood experience whenever it suits our purposes. The elusive continuity between early family experience and how we behave as adults appears to lie in these strategic and context-sensitive links.

ACKNOWLEDGMENTS

For advice in connection with the research on which this article is based, I thank Paul T. Costa, Jr., Bruce Ellis, Eva Dreikurs Ferguson, Lynn Gale, Oliver John, Jerome Kagan, Robert R. McCrae, Lincoln Moses, Michael Shanahan, Phil Starks, and an anonymous peer reviewer. I am also indebted to Percy Rohde for an especially close reading of the text. For their assistance in collecting some of the data discussed in this article, I thank James Brody, Susy Carella, Kevin MacDonald, Ralph Miller, Donna Milmore, Michael Shermer, and David Sloan Wilson. I also thank the following organizations for their help with data collection: the American Academy of Achievement (Washington, D.C.), the American Association of University Women (Malden, Massachusetts), Aubin International (Boston), the Cape Cod Institute (Wellfleet, Massachusetts), the Carnegie Endowment for International Peace (Washington, D.C.), the Milton H. Erickson Foundation (Phoenix), the Institute for Family Enterprise (Montreal), the John D. and Catherine T. MacArthur Foundation (Chicago), the Skeptics Society (Altadena, California), and the Young Presidents Organization (Dallas). This chapter was partially prepared while the author was Frederick Redlich Fellow at the Center for Advanced Study in the Behavioral Sciences. I am grateful for financial assistance furnished by the Center's Foundations Fund for Research in Psychiatry and by National Science Foundation Grant SBR-9022192.

NOTES

[1] As Trivers (1974) has argued, offspring do sometimes have Darwinian reasons to harm their parents, by increasing their own fitness at the expense of their parents' fitness. See also the work of Haig (1993), who has shown how the fetus competes with its mother (and future siblings) for nutritional resources—by secreting hormones, for example, that block the mother's production of insulin and thus increase the amount of sugar that the fetus receives through the mother's blood.

[2] Freud's error about the sources of parent-offspring conflict was part of a

more extensive series of mistakes stemming from his considerable reliance on nineteenth-century biology. Although Freud's overall thinking about human development was deeply imbued with evolutionary ideas and assumptions, his most important theoretical assumptions were non-Darwinian. In particular, Freud's theory of psychosexual development was fundamentally compromised by its reliance on Lamarckian theory and the biogenetic law (Sulloway, 1979, 1982; see also Gould, 1977).

[3] Although haplodiploidy has facilitated the evolution of cooperative behavior in social insects, this genetic peculiarity is neither necessary nor sufficient for the evolution of social organisms. Some bird species (for example, bee eaters) and mammals (for example, naked mole rats) have evolved a high degree of cooperation in their breeding systems without haplodiploidy, and many haplodiploid insects have not done so (Krebs & Davies, 1997).

[4] On the Five Factor Model of personality, see McCrae and Costa (1987), John (1990), Costa and McCrae (1992), McCrae and John (1992).

[5] For critical discussions of the meta-analytic findings on birth order, see Modell (1997), Harris (1998), Townsend (in press), and Sulloway (in press-a).

[6] In my study of CEOs, birth order was significantly correlated with only one variable using unanchored scales—laterborns were more socially liberal than firstborns (r=.12, N=623, p<.005; controlled for age, sex, sibship size, and social class). In addition, birth order was not significantly correlated with a scale score of predicted outcomes for all 11 of the self-reported personality differences (r=.04, N=592; controlled for the same background variables). When I administered direct sibling comparisons with these same subjects, however, the scale score for the same 11 variables yielded a substantially larger partial correlation (r=.20, N=524, p<.0001; controlled for age, sex, sibship size, and social class). The three variables that did not exhibit significant birth-order differences by either method of assessment were being disorganized, sociable, and feminine (predicted to be laterborn traits).

[7] The 6,053 subjects in this study are drawn from 47 different samples from the United States, Canada, Latin America, and Germany (Sulloway, 1999). Thirty-one of these samples comprise people who attended a lecture by the author on the subject of birth order (N=3,269). All questionnaires were filled out before the lecture. Responses from another 2,784 subjects were collected by colleagues at 15 different universities and professional organizations, and through a mailing by the Skeptics Society to 10,000 nationally representative addresses. The two major subsamples make it possible to test for (1) possible inflation of birth-order effects by stereotypes and (2) possible self-selection bias among people who specifically came to hear a lecture on the subject of birth order. To test for these possible effects, I have conducted formal tests of heterogeneity on all of the samples. Overall, there is no significant heterogeneity in the birth-order findings from the individual subsamples. More specifically, there is no significant heterogeneity between the results for subjects who attended a lecture on the subject of birth order and subjects

who did not. In addition, subjects who reported having previously heard about my birth-order research in the news media responded similarly on their surveys compared with other subjects.

Of the 6,053 questionnaires that were collected, 726 (12 percent) contained insufficient data to provide usable responses. Also excluded from the statistical analyses in Table 2 were (1) 71 twins; (2) 112 subjects who reported an age gap of more than nine years between themselves and the sibling they also rated; (3) 86 subjects who reported that their functional birth orders differed from their biological birth orders (owing to adoption, the acquisition of stepsiblings, or the death of a sibling); and (4) 548 only children. As expected, birth-order effects were significantly reduced among the first three groups compared with other subjects. The findings for only children are discussed in note 10.

[8] As McCrae (1994) has shown, Openness to Experience exhibits two general components—openness in the sense of "intellect" and openness in the sense of "nonconformity." The "intellect" component of openness is more strongly correlated with IQ and years of education than is the second, or "nonconformist," component. Because firstborns tend to have higher IQs than laterborns, and because firstborns also tend to excel at academic pursuits, they are expected to score higher than laterborns on openness in intellect, reflected most strongly by the Ideas facet of the Five Factor Model. By contrast, laterborns are expected to score higher on the other facets of this dimension, especially Values. On some questionnaires I included the additional adjective pair "unintelligent/intelligent," which loads highly on Openness to Experience (r=.40). As expected, firstborns were rated as being more intelligent than laterborns (pr=-.15, N=1,641, p<.0001; controlled for age, sex, sibship size, and social class, and based on sibling difference scores). Ratings for this adjective pair also load highly on Conscientiousness (r=.58).

[9] All of the five scale scores except Conscientiousness exhibit significant heterogeneity for their component trait scores. In some cases, these heterogeneities are capable of canceling out birth-order effects, or even reversing them, unless proper adjustments are made. For example, if the trait "assertive (dominant)/unassertive (submissive)" is grouped in my study with the bipolar markers used to define Extraversion (for which it was originally picked as a marker), birth-order effects almost vanish, declining from r=.14 to r=.04 (see Table 2, note a). If the adjective pair "unintelligent/intelligent" is included with Openness to Experience, the scale-score correlation changes from a significant .08 to an equally significant -.08. Hence birth-order effects for this particular dimension can be positive or negative, depending on the degree to which "intellect" is emphasized over "unconventionality." In short, researchers who wish to investigate the influence of birth order on personality should exert caution when combining trait markers within higher-order scales, and should also first test for possible heterogeneity among the individual traits. The absence of birth-order effects in some large studies, such

as that by Ernst and Angst (1983) involving 7,696 Swiss subjects, may be due in part to such scale heterogeneities.

[10] Comparisons involving only children are based on absolute scores for predicted differences rather than on sibling difference scores. On an aggregate measure of these scores, only children are significantly different from laterborns. For the only child/laterborn contrast, $r=.06$ ($N=2{,}955, p<.001$). For the firstborn/only child contrast, $r=.04$ ($N=2{,}185, p<.09$). Both tests are controlled for age, sex, sibship size, and social class.

[11] For parental favoritism and being a middle child, $r=-.07$ ($N=2{,}047, p<.005$; controlled for age, sex, sibship size, and social class).

[12] These results regarding male homosexuality suggest that other aspects of behavior besides sexual orientation might also be influenced by the prenatal environment. If such biological effects exist, they appear to be sufficiently small that they are difficult to detect, even in large populations. In this connection, meta-analysis of the birth-order literature reveals no difference in the frequency of birth-order effects by sex, including for behaviors such as agreeableness, extraversion, and neuroticism that are closely linked to sex differences. In the study I have reviewed in Table 2, birth-order effects are just as frequent and substantial among women as they are among men. For males, moreover, number of elder brothers is not associated with gender-related traits after the analysis is controlled for birth order. In short, the link between number of elder brothers and gender-related traits appears to reflect a developmental outcome of nonpsychological origin that is confined to male homosexuals.

[13] In a canonical correlation model that includes five dependent variables (one for each of the dimensions of the Five Factor Model), the following independent variables are all significant predictors of personality and explain the amount of adjusted variance indicated in parentheses, measured via direct sibling contrasts: birth order, including linear and quadratic trends (4.3 percent of the variance), degree of bossiness toward one's siblings (16.2 percent), acting as a surrogate parent (6.8 percent), parent-offspring conflict (3.2 percent), parental favoritism (2.7 percent), closeness to parents (1.8 percent), and closeness to one's siblings (2.8 percent). These 7 predictors are all controlled for sex (which explains 7.9 percent of the variance), age (2.0 percent), sibship size (0.4 percent), and social class (0.3 percent—$N=3{,}683$ [the harmonic mean]). Note that between-family measures (such as social class and sibship size) contribute very little to the model, whereas predominantly within-family differences (birth order, sex, parent-offspring conflict, and closeness to siblings) collectively explain more than thirty times as much variance.

For 10 of the 11 predictors in this canonical correlation model (all but birth order), we cannot unambiguously infer cause and effect. For example, people who are genetically predisposed to act in a bossy manner during childhood are also likely to express aggressive traits in adulthood. Similarly, adults who are genetically predisposed to be conscientious are likely to have taken on greater responsibilities as surrogate parents during childhood. Behavioral genetic studies are needed to resolve these causal issues.

[14] Nyman (1995) has documented significant differences in the personalities that people associate with differing birth orders, which may be considered "stereotypes." Such stereotypes may also have a basis in reality (as do gender stereotypes) and may also influence behavior in their own right. Not surprisingly, the birth-order differences documented by Nyman resemble those documented in Table 2.

[15] The following formal test suggests that "contrast effects" are modest for the results presented in Table 2. Sibling-difference scores were recoded on a three-step scale, depending on whether subjects rated themselves higher, lower, or the same as a sibling on each given personality attribute. (This procedure truncates the variance at each end of the scales, thereby limiting the expression of contrast effects.) Based on this method of data coding, the partial correlation of birth order with a scale score for all predicted personality differences is reduced from .20 to .19 (controlled for age, sex, sibship size, and social class). Another test for contrast effects is supplied by using absolute scores from self-ratings rather than sibling difference scores. Using this metric, the partial correlation between birth order and a scale score for all predicted personality differences is reduced from .20 to .16 (controlled for age, sex, sibship size, and social class).

[16] Controlled for age, sex, sibship size, and social class, birth order exhibits the following correlation with length of response about "unconventional" or "rebellious" behavior (r=.05, N=2,034, p<.02). For the content of responses, rated by two independent judges, r=.06 (N=2,034, p<.01; the interrater reliability between judges is .92). For both measures combined, the partial correlation with birth order is .07, compared with a partial correlation of .08 for birth order and openness to experience measured via direct sibling comparisons (Table 2).

[17] See Parker (1998); Jefferson, Herbst, and McCrae (1998); Phillips (1998); Paulhus, Chen, and Trapnell (1999); Beer and Horn (2000); and Nicholson (2001). The study by Paulhus et al. (1999) employed direct sibling comparisons for role descriptions, such as the "rebel" of the family, and documented birth-order differences similar in magnitude to those reported here. The other five studies employed unanchored scales and produced mostly null findings.

[18] To avoid plotting some overlapping data points directly on top of one another, I have randomly displaced the data in Figure 1 to a very small degree using SYSTAT's "Jitter" command (Wilkinson & Hill, 1994). All statistical tests are based on the undisplaced data points.

[19] The partial correlation between birth order and a scale score of expected personality differences among close friends is -.02 (N=1,002, p<.50, controlled for age, sex, sibship size, and social class). For the contrast between birth-order effects in ratings of peers and those found in ratings of spouses and roommates, z=2.99 (p<.005) and z=2.14 (p<.05), respectively.

[20] The correlation (.92) between the findings reported in Table 2 and the overall pattern of birth-order outcomes for each dimension of the Five Factor Model of personality, as reported in the birth-order literature, employs the meta-analytic data summarized in Sulloway (in press-a, Table 3). Note: Correlations that are based on aggregate data for each dimension of the Five Factor Model tend to be much larger than correlations based on individual data.

[21] Evidence conflicts regarding whether laterborns are more liberal than firstborns in contemporary populations. In my study of 660 CEOs mentioned earlier, laterborns rated themselves as being significantly more liberal than firstborns (r=.12, N=623, p<.005; controlled for age, sex, sibship size, and social class). Also, in my study of 6,053 subjects from the United States, Latin America, and Germany, I obtained the following partial correlations between birth order and self-reported traits, based on direct sibling contrasts, with a positive correlation denoting higher laterborn scores for the right-hand alternative in the following bipolar adjective pairs: traditional/untraditional, pr=.08 (N=2,357, p<.001); conventional/unconventional, pr=.11 (N=590, p<.01); conservative/liberal, pr=.03 (N=2,317, p<.17). All three correlations are controlled for age, sex, sibship size, and social class. In a study that did not employ direct sibling contrasts, Freese, Powell, and Steelman (1999) found almost no significant results among approximately 1,200 subjects included in the General Social Survey.

[22] The partial correlation between birth order and support for Darwinism in Number's (1998) sample is .27 (N=66, p<.05; controlled for age, social class, and social attitudes). Controlled for social class and sibship size, the partial correlation between birth order and support for evolutionary theory in Ruse's sample is .35 (N=63, p<.005)

[23] The mean-weighted partial correlation I have reported (.09) is controlled for sibship size in all 20 instances and for social class in 17 of the 20 instances. For the 27 studies as a whole (N=14,608), the mean-weighted correlation is .07 (which conservatively assumes a correlation of .00 for all nonsignificant studies that failed to report an effect size). All 27 of these studies are controlled for sibship size, but only 19 of the 27 studies are controlled for social class.

[24] This conclusion about birth order and parent-offspring conflict among radicals is based on a reanalysis of my own data on participants in 28 major scientific controversies (r=.13, N=252, p<.05; for these data the odds ratio is 1.8 to 1 in favor of high parent-offspring conflict having occurred among radical firstborns rather than among radical laterborns, with high conflict being defined as the 75th percentile or higher, based on ratings by independent judges).

[25] These partial correlations are controlled for sibship size (pr=.00) and social class (pr=.01), neither of which variable is statistically significant. For the overall model, the adjusted multiple correlation is .46 (N=1,858 [the harmonic

mean], $p<.0001$). It is important to note that willingness to accept a radical innovation is different from either the inclination or the ability to *initiate* radical intellectual change. Still, those people who are the most likely to support radical changes are also more likely than average to instigate such changes. Hence a willingness to endorse heterodox ways of thinking is a necessary, but not a sufficient, condition for intellectual discovery.

[26] In Figure 3, birth order has been operationalized as the linear trend in relative birth rank, from first to last. The trend for social attitudes is based on expert ratings by 95 historians. Two controversies that span more than a century (preformation/epigenesis and evolutionary theory) have been subdivided into the two major debates that occurred regarding the various theoretical alternatives. The data for two sets of closely related controversies (Hutton/Lyell and special and general relativity) have been combined. In the case of two controversies for which sample size was 20 or less (Devonian theory in geology and preformation theory) I have plotted the effect sizes given by the Estimation-Maximization algorithm, which draws on more information than do bivariate correlations (Sulloway, 1996:389-93; Schafer, 1991). Owing to the use of some new data for birth order and social attitudes, and to an alternative method of operationalizing birth order, the results presented in Figure 3 differ slightly from those presented in Sulloway (1996: Figure 14.1).

[27] For the partial correlation between birth order and ratings of personality among friends, $pr=-.02$ ($N=1{,}002$, $p<.50$; controlled for age, sex, sibship size, and social class; based on the scale score for all predicted attributes). For the interaction between birth order and age, $pr=.08$ ($N=1{,}002$, $p<.01$). For the interaction between birth order and social class, $pr=.06$ ($N=1{,}002$, $p<.05$; both partial correlations are controlled for the specified background factors and main effects). At the personality dimension level, significant interaction effects were found for openness to experience and neuroticism.

[28] For the interaction between birth order and marital status as they relate to perceived personality among friends and spouses, $pr=.05$ ($N=1{,}756$, $p<.05$; controlled for age, sex, sibship size, social class, and both main effects). Longitudinal studies would be particularly useful in providing further tests of these kinds of context-sensitive relationships and hypotheses.

[29] For the relationship between birth order and dominance in marriages, $pr=-.07$ ($N=780$, $p<.05$). Among roommates, firstborns were also more likely than laterborns to be rated high in dominance ($pr=-.16$, $N=165$, $p<.05$). Both partial correlations are controlled for age, sex, sibship size, and social class.

REFERENCES

Adler, Alfred. 1927. *Understanding Human Nature*. New York: Greenberg.

Adler, Alfred. 1928. "Characteristics of the First, Second, and Third Child", *Children*, 3, 14-52.

Beer, Jeremy M. & Horn, Joseph M. 2000. "The Influence of Reading Order

on Personality Development within Two Adoption Cohorts", *Journal of Personality*, 68, 789-819.
Blanchard, Ray. 1997. "Birth Order and Sibling Sex Ratio in Homosexual versus Heterosexual Males and Females", *Annual Review of Sex Research*, 8, 27-67.
Block, Jack. 1995. "A Contrarian View of the Five-Factor Approach to Personality Description", *Psychological Bulletin*, 117, 187-215.
Bragg, Arthur N. 1954. "Further Study of Predation and Cannibalism in Spadefoot Tadpoles", *Herpetologica*, 20, 17-24.
Costa, Paul T., Jr. & McCrae, Robert R. 1992. *NEO PI-R Professional Manual*. Odessa, Fl.: Psychological Assessment Resources.
Daly, Martin & Wilson, Margo. 1988. *Homicide*. New York: Aldine de Gruyter.
Daly, Martin & Wilson, Margo. 1990. "Is Parent-offspring Conflict Sex-linked? Freudian and Darwinian Models", *Journal of Personality*, 58, 163-89.
Darwin, Charles R. 1859. *On the Origin of Species by Means of Natural Selection*. London: John Murray.
Darwin, Charles R. 1887. *The Life and Letters of Charles Darwin,* 3 vols. London: John Murray.
Downey, Douglas B. 2001. "Number of Siblings and Intellectual Development", *American Psychologist*, 56, 497-504.
Drummond, Hugh & García-Chavelas, C. 1989. "Food Shortage Influences Sibling Aggression in the Blue-Footed Booby", *Animal Behavior*, 37, 806-18.
Drummond, Hugh & Osorno, Jose Luis. 1992. "Training Siblings to Be Submissive Losers: Dominance between Booby Nestlings", *Animal Behavior*, 44, 881-93.
Eaves, L.J., Eysenck, H.J., & Martin, N.G. 1989. *Genes, Culture and Personality*. London and San Diego: Academic Press.
Ernst, Cécile & Angst, Jules. 1983. *Birth Order: Its Influence on Personality*. Berlin and New York: Springer-Verlag.
Frank, Laurence G., Glickman, Stephen E., & Licht, Paul. 1991. "Fatal Sibling Aggression, Precocial Development, and Androgens in Neonatal Spotted Hyenas", *Science*, 252, 702-704.
Freese, Jeremy, Powell, Brian, & Steelman, Lala Carr. 1999. "Rebel Without a Cause or Effect: Birth Order and Social Attitudes", *American Sociological Review,* 64, 207-231.
Galton, Francis. 1874. *English Men of Science*. London: Macmillan.
Gould, Stephen Jay. 1977. *Ontogeny and Phylogeny*. Cambridge and London: Harvard University Press.
Haig, David. 1993. "Genetic Conflicts in Human Pregnancy", *Quarterly Review of Biology*, 68, 495-531.
Hamilton, William D. 1964. "The Genetical Evolution of Social Behavior", I and II, *Journal of Theoretical Biology*, 7, 1-16; 17-52.
Hamilton, William D. 1967. "Extraordinary Sex Ratios", *Science,* 156, 477-88.
Harris, Judith Rich. 1995. "Where Is the Child's Environment? A Group So-

cialization Theory of Development", *Psychological Review*, 102, 458-89.
Harris, Judith Rich. 1998. *The Nurture Assumption: Why Children Turn Out the Way They Do*. New York: Free Press.
Hertwig, Ralph, Davis, Jennifer, & Sulloway, Frank J. 2001. "Parental Investment: How an Equality Motive Can Produce Inequality", manuscript submitted for publication.
Hrdy, Sarah Blaffer & Judge, Debra. 1992. "Darwin and the Puzzle of Primogeniture", *Human Nature*, 4, 1-45.
Hunter, John E. & Schmidt, Frank L. 1990. *Methods of Meta-Analysis: Correcting Error and Bias in Research Findings*. Newbury Park, CA: Sage.
Jefferson, Tyrone, Jr., Herbst, Jeffrey H., & McCrae, Robert R. 1998. "Associations between Birth Order and Personality Traits: Evidence from Self-reports and Observer Ratings", *Journal of Research on Personality*, 32, 498-509.
John, Oliver P. 1990. "The 'Big Five' Factor Taxonomy: Dimensions of Personality in the Natural Language and in Questionnaires", in Pervin, L.A., ed., *Handbook of Personality: Theory and Research*. New York: Guilford, pp. 66-100.
Jones, M.B. & Blanchard, Ray. 1998. "Birth Order and Male Homosexuality: Extension of Slater's Index", *Human Biology*, 70, 775-787.
Koch, Helen. 1956. "Attitudes of Young Children Toward Their Peers as Related to Certain Characteristics of Their Siblings", *Psychological Monographs*, 70, 1-41.
Krebs, John R. & Davies, Nicholas B. 1997. *Behavioural Ecology: An Evolutionary Approach,* 4th ed. Oxford: Blackwell.
Krishnamurthy, K.S., Shaanker, R. Uma, & Ganeshaiah, K.N. 1997. "Seed Abortion in an Animal Dispersed Species, *Syzygium cuminii* (L.) Skeels (Myrtaceae): The Chemical Basis", *Current Science*, 73, 869-73.
Lewis, David L. 1970. *King: A Critical Biography*. New York: Praeger.
Loehlin, John C. 1992. *Genes and Environment in Personality Development*. Newbury Park, CA: Sage Publications.
Mayr, Ernst. 1961. "Cause and Effect in Biology", *Science*, 134, 1501-1506.
McCrae, Robert R. 1994. "Openness to Experience: Expanding the Boundaries of Factor V", *European Journal of Personality*, 8, 251-72.
McCrae, Robert R. & Costa, Paul T., Jr. 1987. "Validation of the Five-Factor Model of Personality across Instruments and Observers", *Journal of Personality and Social Psychology*, 52, 81-90.
McCrae, Robert R. & John, Oliver P. 1992. "An Introduction to the Five-Factor Model and Its Applications", *Journal of Personality,* 60:175-215.
Mock, Douglas W. & Parker, G.A. 1998. *The Evolution of Sibling Rivalry*. New York and Oxford: Oxford University Press.
Mock, Douglas W., Drummond, Hugh, & Stinson, Christopher H. 1990. "Avian Siblicide", *American Scientist*, 78, 438-49.
Modell, John. 1996. "Family Niche and Intellectual Bent", review of *Born to Rebel*, by Frank J. Sulloway, *Science*, 275, 624.

Nicholson, Nigel. 2001. "Personality and Birth Order—Is There a Relationship? An Evolutionary Analysis", unpublished manuscript.
Numbers, Ronald. L. 1998. *Darwinism Comes to America*. Cambridge and London: Harvard University Press.
Nyman, Lawrence. 1995. "The Identification of Birth Order Personality Attributes", *Journal of Psychology*, 129, 51-59.
Ozer, Daniel J. 1985. "Correlation and the Coefficient of Determination", *Psychological Bulletin*, 97, 307-315.
Parker, Wayne D. 1998. "Birth Order Effects in the Academically Talented", *Gifted Child Quarterly*, 42, 29-36.
Paulhus, Delroy L., Chen, David, & Trapnell, Paul D. 1999. "Birth Order and Personality within Families", *Psychological Science*, 10, 482-88.
Phillips, Suzanne M. 1998. "Birth Order and Personality—Not Again!", paper presented at the 69th Annual Meeting of the Eastern Psychological Association, February 1998, Boston.
Plomin, Robert & Daniels, Denise. 1987. "Why Are Children in the Same Family So Different from One Another", *Behavioral and Brain Sciences*, 10, 1-60.
Retherford, Robert A. & Sewell, William H. 1991. "Birth Order and Intelligence: Further Tests of the Confluence Model", *American Sociological Review*, 56, 141-58.
Rogers, Joseph Lee. 2001. "What Causes Birth Order-Intelligence Patterns? The Admixture Hypothesis, Revisited", *American Psychologist*, 56, 505-510.
Rodgers, Joseph Lee, Cleveland, Hobart H., van den Oord, Edwin, & Rowe, David C. 2000. "Resolving the Debate Over Birth Order, Family Size, and Intelligence", *American Psychologist*, 55, 599-612.
Rosenthal, Robert. 1987. *Judgment Studies: Design, Analysis, and Meta-analysis*. Cambridge: Cambridge University Press.
Rosenthal, Robert & Rubin, Donald B. 1982. "A Simple, General Purpose Display of Magnitude in Experimenter Effect", *Journal of Educational Psychology*, 74, 166-69.
Salmon, Catherine A. 1998. "The Evocative Nature of Kin Terminology in Political Rhetoric", *Politics and the Life Sciences*, 17, 51-57.
Salmon, Catherine A. 1999. "On the Impact of Sex and Birth Order on Contact with Kin", *Human Nature*, 10, 183-97.
Salmon, Catherine A. & Daly, Martin. 1998. "Birth Order and Familial Sentiment: Middleborns are Different", *Human Behavior and Evolution*, 19, 299-312.
Saudino, Kimberly J. 1997. "Moving beyond the Heritability Question: New Directions in Behavioral Genetic Studies of Personality", *Current Directions in Psychological Science*, 6, 86-90.
Schachter, Frances Fuchs, Gilutz, Gabi, Shore, Ellen, & Adler, Michelle. 1978. "Sibling Deidentification Judged by Mothers: Cross-validation and Developmental Studies", *Child Development*, 49, 543-46.

Schachter, Frances Fuchs. 1982. "Sibling Deidentification and Split-parent Identifications: A Family Tetrad", in Lamb, M.E. & Sutton-Smith, B., eds., *Sibling Relationships: Their Nature and Significance Across the Lifespan*. Hillsdale, NJ: Lawrence Erlbaum, pp. 123-52.

Schafer, Joseph L. 1991. "Algorithms for Multiple Imputation and Posterior Simulation from Incomplete Multivariate Data with Ignorable Nonresponse", Ph.D. Dissertation, Department of Statistics, Harvard University.

Schwabl, Hubert. 1996. "Environment Modifies the Testosterone Levels of a Female Bird and Its Eggs", *Journal of Experimental Zoology*, 276, 157-63.

Schwabl, Hubert, Mock, Douglas W., & Gieg, Jennifer A. 1997. "A Hormonal Mechanism for Parental Favouritism", *Nature*, 386, 231.

Skinner, William G. 1992. "Seek a Loyal Subject in a Filial Son: Family Roots of Political Orientation in Chinese Society", in *Family Process and Political Process in Modern Chinese History*. Taipei, Republic of China: Chiang Ching-kuo Foundation for International Scholarly Exchange, pp. 943-93.

Somit, Albert, Arwine, Alan, & Peterson, Steven A. 1996. *Birth Order and Political Behavior*. Lanham, MD: University Press of America.

Sulloway, Frank J. 1979. *Freud, Biologist of the Mind: Beyond the Psychoanalytic Legend*. New York: Basic Books.

Sulloway, Frank J. 1982. "Freud and Biology: The Hidden Legacy", in Woodward, W.R. & Ash, M.G., eds., *The Problematic Science: Psychology in Nineteenth-Century Thought*. New York: Praeger Publishers, pp. 198-227.

Sulloway, Frank J. 1995. "Birth Order and Evolutionary Psychology: A Meta-analytic Overview", *Psychological Inquiry*, 6, 75-80.

Sulloway, Frank J. 1996. *Born to Rebel: Birth Order, Family Dynamics, and Creative Lives*. New York: Pantheon.

Sulloway, Frank J. 1997. *Born to Rebel: Birth Order, Family Dynamics, and Creative Lives*, revised ed. New York: Vintage.

Sulloway, Frank J. 1999. "Birth Order", in Runco, M.A. & Pritzker, S.R., eds., *Encyclopedia of Creativity*, 1:189-202.

Sulloway, Frank J. (in press-a). "*Born to Rebel* and Its Critics", *Politics and the Life Sciences*.

Sulloway, Frank J. (in press-b). "Sibling-Order Effects", *International Encyclopedia of the Social and Behavioral Sciences*.

Townsend, F. (in press). "Birth Order and Rebelliousness: Reconstructing the Research in *Born to Rebel*", *Politics and the Life Sciences*.

Trivers, Robert L. 1974. "Parent-offspring Conflict", *American Zoologist*, 14, 249-64.

Trivers, Robert L. 1985. *Social Evolution*. Menlo Park, CA: Benjamin/Cummings.

Voland, Ekart. 1988. "Differential Infant and Child Mortality in Evolutionary Perspective: Data from the late 17th to 19th Century Ostfriesland (Germany)", in Betzig, Laura, Mulder, Monique Borgerhoff, & Turke, Paul, eds., *Human Reproduction Behaviour: A Darwinian Perspective*. Cambridge: Cambridge University Press, pp. 253-62.

Voland, Ekart. 1990. "Differential Reproductive Success within the Krummhorn Population (Germany, 18th and 19th Centuries)", *Behavioral Ecology and Sociobiology*, 26, 54-72.

Wilkinson, Leland & Hill, MaryAnn. 1994. *SYSTAT for DOS*, 2 vols. Evanston, Illinois: SYSTAT, Inc.

Williams, Terrance J., Pepitone, Michelle E., Christensen, Scott E., Cooke, Bradley M., Huberman, Andrew D., Breedlove, Nicholas J., Breedlove, Tessa J., Jordan, Cynthia L., & Breedlove, S. Marc. 2000. "Finger Length Ratios and Sexual Orientation", *Nature*, 404, 455-456.

Zajonc, Robert B. 1976. "Family Configuration and Intelligence", *Science*, 192, 227-36.

Zajonc, Robert B. 2001. "The Family Dynamics of Intellectual Development", *American Psychologist*, 56, 490-96.

Zajonc, Robert B. & Mullally, Patricia R. 1997. "Birth Order: Reconciling Conflicting Effects", *American Psychologist*, 52, 685-99.

Zweigenhaft, Richard L. & Von Ammon, Jessica. 2000. "Birth Order and Civil Disobedience: A Test of Sulloway's 'Born to Rebel' Hypothesis", *Journal of Social Psychology*, 140, 624-27.

CHAPTER 3

THE IMPACT OF THE SOCIAL ENVIRONMENT ON THE EVOLUTION OF MIND

DENISE DELLAROSA CUMMINS
University of California at Davis

Abstract. Reasoning about agents constitutes more of a challenge than reasoning about objects. Agents move of their own volition and are motivated by internal states that are essentially private and hidden from view. Social animals must also adhere to social norms that are rarely explicitly represented. Competing or cooperating successfully in such a dynamic environment requires appreciating the social norms that define acceptable behavior and the intentional states that motivate the behavior of others. For these reasons, the social environment constituted a crucible that forged our cognitive capacities during evolution, capacities that color and constrain our thoughts and behaviors from birth through adulthood. These capacities include a biological preparedness to detect, acquire, and reason effectively about the implicit rules that specify what we are permitted, obligated, or forbidden to do, and a capacity for constructing hierarchically embedded mental representations of the minds of others. These capacities are prefigured in nonhuman primates in a way that make certain aspects of human cognition continuous with those of other social species.

> We have the power to defy the selfish genes of our birth...
> —Dawkins, 1976, p. 215

> So although we arrive on this planet with a built-in, biologically endorsed set of biases...we can nevertheless build lives from this base that overthrow those innate preferences. We can tame and rescind and (if need be) repress those preferences in favor of "higher" preferences...
> —Dennett, 1984, p. 45

H.R. Holcomb III (ed.), Conceptual Challenges in Evolutionary Psychology, 85–118.

1. INTRODUCTION

If you are a materialist, then you are committed (at least implicitly) to the view that

> *The mind (collection of cognitive functions) is what the brain does.*

That is, our cognitive processes are instantiated as neurological processes. And unless you are a creationist, you are also committed to the view that

> *The brain was shaped by evolution.*

If you accept these two premises, you are also committed to accepting their logical conclusion, namely, that

> *The mind (collection of cognitive functions) was shaped by evolution.*

It is uncontroversial (in most circles) that visual and motor cognitive functions evolved in response to selective pressures exacted by the physical environment. Eyes, lateral geniculate nuclei, and visual and motor cortexes are the legacies of evolutionary pressures associated with negotiating the physical environment. This paper offers some reasons for thinking that, similarly, several important cognitive functions were shaped (through evolutionary pressure) by the exigencies of the *social* environment. The legacies of this evolutionary history are (a) a *biological preparedness* to detect, acquire, and cognize effectively the structure of *social norms*, that is, the implicit rules that specify what individuals are *permitted*, *obligated*, or *forbidden* to do given their position in their social groups, and (b) a capacity for constructing hierarchically embedded mental representations of what is socially crucial but essentially hidden from view, namely, *the minds of others*. Further, the development of certain cognitive functions is (a) highly canalized, (b) early emerging, and (c) prefigured in nonhuman primates in a way that distinguishes it from many other kinds of reasoning. The implication is that social reasoning is not simply reasoning applied to the social realm, but a relatively independent/autonomous skill that is quite "old" and which can be dissociated from other reasoning skills.

2. THE EXIGENCIES OF THE SOCIAL ENVIRONMENT

2.1 The selective pressure for adhering to social norms

Let me begin by addressing the point that certain social cognitive functions are prefigured in non-human primates. Most non-human primates are social animals who live in groups of varying sizes. Neocortical volume across primate taxa correlates with mean group size, meaning that neocortex volume is related to the number of relationships a primate can keep track of in a complex social environment (Dunbar, 1992 and 1993). As far as primate taxa are concerned, larger societies seem to require larger brains.

Primate societies are characterized by kinship and rank relationships that constrain behavior in particular ways. These constraints are sometimes referred to as *implicit social norms* and they are reflected in virtually every activity, including who is allowed to sit next to, play with, share food with, groom, or mate with whom (Hall, 1964; Aruguete, 1994). Members of such a society of animals need to behave in accord with these implicit social norms in order to avoid agonistic encounters and to remain within the group. Like virtually all social mammals (including humans), this "social order" is best described as a *dominance hierarchy*, meaning that

> Certain individuals have priority of access to resources in *competitive situations.* (Clutton-Brock & Harvey, 1976).

Dominant individuals protect their privileged access to resources by *detecting violations of implicit norms* and punishing them. For example, dominant males monopolize reproduction opportunities by aggressing against or threatening to aggress against females and subordinate males who are caught socializing or consorting (e.g., Cheney & Seyfarth, 1990, p. 227; de Waal, 1982).

It is impossible to ignore dominance rank when theorizing about the evolution of cognition because *dominance rank correlates with reproductive success,* and reproductive success is the engine that drives evolution. Those who dominate resources are less likely to die of predation or starvation (Cheney & Seyfarth, 1990, pp. 33-34), and are more likely to leave living offspring (e.g., Bertram, 1976; Bygott, Bertram, & Hanby, 1979; Dewsbury, 1982; Clutton-Brock, 1988; Ellis, 1995; Fedigan, 1983; Hausfater, 1975; McCann, 1981; Nishida, 1979;

Robinson, 1982; Silk, 1987; Tutin, 1979; de Waal, 1982; Watts & Stokes, 1971). Among some species of primates (e.g., chimpanzees and baboons), the relationship is even more striking because dominance status is unstable; for this reason, the level of reproductive success achieved by any individual is directly related to the length of time during which the individual is high-ranking (Altmann et al., 1996).

If dominance rank correlated simply and strongly with size, it would be difficult indeed to see how it could have much impact on the evolution of higher cognition. But in point of fact, dominance rank does *not* correlate with size in many species of primates (Smuts, 1985; Walters & Seyfarth, 1987). Instead, attaining and maintaining a high-ranking position depends on *forming and maintaining alliances* (Chapais, 1988 and 1992; Datta, 1983a-b; Goodall, 1986; Harcourt, 1988; Harcourt & Stewart, 1987; Harcourt & de Waal, 1992; Packer, 1977; Riss & Goodall, 1977; Seyfarth & Cheney, 1984; Smuts, 1985; Uehara, Hiraiwa-Hasegawa, Hosaka, & Hamai, 1994; de Waal, 1982). These alliances are formed and maintained through *reciprocal obligations*. During contests of rank, individuals typically call for help, and non-kin allies are most likely to supply that help if the individual in question has groomed them, shared food with them, or assisted them in agonistic encounters in the past (Chapais, 1992; Cheney & Seyfarth, 1990, pp. 67-69; Prud'Homme & Chapais, 1993; Seyfarth, 1976; Seyfarth & Cheney, 1984). These reciprocal relationships have the structure of transactions in that rate of intervention by individual A on behalf of B is proportional to the rate of intervention of B on behalf of A (de Waal, 1992). Similarly, chimpanzees retaliate against individuals who are reluctant to share food (i.e., show a low rate of food distribution relative to others) either by directly aggressing against them when they themselves request food (de Waal, 1989) or by misinforming or failing to inform them about the location of food (Woodruff & Premack, 1979).

The important point is that acquiring and maintaining a high-ranking position depends less on brawn than on brain, less on the capacity to out-muscle one's opponents than on the cognitive capacities to remember individuals as well as their kinship and rank, learn what is permitted and what is forbidden given one's position in the group, detect transgressions of social norms, form alliances, monitor transactional imbalances, and recognize when the benefits of reciprocating outweigh its costs. These, I propose, are basic and crucial social cognitive functions. They are, in a very real sense, what the social brain evolved to do. Without the capacity to detect instances of cheating, one cannot dominate resources. Without the capacity to create and maintain alliances based on reciprocal obligations, one cannot acquire and maintain a high-ranking position within the social group. Finally, this analysis also implicates the capacity for detecting what is forbidden and what is permitted given one's rank. Without this capacity, subordinates risk incurring the wrath of their higher-ranking conspecifics, a situation that can (and does) result in ostracism and even death.

2.2 Social norms, cheater detection, and human reasoning

Have the exigencies of living in social dominance hierarchies left their mark on the cognitive functions of *human* primates? Three decades of research have demonstrated conclusively that people are particularly adept at deontic reasoning, that is, reasoning about what one may, ought, or ought not do (see Cummins, 1996b). Whenever one reasons about what one is *permitted*, *obligated*, or *forbidden* to do, one is reasoning deontically. This type of reasoning is distinct from discursive or theoretical reasoning in which the reasoner is required to determine the epistemic status (truth) of a rule or other description of a state of affairs. When reasoning deontically, one is less concerned with what is true than in choosing a correct or prudent course of action.

One of the most striking effects in the literature on human reasoning is what I've termed the *deontic effect:* When reasoning about deontic rules (social norms), people spontaneously adopt a violation detection strategy, that is, *they look for cheaters* (Cheng & Holyoak, 1985 and 1989; Cosmides, 1989; Cosmides & Tooby, 1992; Cummins, 1996a-d, 1997, 1998a-b, 1999; Manktelow & Over, 1991 and 1995). This extremely cogent and crucial reasoning strategy seems to be triggered almost exclusively by problems with deontic content, particularly permissions, obligations, prohibitions, promises, and warnings. The deontic effect has been observed in literally hundreds of experiments using a variety of reasoning tasks (e.g., conditional arguments, categorical syllogisms, equivalence judgments) over the course of nearly thirty years, making it one of the most reliable effects in the psychological literature.

As a simple example, imagine a friend tells you

If John goes to Baltimore, he always travels by car.
(If <p>, then <q>)

Information regarding John's recent travels is recorded on cards. One side of the card indicates where John traveled, and the other side indicates how he got there. You are shown four cards with the following information face up:

Baltimore *Pittsburgh* *Car* *Airplane*

Which card(s) must be turned over to find out whether or not the statement is true?

The typical answer on *hypothesis-testing* (truth-testing) problems like this is p and q (Baltimore and Car).

Now imagine that you are John's boss, and, in order to save money, the company requires that travel to and from Maryland be by car. *Which card(s) must be turned over to find out whether or not John obeyed the rule?*

The typical answer on *deontic* problems like this is p and not-q (Baltimore and Airplane). In other words, when reasoning about the *truth* of a conditional statement (hypothesis), people spontaneously seek to discover whether the antecedent (p) and the consequent (q) did in fact occur together. When reasoning about *social norms*, they instead spontaneously look for possible cheating. But consider this: In the hypothesis-testing case, had the "Airplane" card been turned over and "Baltimore" appeared on the other side, this would have provided incontrovertible proof that the statement was false. Yet it does not occur to us to look for potential violations of the statement in the hypothesis-testing case. The need only seems apparent in the deontic case.

Moreover, the deontic effect emerges early in human development, being in place at least by the third year of life (Cummins, 1996a; Harris & Nunez, 1996). Children spontaneously adopt a violation detection strategy when attempting to determine whether or not a social rule is being followed, but not when attempting to determine whether a conditional utterance is true or false. The magnitude of this effect is equivalent to the magnitude reported in the adult literature. And also like adults, they find it easier to recognize instances of cheating than instances that prove a rule false.

Recent work by Stanovich and West (in press) also indicates that performance on the truth-testing version of the card selection task correlates with overall cognitive ability (as measured by SAT scores), but the deontic version does not. Students in these studies performed well on deontic versions of the card selection task regardless of cognitive ability, but only those with high SAT scores performed well on the truth-testing version of the task.

Of even greater importance to the claims made here, consideration of relative social rank has a profound effect on the likelihood of adopting a cheater detection strategy. Cummins (1999) found that reasoners are far more likely to look for cheaters when checking on individuals who are lower ranking than themselves (65%) than when checking on individuals of equally high (20%), equally low (18%) or higher rank (20%) than themselves (Cummins, 1999). Social rank had no effect on the likelihood of adopting a violation detection strategy in a lie detection condition. (The percentage "p and not-q" responses ranged from 15% to 18% across the rank manipulations in the lie-detection condition). Similarly, Mealey and her colleagues found that subjects were far better at remembering low-status cheaters than high-status cheaters or non-cheaters of any rank (Mealey, Daood, & Krage, 1996). Other studies have found that increases in blood pressure associated with anger or frustration in social situations can be eliminated if the individual is given an opportunity to aggress against the person who caused their distress (target), *but only if the target is of lower status than the retaliator*; if the target is of higher status, blood pressure remains at the frustration-induced elevated level (Hokanson, 1961; Hokanson & Shetler; 1961).

Results like these make it difficult to avoid the conclusion that social dominance hierarchies constituted the crucible in which our cognitive and emotional functions were forged. So strongly does social rank influence our decision processes that a good deal of our legal code is devoted to ensuring equal access to shared resources, to discourage cronyism, and to protect the socially disadvantaged by exploitation from those of higher rank.

3. THE SELECTIVE PRESSURE FOR THWARTING SOCIAL NORMS THROUGH GUILE

3.1 Truth, deception, and reasoning about other minds in non-human primates

If it is in the interest of high-ranking individuals to detect acts of cheating in order to dominate available resources, it is also in the interest of low-ranking individuals to garner a larger share of resources by eluding detection when they cheat. This produces a fertile environment for the development of *deception*. And this is precisely what one finds in the primatology literature.

Consider the following excerpt from Menzel (1974). In this excerpt, a young female chimpanzee (Belle) was shown the location of hidden food and then was taken back to the troop. All of the chimps were then released. Over a series of trials, Belle was observed leading the others to the food, until a certain male (Rock) began taking all of the food himself. Menzel recorded Belle's responses to this situation, as follows:

> Belle, accordingly stopped uncovering the food if Rock was close. She sat on it until Rock left. Rock, however, soon learned this, and when she sat in one place for more than a few seconds, he came over, shoved her aside, searched her sitting place, and got the food. Belle next stopped going all the way (to the food). Rock, however, countered by steadily expanding the area of his search through the grass near where Belle had sat. Eventually, Belle sat farther and farther away, waiting until Rock looked in the opposite direction before she moved toward the food at all, and Rock in turn seemed to look away until Belle started to move somewhere. On some occasions Rock started to wander off, only to wheel around suddenly precisely as Belle was about to uncover some food...on a few trials, she actually started off a trial by leading the group in the opposite direction from the food, and then, while Rock was engaged in his search, she doubled back rapidly and got some food.

It is clear that Belle is using whatever cognitive capacities she has to elude detection by Rock, and that Rock counters by using whatever cognitive capacities he has to forecast her behavior (Krebs & Dawkins, 1984), thereby thwarting her attempts to deceive him.

Deception attempts such as these are not at all unusual in the primatology literature. For example, subordinate males attempting to mate with forbidden females will hide their erections, females will suppress their copulation cries, and both parties will attempt to move their clandestine trysts out of line of sight of dominant individuals (Kummer, 1988; de Waal, 1988). These clandestine trysts do indeed improve the deceivers' reproductive opportunities. Gagneux, Woodruff, & Boesch (1997) report that over 50 percent of the offspring born to females in their study group were fathered by dominant males from *other troops*. The females in question had surreptitiously disappeared around the times of their estrus and reappeared a few days later. During these times, they had apparently engaged in clandestine matings (with high ranking males from other troops).

Similar observations have been made with respect to illicit grooming and food sharing (see Whiten & Byrne, 1988b for numerous examples.) In the majority of these cases, *the deception serves the very useful purpose of allowing lower-ranking individuals to hide their forbidden activities from higher-ranking individuals.* Not only does this allow subordinates to garner a larger share of resources, it also allows them to form illicit alliances with forbidden individuals, which is crucial for changing rank.

3.2 What deceptions reveal about mental capacities

Deceptions are particularly useful data for determining the mental capacities of organisms because they can be used to infer the minimum level of *intentionality* required to perpetrate the deception (Dennett, 1988).

A zero-order intentional system is incapable of beliefs and desires, but may be capable of deception. Using an example from Byrne (1995), the hawk moth flicks open its hind wings in response to looming objects. The spots on its hind wings look strikingly similar to the eyes of a large hawk. The deception is strategic in that it is used only when something is looming at (threatening) the moth. But this sort of "deception" requires no capacity for mental representation. In contrast, a first-order intentional system is capable of intentional mental states such as

"I *believe*
that the berries are ripe."

or

"I *want*
the ripe berries."

but cannot reflect on its own mental states nor the mental states of others. To do so requires second order intentionality, a capacity to represent states such as

"I *want*
you to *believe*
that I am hungry."

or

"I *know*
that I *want*
your food cache."

A third-order system is capable of intentional states such as

"I *want*
you to *believe*
that I *know*
that the food is in location A."

or

"I *want*
you to *believe*
that I *believe*
that the food is in location A."

and so on.

Note that as the level of intentionality increases, *so does capacity for hierarchical embeddings of mental representations*. Further, as the level of intentionality increases, so does the capacity to represent the mental states of others and counterfactual representations (representations that are contrary to the reality of the situation) that can take on the roles of goal states, setting the stage for suppositional reasoning and forward planning.

3.3. Intentionality in non-human primates

Analyses of deceptions from the viewpoint of intentionality are not without controversy. For example, moving one's clandestine behaviors out of line of sight of dominant individuals (or waiting until the dominant individual looks away, as Belle does) suggests an appreciation of the seeing-knowing relation, that is, that the dominant individual cannot *know* what I'm doing if he or she can't *see* what I'm doing. So does another frequently reported observation, *misdirected attention*. Here, one refrains from looking at a desired object until another individual looks away. An intentional explanation of this phenomenon is that the deceiver inhibits his or her gaze because he or she *knows* that gazing at the *desired* object will attract the other individual's attention to the object and produce an unwanted competitive situation. An alternative and strictly behavioral explanation of these observations, however, is that the animal has learned that negative consequences occur when certain actions are performed in view of other individuals, and so those actions are inhibited while others are around and/or looking in their direction.

In order to test the capacity of non-human primates for representing the mental states of others, some researchers have observed the performance of various species of primates on carefully controlled, laboratory-conducted tasks. The results so far have been mixed. Many species of primates have been found to perform well on tasks that require facility with relational categories, that is, those that require an understanding of how objects, events, and social partners relate to each other. Examples include oddity, transitivity, relational matching, and sorting out third party dominance and kinship relationships (Tomasello, in press; Tomasello & Call, 1998).

Less clear, however, is evidence derived from tasks that require mental representation of others' mental states. For example, Povinelli and colleagues have demonstrated convincing evidence of self-recognition (when looking in mirrors) among chimpanzees but not monkeys (Povinelli, 1994). They also demonstrated unique appreciation of the seeing-knowing relation among chimps by requiring them to choose which of two hiding places to look for food, one pointed to by a human observer who had watched while the food was hidden and one who had not observed the hiding event; however, it took them from 100 to 250 trials to learn this task (Povinelli, Nelson, & Boysen, 1990). No convincing demonstration of false belief reasoning (see below) has yet been demonstrated among any species of non-human primate.

The jury therefore is still out on the capacity of non-human primates for higher-order intentionality (see, e.g., Heyes, 1998). Field studies of wild monkeys interacting with conspecifics in their natural social and physical environments present a very different picture of their capacity for higher-order intentionality than do laboratory studies of captive primates living in artificially sparse social and physical environments. Despite these differences, however, researchers on both sides of the question seem to agree on one thing, and that is the social origins of primate intelligence. As Tomasello (in press) puts it

> ...what distinguishes primate cognition is not the way individuals understand and categorize objects, events, and social partners in the environment directly, but rather the way they understand and categorize how outside objects, events, and social partners relate to one another—the understanding of tertiary relations in Piaget's (1952) terminology. Our evolutionary hypothesis is that

this ability evolved first to deal with social relationships, such as kinship and dominance among group mates, and only in special instances is it used with inanimate objects.

3.4 Truth, deception, and reasoning about other minds in human development

In contrast to the early emergence of cheater detection, the capacity for mental state attribution exhibits a more gradual developmental pace, with separate subskills developing at different rates. This type of reasoning is called *theory of mind reasoning* in the developmental literature, and it comprises a number of separate components.

For example, even two year olds can succeed at tasks that require them to grasp another's goals, desires, or preferences (e.g., Bartsch & Wellman, 1989; Flavell, Flavell, Green, & Moses, 1990; Melztoff, 1995). By three years of age, children readily succeed at tasks that require sabotage (in order to prevent an agent from achieving a goal) (Perner, 1991; Sodian, 1991). Prior to the fourth or fifth year of life, however, they fail at tasks that require intentional deception in order to manipulate the behavior of others (Perner, 1991; Sodian, 1991). In fact, children cannot reliably label an utterance as the truth, a mistake, or a lie prior to about age 3; often, they cannot even respond coherently to the question. Somewhere in the third year of life, they begin to distinguish between utterances based on their factuality; inaccurate statements are labeled as lies, regardless of whether the speaker had reason to believe the statement to be true. It is not until about the seventh or eighth year life that they can reliably take into account what a speaker believed or knew in order to distinguish lies from simple falsehoods. (See Haugaard & Reppucci, 1992 and Bussey, 1992 for reviews of this literature.)

Spontaneous self-exploration in mirror self-recognition tasks emerges quite early in human development (between 18 to 24 months of age), which is generally taken as evidence of first-order intentionality. But full emergence of even this cognitive function shows a protracted developmental pace. Povinelli (1994) videotaped 2, 3, and 4-year-olds playing a game with an adult. During the game, the experimenter surreptitiously placed a sticker in the child's hair just above the forehead. They were then allowed to watch the videotape. None of the 2-year-olds and only 25% of the 3-year-olds spontaneously reached up to search for the sticker in their hair, but 75% of the 4-year-olds did. These developmental differences occurred even though all of the children readily identified themselves as the child in the videotape.

The fourth year of life also marks the emergence of a full appreciation of the seeing-knowing relation. Two-year-olds appreciate that in order for someone else to see something, their view must not be obstructed, they must have their eyes open, and they must be looking at it; they also don't assume that others can see what they can see (Flavell, Abraham, Croft, & Flavell, 1981). But they fail at tasks that require full appreciation of the seeing-knowing relation. For example, if they are shown part of a drawing and can't guess what the full drawing depicts, and then are shown the entire drawing and allowed to identify it, children younger than about 4 will insist that a new person who is shown only part of the drawing knows what the entire drawing depicts (Taylor, 1988; Taylor, Cartwright, & Bowden, 1991).

Prior to about age 4, children also perform inconsistently on tasks that require attributing false beliefs to others. In the standard false belief task, children watch while a puppet hides a toy in location A and then leaves. A second puppet then appears, finds the toy, and hides it in location B. The child is then asked where the first puppet left the toy, where the toy is now, and where the first puppet thinks the toy is. Prior to about age 4, children answer the first two questions correctly, yet believe the first puppet thinks the toy is in the current location (location B) (see Gopnik, 1993; Gopnik & Wellman, 1994; and Leslie, 1987 and 1994 for reviews of this literature.)

Finally, it is not until about ages 5 to 7 that children can reliably perform tasks that require understanding statements that express second-order intentionality such as "John wants his mother to think that..." (Perner & Wimmer, 1985) or require appreciating a speaker's intention when distinguishing irony from "white lies" (Winner & Leekam, 1991).

The results of these diverse literatures all seem to point to the same conclusion, namely, that representing the internal states of others—a skill that requires deep embedding of mental representations—emerges later in human development than does cheater detection—a skill that depends only on inducing social norms and observing overt behavior. Further, deep embedding of mental representations is a skill at which we as humans excel compared to other species, while cheater detection is a more primitive cognitive capacity that we share with other social mammals. *The cognitive capacities that emerge earliest are also those that have the deepest evolutionary roots.*

Could the selective pressure exerted by the exigencies of surviving in a social dominance hierarchy help to explain the rise of reasoning about the mental states of others? Reasoning about agents provides more of a challenge than reasoning about objects. Agents move of their own volition, requiring constant updating of one's representations of their current locations. Their behavior is internally motivated and goal driven, and competing or cooperating successfully requires some appreciation of these motives and goals. As this discussion on deception shows, the capacity to cognitively construct an agent model is clearly adaptive in a dominance hierarchy, and is directly related to reproductive success. Such a capacity is required for sophisticated social reasoning, but not for discursive reasoning about the physical environment. Thus, we should look to the pressures of social living to discern the evolutionary origins of theory of mind reasoning. And recent studies showing that theory of mind reasoning emerges earlier in later borns than first borns suggest quite strongly that the capacity to mentally represent the internal states of others can serve the very useful purpose of thwarting dominance through guile (Jenkins & Astington, 1996; Lewis, Freeman, Kyriakidou, Maridaki-Kassotaki, & Berridge, 1996; Perner, Ruffman, & Leekam, 1994; Ruffman, Perner, Naito, Parkin, & Clements, 1998).

4. THEORETICAL EXPLANATIONS AND IMPLICATIONS

Numerous theories have been developed to explain the deontic effect (e.g., Cheng & Holyoak, 1985 and 1989; Cosmides, 1989; Cosmides & Tooby, 1992; Cummins, 1996a-d; Manktelow & Over, 1991; Oaksford & Chater, 1994) and developmental effects in theory of mind reasoning (e.g., Gopnik, 1993; Leslie, 1987; Perner, 1991). Broadly speaking, the theoretical mechanisms invoked to explain them fall into either the empiricist or nativist camp. Empiricist theories appeal to knowledge acquisition effects, the idea being that fundamental changes occur in the reasoner's representational schemes as a result of acquiring knowledge about the world (e.g., Cheng & Holyoak, 1985; Gopnik, 1993). Nativist theories instead appeal to neural or cognitive architecture (as opposed to knowledge acquisition) to explain these effects (e.g., Cosmides, 1989; Cummins, 1996a-d).

4.1 Empiricist theories

Pragmatic Reasoning Schema Theory explains the deontic effect in terms of general learning mechanisms (Cheng & Holyoak, 1985 and

1989). People induce special sets of reasoning rules through experience with frequently occurring, highly important classes of situations, such as permissions, obligations, prohibitions, causations, and the like. These learned rule sets are called reasoning schemas, and they are evoked whenever a problem with the appropriate content is encountered. The deontic effect is therefore a simple reflection of regularities in the environment, regularities that lead to the induction of domain-specific rules. No special-purpose cognitive architecture is needed to explain the effect; *induction from experience* is sufficient.

Oaksford and Chater (1994) argue that differential performance on the deontic and non-deontic versions of the card selection task can be explained in terms of a single reasoning strategy: optimal data selection. They argue that human reasoning performance should be judged rational insofar as it is optimally adapted to the environment. The truth-testing version of the card-selection task constitutes an environment in which one must decide which experiments to perform (cards to inspect) in order to choose between rival hypotheses. In such an environment, an optimal strategy is to choose those experiments (cards), which will be most informative, that is, the experiments that will maximize expected information gain. They demonstrate that under certain circumstances, the "p and q" card selection pattern constitutes optimal data selection, and that, more generally, observed selection frequencies across a variety of experimental conditions are monotonically related to expected information gain. The deontic version of the task, however, is handled differently. They refocus their probabilistic model away from truth-testing and toward rule-compliance, and emphasize maximizing expected utility through discovery of rule-violations (e.g., cheating).

Manktelow and Over (1991) also argue that people adopt different strategies when performing the deontic and non-deontic versions of the selection task. When reasoning about the deontic version, they seek to maximize subjective utility, choosing those cards that are pertinent to their interests. For example, given the rule "If you tidy your room, you may go out to play," the promisor (parent) seeks to ensure that the condition (tidying the room) has been satisfied before the action (going out to play) has been taken, while the promisee (child) seeks to ensure that the parent has made good on the promised action (allowing him or her to go out to play) once the condition has been satisfied (room has been tidied). When reasoning about the truth-testing version of the selection task, people seek to verify truth, adopting a confirmation-seeking strategy.

As is apparent from these summaries, all theorists agree that people adopt different, domain-specific strategies on the two versions of the task, and they even agree about which strategies those are: When testing truth, seek to confirm that p and q did in fact occur together, and when testing compliance, seek to identify cheaters. Essentially, there must be agreement on these points if one is to pay any attention at all to the data. The empiricist view is that people exhibit these performance patterns because they've induced optimal, domain-specific reasoning strategies through experience with these types of situations, and that in the deontic case, these strategies are intimately connected to the reasoner's goals or subjective utility.

But this concedes the major points of the nativist view, namely that social reasoning is dissociable from other types of reasoning, and the skills involved in social reasoning are fitness enhancing. They do NOT explain, however, (a) why performance on the non-deontic version correlates with cognitive ability while performance on the deontic version does not, (b) why effective deontic reasoning strategies emerge earlier than effective truth-testing and false belief- testing strategies, c) why the optimal strategy for the deontic version of the card selection task is patently obvious even to 3-year-olds, yet even philosophers of science disagree (sometimes violently) on what constitutes an optimal hypothesis-testing (truth-testing) strategy, and (d) why social reasoning performance is selectively impaired in various neurological syndromes, but (as far as I know), truth-testing is not. For example, Turner syndrome is a genetic abnormality in which a female lacks all or part of one X-chromosome. Individuals with only a maternally-inherited X-chromosome show marked social difficulties, particularly on measures of social insight and adeptness (McGuffin & Scourfield, 1997). Autism is a neurodevelopmental syndrome whose most vivid impact at the cognitive level is an impaired ability to reason about social stimuli (e.g., Baron-Cohen, 1995; Leslie & Roth, 1993). Prefrontal lobe syndrome is a pattern of impaired reasoning performance that results from bilateral damage to the ventromedial prefrontal cortical lobes (Damasio, 1994). In humans, this syndrome is characterized by an impaired capacity to reason effectively about socio/emotional stimuli, while leaving other types of intelligent reasoning virtually untouched. The selective impairment of social reasoning in each of these syndromes suggests that social reasoning may be neurologically separable from other types of intelligent reasoning, and the congenital and genetic components of autism and Turner's syndrome further suggest that we are genetically "wired" (in the normal case) to develop specialized neurological circuitry for cognizing social stimuli.

4.2 Nativist theories

I've explained these phenomena here and elsewhere (Cummins, 1996a-d, 1997, 1999; Cummins & Cummins, 1999) in terms fitness enhancing adaptations to social dominance hierarchies, an explanation I call *Dominance Theory*. The major competing nativist theory offered to explain the same phenomena is *Social Contract Theory* (Cosmides, 1989; Cosmides & Tooby, 1994). Social Contract Theory also analyzes deontic reasoning in terms of cost/benefit analysis and cheater detection. These strategies are proposed to be innate, having been selected for during the evolution of our species in order to reason effectively about social exchange (cooperative action for mutual benefit). Both Dominance Theory and Social Contract Theory explain better performance on social versions of the card selection tasks (more accurate and faster) in terms of cognitive/neural adaptations in response to evolutionary pressures, but there are important differences between the two:

(a) Unlike Social Contract Theory, Dominance Theory explains the evidence from ethology, making it a primate (or social species) adaptation, not a human adaptation. The human capacity for deontic reasoning is rooted in primitive concepts of permission, obligation, and prohibition that are apparent in the cognition of other social mammals. With respect to this aspect of social reasoning, the difference between human and non-human cognition is one of degree and not kind.

(b) Dominance Theory gives a better explanation of the selective pressure, since dominance relates directly to reproductive success. To say that a cognitive or behavioral trait was selected essentially means that the trait bestowed a reproductive advantage on those possessing it. By showing that dominance (and the thwarting of dominance) has a directly measurable impact on reproductive success, and that dominance (and the thwarting of it) in higher primates depends more on social acumen than on physical size, Dominance Theory provides a better explanation of the deontic effect than Social Contract Theory, which instead rests on conjectures about the evolutionary pressures faced by our ancestors during the Pleistocene.

(c) Dominance Theory gives a better explanation of the experimental evidence, such as the fact that differential costs don't seem to influence card selection performance (Cheng & Holyoak, 1989), but that consideration of social rank does (Cummins, 1999). According to Dominance Theory, it is not exchange of goods or services that lies at the root of the deontic effect but the concept of a *social norm*, that is, the constraints placed on one's behavior by the social group. Social animals are primed to detect deviations from social norms, particularly when they disturb the "social order." The capacity for reciprocity (as in social exchange) is the function that allows one to build and maintain the alliances that are required to improve (and maintain) one's rank. The capacity for reciprocity, therefore, is also part of the cognitive architecture that we share with other species, and is not an adaptation that is specific to humans.

(d) Dominance Theory explains natural selection effects on neurological architecture in terms of *biological preparedness* and *canalization*, which are more consistent with the developmental plasticity of the brain than are Social Contract Theory's innate cognitive modules. According to Social Contract Theory, the brain is like a Swiss Army knife, a collection of special-purpose modules that constitute separate adaptations to evolutionary pressures. These modules are apparently innate and fully developed, requiring only environmental triggers to bring them into play. As Tooby and Cosmides (1995) put it:

> [O]ur cognitive architecture resembles a confederation of hundreds or thousands of functionally dedicated computers (often called modules) designed to solve adaptive problems endemic to our hunter-gatherer ancestors. Each of these devices has its own agenda and imposes its own exotic organization on different fragments of the world. There are specialized systems for grammar induction, for face recognition, for dead reckoning, for construing objects and for recognizing emotions from the face. There are mechanisms to detect animacy, eye direction, and cheating. There is a "theory of mind" module ... a variety of social inference modules ... and a multitude of other elegant machines.

One alternative explanation is not to think in terms of modules but in terms of "fast tracked" development of social cognitive functions. In other words, rather than assume that early emerging and specialized cognitive capacities are either innate or learned, we may suppose instead that organisms do not inherit modules fully formed, but have a *biological preparedness* (Seligman, 1971) to very quickly develop specialized cognitive functions for solving classes of problems that were critical to the survival and reproductive success of their ancestors. Conceiving of cognitive functions in this way puts them on a par with other biological traits that can differ in their degree of *canalization*, that is, in the degree to which the environment plays a role in their expression (Ariew, 1996; McKenzie & O'Farrell, 1993; Waddington, 1975). (See Cummins & Cummins, 1999 for a complete exposition of this view.)

4.3 Biological preparedness and evolutionary explanation

The concept of biological preparedness is not one that psychologists readily accept; yet it is one that we invariably find ourselves having to incorporate in our theories. Early learning theories rested on the assumption that an association could be made between any two stimuli through repeated pairings, yet it soon became apparent that some associations were learned more readily than others. Further, this "fast-tracked" learning typically involved contingencies that had significant survival advantages. For example, food aversions are notoriously easy to acquire, even when the time interval between ingestion and illness is quite long (Bernstein & Borson, 1986; Etscorn & Stephens, 1973; Garcia, Brett, & Rusiniak, 1989; Logue, 1988). Primates (including humans) also seem biologically predisposed to acquire fear responses to certain classes of animals (such as snakes and spiders) rather than others (Cook & Mineka, 1989 and 1990; Öhman, 1986; Seligman, 1971). As Hilgard and Bower (1975, p. 574) put it: "One might say that the animal is innately preprogrammed to see certain cues and responses as 'naturally fitting' together, so that they are readily learned."

Similarly, the explosion of data on infant cognition that has come about since the introduction of the "looking time" paradigm has forced developmental psychologists to re-examine their assumptions about the infant mind. Rather than Piagetian sensory-motor systems, infants appear to be cognitively predisposed to interpret the world in terms of agents and objects whose behaviors are constrained by different sets of principles (Leslie, 1994; Shultz, 1982; Spelke, 1996). These predispositions "fast track" certain kinds of learning and reasoning processes throughout childhood and into adulthood.

Infants show not only a preference for looking at faces over any other stimuli; they also can distinguish among emotional facial expressions (Entremont & Muir, 1997). They also engage in *social referencing*, looking at their caregivers' reactions to novel stimuli (e.g.

Stenberg & Hagekull, 1997). It seems likely that it is just this predisposition for attending to social stimuli that "fast tracks" learning about social norms, evokes early cheater detection reasoning strategies, and facilitates the development of agent models. A biological predisposition to entrain one's attention on agents and their facial expressions ensures that infants are provided ample opportunity to notice contingencies between agents' actions and their consequences, and to learn appropriate social behavior, that is, the social norms that facilitate and constrain social interactions.

4.4 Predispositions versus innate, intact modules

Biological preparedness as an explanans (explanatory factor) is also more consistent with contemporary neuroscience than are innate modules. I have discussed this issue in some detail elsewhere (Cummins & Cummins, 1999), but will briefly summarize the discussion here. The picture of the brain that is emerging from decades of neuroscience research is one in which precious little is fixed at birth. Instead, innate processing "biases" are elicited or modified through interactions with the environment (Figueredo, 1992). For example, binocular columns (used in depth perception) are not present at birth, but appear in the visual cortex during a critical period after the infant has received visual input (Banich, 1997, p. 472). Other visual cortical cells show diffuse line orientation "preferences" at birth, firing maximally to lines of a particular orientation (e.g., vertical), but responding to lines of other orientations as well, albeit to a lesser degree (Hubel, 1988). After receiving visual input, however, these cell preferences are sharpened so that they respond maximally *only* to lines of a particular orientation (Blakemore, 1970). Further, if visual input is restricted to only a single orientation (e.g., the animal is exposed only to lines of vertical orientation), the majority of cells will shift their preferences to match their visual experiences, responding maximally to lines of vertical orientation even if their initial preferences were for lines of other orientations (Blakemore & Cooper, 1970; Hirsh & Spinelli, 1970). The animal, in short, is blind to all line orientations except that to which it was exposed during this critical period.

Like vision, language development also shows a complex pattern of interplay between innate biases and environmental input. Deaf babies will begin to babble vocally just as hearing babies do, but their babbling declines and eventually ceases, presumably because they don't receive the auditory feedback hearing babies do (Oller & Eiler,

1988). Babbling deaf babies are practicing sounds that they have never heard, a phenomenon that is perhaps best explained as the unfolding of a biological program that requires environmental feedback to fully develop. Infants are also born with the capacity to hear all phonetic contrasts that occur in human communicative systems, yet lose the capacity to distinguish among phonemes that are not marked in their language community within the first year of life (Kuhl, 1987; Eimas, 1975). There also appears to be a critical period for language acquisition that ends approximately at puberty; children who do not acquire their first language during this critical period fail to acquire the rules governing the use of grammatical morphemes and the syntactic constraints necessary for forming grammatical sentences (Curtiss, 1977). Further, this same inability to extract the grammatical rules of a natural language is selectively impaired in certain genetic disorders (M. Gopnik, 1990a and b).

As pointed out earlier, social learning (like language) also seems to depend on having the right genetically- or congenitally-wired neurological substrates. The selective impairment in social reasoning seen in Turner's syndrome and autism may occur because the neurological substrates necessary for detecting and attending to social stimuli are congenitally absent or fail to develop normally.

4.5 Biological predispositions and canalization

Biological preparedness comes in degrees, and is probably best explained in terms of *canalization* (Ariew, 1196; McKenzie & O'Farrell, 1993; Waddington, 1975). A trait is said to be more or less canalized, as its expression is more or less independent on environmental influence. A combination of genetic and environmental factors causes development to follow a particular pathway, and once begun, development is bound to achieve a particular end-state. Limb development is highly canalized in humans (humans everywhere grow limbs in the same way) but not perfectly so, as the example of Thalidomide shows. As we saw, language is highly canalized, though not so highly as limb development.

The environment can influence trait development in many different ways. The most interesting of these to the psychologist is learning. It is not often fully appreciated that learning can affect the development of even highly canalized traits. Thus language, though highly canalized, is still learned. Biology puts strong constraints on what properties a language must have in order to be learnable (as a first language), and it virtually guarantees that language will be learned in a huge variety of environments. This is what is meant by the claim that there is a specific biological preparedness for language acquisition.

Similarly, the claim that there is a biological preparedness for the acquisition of social reasoning skills is the claim that such skills are relatively highly canalized compared to, e.g., truth-testing. The claim is not that these skills are not learned, nor is it the claim that learning or other environmental influences cannot utterly undermine or overwrite them. It is rather the claim that, like language and face recognition, social reasoning skills tend to develop in a way that is remarkably invariant with respect to life experiences. Effective and sophisticated reasoning about physical objects, for example, is highly sensitive to specialized training in a way that comparable social reasoning is not.

When it comes to nativist theses about cognition, there is a temptation to ask which information (rules, theories, etc) is innate, and which is learned. Couching the issue in terms of canalization or biological preparedness, however, allows us to see things quite differently. Consider a jointly authored paper. We might ask who authored which sections or paragraphs or even sentences. This is how people tend to think of nature versus nurture in the cognitive realm. But it could also happen that both authors are responsible for every sentence, with the *degree* of responsibility varying from sentence to sentence, or section to section. The suggestion is that we should think of our cognitive abilities as all thoroughly co-authored. From this perspective the question is not which ideas are contributed by the genes, and which by learning, but rather how canalized the development of a given idea (or concept or ability) is: how much variability in the learning environment will lead to the same developmental end-state. An advantage of this way of thinking is that we see at once that nothing, not even limb-development, is inevitable. And when we investigate things in this light, we are led to ask which variations in the learning environment will divert the stream into a different and perhaps preferable canal.

5. NATURE AND NURTURE: THE ROLE OF THE SOCIAL ENVIRONMENT

One consequence of adopting a biological preparedness and canalization view of social norm reasoning is that *the environment plays a very large role in how and whether certain biological predispositions get expressed.* I have argued here that humans (like most other social mammals) have a propensity to form dominance hierarchies and to use dominance-centered strategies for controlling social interactions. In fact, dominance hierarchies are apparent in the social groups formed by children as young as 3 years of age (Smith, 1988). In their most benign form, such hierarchies reflect nothing more than leadership qualities among a few individuals. In their more malevolent form, such hierarchies can constitute rigid and profound socioeconomic stratification, as in the caste system in India, the warlords of Somalia, or monarchies that characterized human societies for millennia. *Which environmental factors play a role in determining whether our biological predisposition for dominance-centered strategies gets expressed?* In the sections below, I offer some data that hopefully illuminate that question.

5.1 Exposure to dissenting voices and non-conforming individuals

Perhaps the most famous examples of this are the Milgram experiments (Milgram, 1974). Sixty-five percent of the subjects participating in these studies obeyed a malevolent authority's orders to deliver a potentially fatal electric shock to an innocent victim. (In fact, no shocks were delivered and the "victim" was a confederate of the experimenter, but the subjects were not made aware of these facts until after they agreed or refused to obey the order.) These experiments (along with countless others in the social psychology literature) indicate that our predispositions to acquire social norms and detect violations of them also predispose us to value conformity, compliance, and submission to authority (dominant individuals)—even malevolent authority.

But these studies also show that these predispositions can be overcome by exposure to a single dissenting individual. Milgram reported that when a second confederate was included who refused to obey the order, obedience rates among the real subjects plummeted to less than 10 percent.

5.2 Exposure to social tolerance models

Rhesus monkeys are notorious for their aggressive natures, while stump-tail monkeys typically are characterized by cohesive group life, high social tolerance, and frequent reconciliation after fights. Co-housing juveniles between the two species, however, produces a dramatic shift in social interaction strategies among the rhesus (deWaal and

Johanowicz, 1993.) Those co-housed with stump-tailed macaques adopt many of the cooperative and conciliative behaviors typically seen only in stump-tails.

As this example shows, social mammals are biologically predisposed to acquire the social norms that exist within their troops. That is the aspect of their cognition that is canalized. Which norms are acquired, however, depends on the social environment they find themselves in.

5.3 Availability of resources

The relative availability of resources in the environment has a profound effect on the type of social structure that emerges within a group. In many species, the role of dominance is most pronounced in situations characterized by high levels of competition for resources, such as high population density or the onset of breeding season (Clutton-Brock & Harvey, 1976). Predation rates also affect the social order. Among feral horses, strong dominance hierarchies are observed in herds that live in regions characterized by high predation rates from mountain lions (such as the Montgomery Pass region near the California-Nevada border). In these herds, every group movement is initiated by the alpha mare and reflects social order and status, and within-group conflict is relatively uncommon. In contrast, the dominance hierarchies formed by feral horses in predator-free environments (such as the Pryor Mountains between the Montana-Wyoming border) are far more fluid and dynamic, with a good deal of jockeying for position and territory breaches occurring (Coates-Markle, 1997).

5.4 Gender and kinship characteristics in the social environment

The relative distribution of power across the two sexes also has a significant impact on the type of social order that develops in a group. As in human societies, males are dominant to females in the majority of primate species (the Bonobos being a notable exception), but the rate of sexual aggression of males against females varies significantly across species (Smuts & Smuts, 1993). A major factor is the migration pattern that occurs when juveniles reach sexual maturity. Among most monkey species, the males migrate out so that females remain among kin throughout their lifespan. Among most ape species, the females migrate out so that females find themselves among non-kin during their adult years. Sexual aggression is far more rampant among apes than monkeys because aid is more likely to be given to kin than non-kin in agonistic encounters. Female apes, therefore, are less likely to obtain help when under sexual attack by males than are female monkeys (Smuts, 1992). Recently, Figueredo (1998) undertook an analysis of factors that predicted a woman's likelihood of ending up in a battered women's shelter in a variety of cultures, including the U.S. A significant predictor was geographic distance from kin. Other researchers have noted that sexual aggression against females is higher in cultures where females marry outside of their clans or villages than in cultures where females stay close to kin (Smuts, 1992).

With respect to gender differences, some primatologists have argued that anything that increases female vulnerability will also increase the rate of sexual aggression of males against females. For example, orangutans in the wild mate approximately three times a month. But when male orangutans were housed with females who could not escape, mating occurred every day, with the majority involving forced copulation (Nadler, 1988). The aggression stopped immediately (and mating resumed the frequency observed in the wild) when the females were given a way to escape. This was done by making a door in the female's enclosure that was too small for the male to fit through. The researchers concluded

> ...the probability of a human female encountering sexual aggression is directly related to conditions—social, cultural, and environmental—that permit the male's dominance over the female to assume the predominant role in the regulation of sexual interactions and thereby reduce the female's prerogatives for refusing sexual overtures. They also point out that ... *such aggression is likely to be expressed under a variety of conditions unless society and individuals take specific measures to preclude it.* (italics mine)

Perhaps the recent scandals surrounding sexual harassment and exploitation of young, low-ranking female recruits in the military could have been avoided had these words been heeded.

As an interesting aside, the typical difficulty and displeasure adolescent females experience in science and math classes seem to disappear when they learn these subjects in female-only class settings, away from adolescent males who are (probably unwittingly) exercising strategies of domination and intimidation to control the classroom social environment (Abcarian, 1993; Gross, 1993).

6. CLOSING COMMENTS

Like all social animals, humans live in social environments that exert extraordinary cognitive and socio-emotional pressures. Extracting the social norms that implicitly (or explicitly) regulate our behavior and allow continued group membership is crucially important, as is developing the capacity to read the intentions, desires, and beliefs of others. As an evolutionary scientist, I would find it astonishing if this crucially important knowledge and these crucially important skills had to be learned from scratch by each succeeding generation in the same way we learn to play tennis or chess. Instead, I think the majority of comparative, developmental, and neuroscientific evidence weighs in on the side of fast-track learning through biological "biases" or predispositions that entrain the focus of our attention on the environmental stimuli and contingencies that really matter. Our biological predispositions impose the framework that is necessary to learn the things most vital for survival in a complex social environment, while neurological plasticity allows our actual environmental experiences the final say in whether and how those predispositions are expressed.

REFERENCES

Abcarian, R. 1993. "Answer for Girls is Math, Science Classes Minus Boys", *Los Angeles Times*, 112, E1.

Altmann, J., Alberts, S.C., Haines, S.A., Dubach, J., Muruth, P., Coote, T., Geffen, E., Cheesman, D.J., Mututua, R.A., Saiyalel, S.N., Wayne, R.K., Lacy, R.C., & Bruford, M.W. 1996. "Behavior Predicts Genetic Structure in a Wild Primate Group", *Proceedings of the National Academy of Sciences*, 93, 5795-5801.

Ariew, A. 1996. "Innateness and Canalization", *Philosophy of Science*, 63, S19-S27.

Aruguete, M. 1994. "Cognition, Tradition, and the Explanation of Social Behavior in Non-Human Primates", [Review of *Social Processes and Mental Abilities in Non- Human Primates*], *American Journal of Primatology*, 33, 71-74.

Banich, M.T. 1997. *Neuropsychology: The Neural Bases of Mental Function*. Boston: Houghton-Mifflin.

Baron-Cohen, S. 1995. *Mindblindness: An Essay on Autism and Theory of Mind*. Cambridge, MA: MIT Press.

Bartsch, K. & Wellman, H. M. 1989. "Young Children's Attribution of Action to Beliefs and Desires", *Child Development*, 60, 946-964.

Bernstein, I.L. & Borson, S. 1986. "Learned Food Aversion: A Component of Anorexia Syndromes", *Psychological Review*, 93, 462-472.

Bertram, B.C.R. 1976. "Kin Selection in Lions and Evolution", in Bateson, P.P.G. & Hinde, R.A., eds., *Growing Points in Ethology*. Cambridge: Cambridge University Press, pp. 281-302.

Blakemore, C.1974. "Developmental Factors in the Formation of Feature Extracting Neurons", In Worden, F.G. & Smith, F.O., eds., *The Neurosciences, 3rd Study Program*. Cambridge, MA: MIT press.

Blakemore, C. & Cooper, G.F. 1970. "Development of the Brain Depends on Visual Environment", Nature, 228, 477-478.

Bygott, J.D., Bertram, B.C.R., & Hanby, J.P. 1979. "Male Lions in Large Coalitions Gain Reproductive Advantage", *Nature*, 282, 839-841.

Byrne, R. 1995. *The Thinking Ape: Evolutionary Origins Of Intelligence*. Oxford: Oxford University Press.

Byrne, R.W. & Russon, A.E. (in press). "Learning by Imitation: A Hierarchical Approach", *Behavioral and Brain Sciences*.

Bussey, K. 1992. "Children's Lying and Truthfulness: Implications for Children's Testimony", in Ceci, S.J., De Simone Leichtman, M., & Putnick, M., eds., *Cognitive and Social Factors in Early Deception*. Hillsdale, NJ: Erlbaum, pp. 89-110.

Chapais, B. 1988. "Rank Maintenance in Female Japanese Macaques: Experimental Evidence For Social Dependency", *Behavior*, 104, 41-59.

Chapais, B. 1992. "Role of Alliances in the Social Inheritance of Rank Among Female Primates", in Harcourt, A. & De Waal, F.B.M., eds., *Cooperation in Contests in Animals and Humans*. Oxford: Oxford University Press, pp. 29-60.

Cheney, D.L. & Seyfarth, R.M. 1990. *How Monkeys See the World*. Chicago: University of Chicago Press.

Cheng, P.W. & Holyoak, K.J. 1985. "Pragmatic Reasoning Schemas", *Cognitive Psychology*, 17, 391-416.

Cheng, P.W. & Holyoak, K.J. 1989. "On the Natural Selection of Reasoning Theories", *Cognition*, 33, 285-313.

Clutton-Brock, T.H. 1988. "Reproductive Success", in Clutton-Brock, T.H., ed., *Reproductive Success*. Chicago: University of Chicago Press.

Clutton-Brock, T.H. & Harvey, P.H. 1976. "Evolutionary Rules and Primate Societies", in Bateson, P.P.G. & Hinde, R.A., eds., *Growing Points in Ethology*. Cambridge: Cambridge University Press, pp. 195-238.

Coates-Markle, L. 1997. "Choosing to Survive", *Equus*, 231, 34-41.

Cook, L.M. & Mineka. S. 1989. "Observational Conditioning of Fear to Fear-Relevant versus Fear-Irrelevant Stimuli in Rhesus Monkeys", *Journal of Abnormal Psychology*, 98, 448-459.

Cook, L.M. & Mineka, S. 1990. "Selective Associations in the Observational

Conditioning of Fear in Rhesus Monkeys", *Journal of Experimental Psychology: Animal Behavior Processes*, 16, 372-389.

Cosmides, L. 1989. "The Logic of Social Exchange: Has Natural Selection Shaped How Humans Reason? Studies with the Wason Selection Task", *Cognition*, 31, 187-276.

Cosmides, L. & Tooby, J. 1992. "Cognitive Adaptations for Social Exchange", in Barkow, J., Cosmides, L., & Tooby, J., eds., *The Adapted Mind: Evolutionary Psychology and the Generation of Culture*. New York: Oxford University Press.

Cosmides, L. & Tooby, J. 1994. "Origins of Domain Specificity: The Evolution of Functional Organization", in Hirshfeld, L.A. & Gelman, S., eds., *Mapping the Mind: Domain Specificity in Cognition and Culture*. Cambridge: Cambridge University Press, pp. 85-116.

Cummins, D.D. 1996(a). "Evidence of Deontic Reasoning in 3- and 4-Year-Olds", *Memory & Cognition*, 24, 823-829.

Cummins, D.D. 1996(b). "Evidence for the Innateness of Deontic Reasoning", *Mind & Language*, 11, 160-190.

Cummins, D.D. 1996(c). "Dominance Hierarchies and the Evolution of Human Reasoning". *Minds & Machines*, 6, 463-480.

Cummins, D.D. 1996(d). "Human Reasoning from an Evolutionary Perspective", *Proceedings of the 18th annual meeting of the Cognitive Science Society*, 18, 50-51.

Cummins, D.D. 1997. "Rationality: Biological, Psychology, and Normative Theories", *Cahiers de Psychologie Cognitive (Current Psychology of Cognition)*, 16, 78-86.

Cummins, D.D. 1998(a). "Social Norms and Other Minds: The Evolutionary Roots Of Higher Cognition", in Cummins, D.D. & Allen, C.A., eds., *The Evolution Of Mind*. New York: Oxford University Press, pp. 30-50.

Cummins, D.D. 1998(b) "Can Humans Form Hierarchically Embedded Mental Representations?", commentary on Byrne, R.W. & Russon, A.E., "Learning By Imitation: A Hierarchical Approach", *Behavioral and Brain Sciences*, in press.

Cummins, D.D. 1999. "Cheater Detection is Modified by Social Rank, The Impact of Dominance on the Evolution of Cognitive Functions". *Evolution and Human Behavior*, 20, 229-248.

Cummins, D.D. & Cummins, R.C. 1999. "Biological Preparedness and Evolutionary Explanation", *Cognition*, 73, B37-B53.

Curtiss, S. 1977. *Genie: A Psycholinguistic Study of a Modern Day "Wild Child"*. New York: Academic Press.

Damasio, A.R. 1994. *Descartes' Error: Emotion, Reason, and the Human Brain*. New York: Grosset/Putnam.

Datta, S.B. 1983(a). "Relative Power and the Acquisition of Rank", in Hinde, R.A., ed., *Primate Social Relationships*. Oxford: Blackwell, pp. 93-102.

Datta, S.B. 1983(b). "Relative Power and the Maintenance of Rank", in Hinde, R.A., ed., *Primate Social Relationships*. Oxford: Blackwell, pp. 103-111.

Dawkins, R. 1976. *The Selfish Gene*. Oxford: Oxford University Press.
Dennett, D. 1984. *Elbow Room*. Cambridge, MA: Bradford/MIT Press.
Dennett, D. C. 1988. "The Intentional Stance in Theory and Practice", in Byrne, R.W. & Whiten, A., eds., *Machiavellian Intelligence*. Oxford: Oxford University Press, pp. 180-202.
Dewsbury, D.A. 1982. "Dominance Rank, Copulatory Behavior and Differential Reproduction", *Quarterly Review of Biology*, 57, 135-159.
Dunbar, R.I.M. 1992. "Neocortex Size as a Constraint on the Behavioral Ecology of Primates", *Journal of Human Evolution*, 20, 469-493.
Dunbar, R.I.M. 1993. "Coevolution of Neocortical Size, Group Size, and Language in Humans", *Behavioral and Brain Sciences*, 16, 681-735.
Eimas, P.D. 1975. "Speech Perception in Early Infancy", in Cohen, L.B. & Salapafek, P., eds., *Infant Perception*. New York: Academic Press.
Ellis, L. 1995. "Dominance and Reproductive Success Among Nonhuman Animals: A Cross-Species Comparison", *Ethology & Sociobiology*, 16, 257-333.
Entremont, B. & Muir, D.W. 1997. "Five-Month-Olds Attention and Affective Responses to Still-Faced Emotional Expressions", *Infant Behavior & Development*, 20, 563-568.
Etscorn, F. & Stevens, R. 1973. "Establishment of Conditioned Taste Aversions with a 24-Hour CS-US Interval", *Physiological Psychology*, 1, 251-253.
Fedigan, L. 1983. "Dominance and Reproductive Success in Primates", *Yearbook of Physical Anthropology*, 26, 91-129.
Figueredo, A.J. 1998. "Family Deterrence of Domestic Violence", unpublished manuscript, Department of Psychology, University of Arizona.
Figueredo, A.J. 1992. "Preparedness And Plasticity: A Stochastic Optimality Theory", presented at the 4th Annual Meeting of the Human Behavior and Evolution Society, Albuquerque.
Flavell, J.H., Abraham, B., Croft, K., & Flavell, E.R. 1981. "Young Children's Knowledge About Visual Perception: Further Evidence for a Level 1-Level 2 Distinction", *Developmental Psychology*, 17, 99-103.
Flavell, J.H., Flavell, E.R., Green, G.L., & Moses, L.J. 1990. "Young Children's Understanding of Fact Beliefs Versus Value Beliefs", *Child Development*, 61, 915-928.
Gagneux, P., Woodruff, D.S., & Boesch, C. 1997. "Furtive Mating in Female Chimpanzees", *Nature*, 387, 358-369.
Garcia, J., Brett, L.P., & Rusiniak, K.W. 1989. "Limits of Darwinian Conditioning", in Klein, S.B. & Mowrer, R.R., eds., *Contemporary Learning Theories: Instrumental Conditioning Theory and the Impact of Biological Constraints on Learning*. Hillsdale, NJ: Erlbaum.
Goodall, J. 1986. *The Chimpanzees of Gombe*. Cambridge: Belknap Press.
Gopnik, A. 1993. "How We Know Our Minds: The Illusion of First-Person Knowledge of Intentionality". *Behavioral and Brain Sciences*, 16, 1-15.
Gopnik, A. & Wellman, H.M. 1994. "The Theory Theory", in Hirshfeld, L.A.

& Gelman, S.A., eds., *Mapping the Mind: Domain Specificity in Cognition and Culture*. Cambridge: Cambridge University Press, pp. 257-293.

Gopnik, M. 1990a. "Feature Blindness: A Case Study", *Language Acquisition: A Journal of Developmental Linguistics*, 1, 139-164.

Gopnik, M. 1990b. "Feature Blind Grammar and Dysphasia", *Nature*, 344, 715.

Gross, J. 1993. "To Help Girls Keep Up, Girls-Only Math Classes", *New York Times*, 143, A1(L).

Hall, K.R.L. 1964. "Aggression in Monkey and Ape Societies", in Carthy, J. & Ebling, F., eds., *The Natural History of Aggression*. London: Academic Press, pp. 51- 64.

Harcourt, A.H. 1988. "Alliances in Contests and Social Intelligence", in Byrne, R.W. & Whiten, A., eds., *Machiavellian Intelligence*, Oxford: Oxford University Press, pp. 131-152.

Harcourt, A.H. & Stewart, K.J. 1987. "The Influence of Help in Contests on Dominance Rank in Primates: Hints From Gorillas", *Animal Behaviour*, *35*, 182-190.

Harcourt, A.H. & De Waal, F.B.M, eds., 1992. *Coalitions and Alliances in Humans and Other Animals*. Oxford: Oxford University Press.

Harris, P.L. & Nuñez, M. 1996. "Understanding of Permission Rules by Preschool Children", *Child Development*, 67, 1572-1591.

Haugaard, J. & Reppucci, N.D. 1992. "Children and the Truth", in Ceci, S.J., De Simone Leichtman, M., & Putnick, M., eds., *Cognitive and Social Factors in Early Deception*. Hillsdale, NJ: Erlbaum, pp. 29-46.

Hausfater, G. 1975. "Dominance and Reproduction in Baboons (*Papio Cynocephalus*): A Quantitative Analysis", *Contributions in Primatology*, 7, 1-150.

Heyes, C. M. 1998. "Theory Of Mind In Nonhuman Primates", *Behavioral & Brain Sciences*, 21, 101-134.

Hilgard, E.R. & Bower, G.H. 1975. *Theories of Learning*. Englewood Cliffs, NJ: Prentice-Hall.

Hirsh, H.V.B. & Spinelli, D.N. 1970. "Visual Experience Modifies Distribution of Horizontally and Vertically Oriented Receptive Fields in Cats", Science, 168, 869-871.

Hokanson, J.E. 1961. "The Effect of Frustration and Anxiety on Overt Aggression", *Journal of Abnormal and Social Psychology*, 62, 346- 351.

Hokanson, J.E. & Shetler, S. 1961. "The Effect of Overt Aggression on Physiological Arousal", *Journal of Abnormal and Social Psychology*, 63, 446-448.

Hubel, D.H. 1988. *Eye, Brain, and Vision*. New York: W.H. Freeman.

Jenkins, J. & Astington, J.W. 1996. "Cognitive Factors and Family Structure Associated with Theory of Mind Development in Young Children", *Developmental Psychology*, 32, 70-78.

Krebs, J.R. & Dawkins, R. 1984. "Animal Signals: Mind-Reading and Ma-

nipulation", in Krebs, J.R. & Davies, N.B., eds., *Behavioral Ecology: An Evolutionary Approach*, 2nd ed. Oxford: Blackwell, pp. 380-402.
Kuhl, P.K. 1987. "Perception of Speech and Sound in Early Infancy", in Cohen, L.B. & Salapafek, P., eds., *Infant Perception*. New York: Academic Press.
Kummer, H. 1988. "Tripartite Relations in Hamadryas Baboons", in Byrne, R.W. & Whiten, A., eds., *Machiavellian Intelligence*. Oxford: Oxford University Press, pp. 113-121.
Leslie, A.M. 1987. "Pretense and Representation: The Origins of 'Theory of Mind'", *Psychological Review*, 94, 412-426.
Leslie, A.M. 1994. "ToMM, ToBY, and Agency: Core Architecture and Domain Specificity", in Hirshfeld, L.A. & Gelman, S.A., eds., *Mapping the Mind: Domain Specificity in Cognition and Culture*. Cambridge: Cambridge University Press, pp. 119-148.
Leslie, A.M. & Thaiss, L. 1992. "Domain Specificity in Conceptual Development: Neuropsychological Evidence from Autism", *Cognition*, 43, 225-251.
Lewis, C., Freeman, N., Kyriakidou, C., Maridaki-Kassotaki, K., & Berridge, D. 1996. "Social Influences on False Belief Access: Specific Contagion or General Apprenticeship?", *Child Development*, 67, 2930-2947.
Logue, A.W. 1988. "A Comparison of Taste Aversion Learning in Humans and Other Vertebrates: Evolutionary Pressures in Common", in Bolles, R.C. & Beecher, M.D., eds., *Evolution and Learning*. Hillsdale, NJ: Erlbaum.
Manktelow, K.I. & Over, D.E. 1991. "Social Roles and Utilities in Reasoning With Deontic Conditionals", *Cognition*, 39, 85-105.
Manktelow, K.I. & Over, D.E. 1995. "Deontic Reasoning", in Newstead, S.E. & Evans, J. St. B., eds., *Perspectives On Thinking and Reasoning*. Englewood Cliffs, NJ: Erlbaum.
McCann, T.S. 1981. "Aggression and Sexual Activity of Male Southern Elephant Seals (*Mirounga Leonina*)", *Journal of Zoology*, 195, 295-310.
McGuffin, P. & Scourfield, J. 1997. "A Father's Imprint on His Daughter's Thinking", *Nature*, 387, 652-653.
McKenzie, J.A. & O'Farrell, K. 1993. "Modification of Developmental Instability and Fitness: Malathion-Resistance in the Australian Sheep Blowfly", *Genetica*, 89, 67-76.
Mealey, L., Daood, C., & Krage, M. 1996. "Enhanced Memory for Faces of Cheaters", *Ethology and Sociobiology*, 17, 119-128.
Meltzoff, A.N. 1995. "Understanding the Intentions of Others: Re-enactment of Intended Acts by 18-Month-Old Children", *Developmental Psychology*, 31, 838-850.
Menzel, E.W. 1974. "A Group of Chimpanzees in a 1-Acre Field: Leadership and Communication", in Schrier, A.M. & Stollnitz, F., eds., *Behavior of Nonhuman Primates*. New York: Academic Press, pp. 83-153.
Milgram, S. 1974. *Obedience to Authority*. New York: Harper & Row.
Morris, J.S., Frith, C.D., Perrett, D.L., & Rowland, D.A. 1996. "Differential

Neural Response in the Human Amygdala to Fearful and Happy Facial Expressions", *Nature*, 383, 812-815.

Nadler, R. 1988. "Sexual Aggression in the Great Apes", *Annals of the New York Academy of Sciences*, 528, 154-162.

Nishida, T. 1983. "Alpha Status and Agonistic Alliance in Wild Chimpanzees (*Pan Troglodytes Schweinfurthü*)", *Primates*, *24*, 318-336.

Oaksford, M. & Chater, N. 1994. "A Rational Analysis of the Selection Task As Optimal Data Selection", *Psychological Review*, 101: 608-631.

Öhman, A. 1986. "Face the Beast and Fear the Face: Animal and Social Fears as Prototypes for Evolutionary Analysis of Emotion", *Psychophysiology*, 23, 123-145.

Oller, D.K. & Eilers, R.E. 1988. "The Role of Audition in Infant Babbling", *Child Development*, 59, 441-449.

Packer, C. 1977. "Reciprocal Altruism in Olive Baboons", *Nature*, 265, 441-443.

Perner, J. 1991. *Understanding the Representational Mind.* Cambridge, MA: MIT Press.

Perner, J., Ruthman, T., & Leekam, S.R. 1994. "Theory of Mind is Contagious: You Catch It from Your Sibs", *Child Development*, 65, 1228-1238.

Perner, J. & Wimmer, H. 1985. "John Thinks That Mary Thinks That...: Attribution of Second-Order Beliefs by 5-10-Year-Old Children", *Journal of Experimental Child Psychology*, 38, 437-471.

Piaget, J. 1952. *The Origins of Intelligence in Children.* New York: International University Press.

Povinelli, D.J. 1994. "What Chimpanzees (Might) Know about the Mind", in Wrangham, R.W., McGrew, W.C., de Waal, F.B.M., & Heltne, P.G., eds., *Chimpanzee Cultures*. Cambridge, MA: Harvard University Press, pp. 285-300.

Povinelli, D.J., Nelson, K.E., & Boysen, S.T. 1990. "Inferences about Guessing and Knowing by Chimpanzees (*Pan troglodytes*)", *Journal of Comparative Psychology,* 104, 203-210.

Prud'homme, J. & Chapais, B. 1993. "Aggressive Interventions and Matrilineal Dominance Relations in Semifree-Ranging Barbary Macaques", *Primates*, 34, 271-283.

Riss, D.C. & Goodall, J. 1977. "The Recent Rise to the Alpha-Rank in a Population of Free-Living Chimpanzees", *Folia Primatologica*, 27, 134-151.

Robinson, J.G. 1982. "Intrasexual Competition and Mate Choice in Primates", *American Journal of Primatology*, 1 Supplement, 131-144.

Ruffman, T., Perner, J., Naito, M., Parkin, L., & Clements, W.C. 1998. "Older (But Not Younger) Siblings Facilitate False Belief Understanding", *Developmental Psychology*, 34, 161-174.

Seligman, M.E.P. 1971. "Phobias and Preparedness", *Behavior Therapy*, 2, 307-320.

Seyfarth, R.M 1976. "Social Relationships among Adult Female Monkeys", *Animal Behavior*, 24, 917-938.

Seyfarth, R.M. & Cheney, D.L. 1984. "Grooming, Alliances, and Reciprocal Altruism in Vervet Monkeys", *Nature*, 308, 541-543.

Shultz, T.R. 1982. "Causal Reasoning in the Social and Nonsocial Realms", *Canadian Journal of Behavioral Science*, 14, 307-322.

Silk, J.B. 1987. "Social Behavior in Evolutionary Perspective", in Smuts, B.B., Cheney, D.L., Seyfarth, R.M., Wrangham, R.W., & Struhsaker, T.T., eds., *Primate Societies*. Chicago: University of Chicago Press, pp. 318-329.

Smith, P.K. 1988. "The Cognitive Demands of Children's Social Interactions With Peers", in Byrne, R.W. & Whiten, A., eds., *Machiavellian Intelligence: Social Expertise and the Evolution of Intellect in Monkeys, Apes, and Humans*. Oxford: Oxford University Press, pp. 94-110.

Smuts, B. 1985. *Sex and Friendship in Baboons*. Hawthorne: Aldine Press.

Smuts, B.B. 1992. "Male Aggression against Women: An Evolutionary Perspective", *Human Nature*, 3, 1-44.

Smuts, B.B. & Smuts, R.W. 1993. "Male Aggression and Sexual Coercion of Females in Nonhuman Primates and Other Mammals Evidence and Theoretical Implications", in Slater, P.J.B. et alia, eds., *Advances in the Study of Behavior*, Vol. 22. San Diego, CA: Academic Press, pp. 1-63.

Sodian, B. 1991. "The Development of Deception in Young Children", *British Journal of Developmental Psychology*, 9, 173-188.

Spelke, E. 1994. "Initial Knowledge: Six Suggestions", *Cognition*, 50, 431-445.

Stanovich, K.E. & West, R.F. (in press). "Cognitive Ability and Variation in Selection Task Performance", *Thinking & Reasoning*.

Stenberg, G. & Hagekull, B. 1997. "Social Referencing and Mood Modification in 1-year-olds", *Infant Behavior & Development*, 20, 209-217.

Taylor, M. 1988. "Conceptual Perspective Taking: Children's Ability to Distinguish What They Know from What They See", *Child Development*, 59, 703-718.

Taylor, M., Cartwright, B.S., & Bowden, T. 1991. "Perspective Taking and Theory Of Mind: Do Children Predict Interpretive Diversity as a Function of Differences in Observers' Knowledge?", *Child Development*, 62, 1334-1351.

Tomasello, M. (in press). "Uniquely Primate, Uniquely Human", *Developmental Science*.

Tomasello, M. & Call, J. 1997. *Primate Cognition*. Oxford: Oxford University Press.

Tooby, J. & Cosmides, L. 1995. Foreword, in Baron-Cohen, S., *Mind-blindness*. Cambridge, MA: MIT Press, xi-xviii.

Tutin, C.E.G. 1979. "Mating Patterns and Reproductive Strategies in a Community of Wild Chimpanzees (*Pan Troglodytes Schweinfurtii*)", *Behavioral Ecology and Sociobiology*, *6*, 29-38.

Uehara, S., Hiraiwa-Hasegawa, M., Hosaka, K., & Hamai, M. 1994. "The Fate of Defeated Alpha Male Chimpanzees in Relation To Their Social Networks", *Primates*, *35*, 49-55.

de Waal, F. 1982. *Chimpanzee Politics*. Baltimore. Johns-Hopkins University Press.

de Waal, F. 1988. "Chimpanzee Politics", in Byrne, R.W. & Whiten, A., eds., *Machiavellian Intelligence*. Oxford: Oxford University Press, pp. 122-131.

de Waal, F. 1989. " Food Sharing and Reciprocal Obligations among Chimpanzees", *Journal of Human Evolution, 18*, 433-459.

de Waal, F. 1992. "Coalitions As Part of Reciprocal Relations in the Arnhem Chimpanzee Colony", in Harcourt, A.H. & de Waal, F., eds., *Coalitions and Alliances in Humans and Other Animals*. Oxford: Oxford University, pp. 233-258.

de Waal, F.B. & Johanowicz, D.L. 1993. "Modification of Reconciliation Behavior through Social Experience: An Experiment with Two Macaque Species", *Child Development*, 64, 897-908.

Waddington, C.H. 1975. "*The Evolution of an Evolutionist*. Ithaca, NY: Cornell University Press.

Watts, C.R. & Stokes, A.W. 1971. "The Social Order of Turkeys", *Scientific American*, 224,112-118.

Whiten, A. & Byrne, R.W. 1988. "The Manipulation of Attention in Primate Tactical Deception", in Byrne, R.W. & Whiten, A., eds., *Machiavellian Intelligence*. Oxford: Oxford University Press, pp. 211-224.

Winner, E. & Leekam, S.R. 1991. "Distinguishing Irony From Deception: Understanding the Speaker's Second-Order Intention", *British Journal of Developmental Psychology*, 9, 257-270.

Woodruff, G. & Premack, D. 1979. "Intentional Communication in the Chimpanzee: The Development of Deception", *Cognition*, *7*, 333-362.

Wooldridge, D. 1963. *The Machinery of The Brain*. New York: McGraw-Hill.

PART TWO

SEXUALLY SELECTED DECISION MAKING IN MATING AND PARENTING

INTRODUCTION TO PART II

HARMON R. HOLCOMB III

Evolutionary psychology regards the human mind as our most distinctive adaptive instrument. In Part One we examined research strategies for achieving the aim of evolutionary psychology to reveal a domain-specific collection of well-adapted tools in human mental capacities in general. In Part Two we examine the area in the field that is most well-developed, namely, mating and parenting. What is the nature of the psychological mechanisms in human (and primate) minds? I shall provide a philosophical analysis of the reasoning in these four chapters that has three aims: to state the questions they ask and the answers they give, to exhibit the logical structure of the case they make for their main theses, and to highlight the larger significance of each chapter by critiquing the likely ways people's preconceptions will guide their evaluations of them. For instance, critics who wrongly presume that evolutionary psychology tries to explain everything are unaware that hypotheses are being tested about topics involving issues that pose current adaptive problems similar to those that existed in the environments of our evolution, such as mate choice, not evolutionary byproducts like reading and writing.

Gangestad's chapter exemplifies the fact that evolutionists seek to understand sexual selection as a process and the imprint of sexual selection on human nature. As he says, the major theoretical debate about sexual selection during the past two decades concerns what accounts for mate preferences that drive intersexual selection. In view of extensive work on preferences selected because of material benefits (resources) passed onto offspring and less work on preferences that exploit sensory biases of the mate chooser, he argues that preferences for mates who possess markers of genetic quality have had an important role to play. One significant aspect of his chapter lies in its

H.R. Holcomb III (ed.), Conceptual Challenges in Evolutionary Psychology, 121–141.

use as an illustration of how selectionist theories can be applied to a study of human mating to answer questions like, "With regard to the sexual selection operating on any specific species, what forms of sexual selection account for the selection of which traits?" Evolutionary psychology deals with questions intrinsically related to general concerns for all evolutionary theorizing about behavior. Contrary to many critics, who hold that evolutionary psychology of humans is illegitimate but evolutionary biology or psychology of nonhumans is legitimate, it is wrongheaded to separate the formulation and application of evolutionary thought to humans from that which involves nonhumans. Instead, although individual studies have to be evaluated on their own merits, evaluation of evolutionary psychology must treat it as woven from the same fabric as evolutionary theory and related empirical work as a whole. Successful application of evolutionary theories, models, and principles of individual behavior and social behavior not only inspire their extension to the human species, but also provide evidence independent of the human studies for the theories, models, and principles they use (as found in Sulloway's chapter).

In point of historical fact, during the 1970s and 1980s, criticisms of adaptationism in sociobiology were made as if they uniquely revealed the illegitimacy of sociobiology and left evolutionary work on nonhuman animals untouched, but debate over adaptationism spilled over into evolutionary ecology. No matter what species or taxon (taxonomic group) an evolutionary ecologist studies, each researcher by now has a stand on the issues surrounding adaptationism. Gangestad's response to adaptationism represents general concerns about adaptationist and selectionist methodology. These concerns, discussed in Part Three on "*Adaptationist Research Strategies and Evolutionary History,*" are relevant to specialized studies that constitute theoretical and empirical advances in evolutionary psychology.

Gangestad documents and tests the effects of "good genes sexual selection" (mate preferences based on the potential mate's genetic endowment) on human mating in the following manner. He begins by exploring the power and limits of two major forms of explanatory argument: tests of optimization models and arguments from design. Here he expresses his stand on adaptationism. Then, he addresses two concerns about the viability of accounts in terms of good genes sexual selection. First, do natural populations contain sufficient genetic variation in fitness to fuel the process? His argument, which connects life-history theory to data on heritabilities in the spirit but not letter of Mealey's proposed methodology, concludes that genetic variation in fitness between potential mates remains substantial. This argument is interesting in terms of the rationality of scientific change, as it illustrates an episode in which new ways of thinking about "the things we thought we knew," together with new evidence, serves to rationally change minds very much committed to a negative answer.

Second, could genetic quality possibly be important relative to material benefits in species characterized by substantial biparental care, such as humans? Gangestad answers by considering various possible interactions of the good genes process with the material benefits process. In conditions when females provide males a pool of multiple mating opportunities, males with good genes may provide fewer material benefits, even if they have potential to provide more, and female preference for such males would still be selected for. One thinks here of extra pair copulations (EPC's) in birds as well as infidelity by human females: A female might pick a long-term partner for material benefits and yet pick a short-term partner for his good genes. Recent work shows that human females make sexual displays and commit adultery far more often when they are ovulating than at other times of their menstrual cycle. Studies of sperm competition between mated partners and other males within the female's reproductive organs add much behavioral and physiological detail to these sorts of findings.

Evaluating Gangested's chapter provides a nice opportunity to stress the way scientific progress arises through the point and counter-point of scientific discussion. The emerging picture is that human and nonhuman females control their reproductive behavior and physiology to enhance their fitness by seeking the best possible mix of males with good looks, good behaviors, good material benefits, and good genes, given the constraints of their internal and external circumstances. This is not anthropomorphically to compare nonhuman with human behavior, but to subsume both human and nonhuman behavior under the same evolutionary principles as special cases. Critics unaware of the body of biological evidence are prone to call such human-nonhuman comparisons "just-so stories," but such pejorative attributions of baseless speculation usually only reveal that the critic is uninformed. For instance, Gangestad cites empirical tests of the good genes hypothesis in terms of genetic variation that leads to developmental morphological asymmetries. Such tests do achieve standards of rigor normal in evolutionary studies of nonhuman behaviors, a necessity for human evolutionary research to become "good science" in the sense of a final product. Improvements in work toward these standards are necessary for evolutionary research to be counted as "good science" in the sense of bringing it to a more advanced, effective, or usable state.

Once sexually mature, we search for, attract, and retain appropriate mates to start our own families. Gangestad argues that the imprint of natural selection on human mating can be seen in a preference for mates having good genes, that is, genes they will pass on to offspring that enhance offspring survival and reproductive success. This type of argument illustrates the need for the methodology suggested by Mealey. Good genes mate selection is presumed to be a species-typical strategy, which is part of a repertoire of mating strategies that is subsumed under the study of human nature. Yet his argument also requires that humans have general traits that exhibit genetic differences (for which study of heritabilities using behavior genetics would be helpful). When we select a mate we respond to environmental cues that tell us whether a potential mate has better genes than another potential mate (our preference is environmentally caused, but the ability to respond to environmental inputs with preferences that enhance reproductive success is a genetically-based evolved type of strategy). Again, evolutionary psychology need not restrict itself, as it usually has, to well-known theories of mate selection for good resources or for good looks (Buss, 1994).

Let's recast Gangested's argument in a way that counters the objection that adaptationist research strategies yield only speculative results by showing how he advances the current research situation. Gangestad argues that good genes sexual selection has played an important role in the evolution of mating. This is part of a large number of evolutionary studies that attempt to understand who mates with whom and why (Buss, 1994). Most studies of mate preferences in humans study female preferences for male mates for their resources (material benefits) and male preferences for female mates for their childbearing aptitude (youth, health, attractiveness). Gangestad is proposing that we also consider preferences for a mate of the opposite sex for their "good genes." The two types of selection are closely related, because good genes are what give us good looks, good health, and so forth. The difference is that preferences for "resources" or "childbearing aptitude" revolve around the behavior of mates, whereas preferences for mates with "good genes" are limited to genetic inheritance from one's mate to one's offspring. Potential mates with good genes have greater access to mates than potential mates with bad genes, and are therefore favored by sexual selection. Although good genes natural selection has long been considered a possible selection process that shapes families, Gangestad marshals a variety of arguments to show that this hypothetical possibility can have (according to theory) and has had (according to data) a significant role to play in humans. Gangestad's contribution speaks to a persistent failing on the part of evolutionists and their critics: When it comes to human evolution, both the researchers and the critics are led into a variety of needless mistakes and objections because they do not know evolutionary theory well enough. In a field where most explanations work by appeal to differential reproductive success, one needs to know natural selection theory in far greater depth, detail, and subtlety than the basic understanding that drives the typical reader's criticisms. This understanding is evident in the way Gangestad concurs with Mealey's (this volume) admonition to establish, rather than merely assume, that a trait under study is an adaptation, possesses genetic variance, and is universal for the human species.

Specifically, Gangestad's understanding of natural selection theory draws on recent theoretical and empirical advances in evolutionary ecology and behavioral ecology. After distinguishing good genes selection from three other selection processes that drive intersexual selection, he speaks to general issues concerning criteria for evaluating adaptationist accounts, particularly, tests of fitness optimization models and arguments from adaptive design. He is well aware that most selectionist arguments are incomplete (because we do not know many of the time-dependent and system-dependent constraints). Even so, he refuses to draw the critics' conclusion that selectionist accounts are "merely speculative," on the grounds of an inference to the best explanation: "Given what we know in terms of general phenomena responsible for evolution, no other evolutionary account can do as good a job at fitting the facts." Optimization models are falsifiable, and when falsified, researchers know which possible sources of error to look into. Traits can be established as adaptations, in the absence of a dynamically sufficient model or optimal adaptation, because the criteria of specificity, efficiency, or economy (i.e., design features) in producing an effect can show that the machinery behind the behavior was fashioned by natural selection for a fitness-related goal. This is a typical way evolutionary psychologists take into account the critique of adaptationist methodology, just as many biologists do when considering the legitimacy of adaptationism in studying nonhumans, without the damaging implications that the field is "unscientific" because "speculative" or "unfalsifiable." Neither animal nor human evolutionary behavioral analyses are "unconstrained speculations" because they are constrained in many ways even if they are not constrained specifically by detailed comparative and phylogenetic methods focusing on evolutionary dynamics (this is a counter-point to the critiques of adaptationism by Griffiths and Richardson in Part Three).

Gangestad does not shy away from theoretical reasons to doubt the efficacy of good genes natural selection. Against the doubt that natural populations contain too little genetic variation in fitness to fuel the process, Gangestad shows that the source of the doubt in R.A. Fisher's fundamental theorem in population genetics was mistaken. Even though Fisher was correct in stating that natural selection tends to exhaust the additive genetic variance in fitness, abundant genetic variance in fitness, which might lead to preferences for mates with good genes, arises from mutation, gene flow, fluctuating selection, and degradation of the environment. In fact, fitness-relevant life history traits such as fertility, fecundity and survivorship have two to four times the additive genetic coefficient of variation as do ordinary morphological traits in nonhuman species. In order to rebut the doubt that genetic quality could not be important relative to material benefits in species having substantial parental care (e.g., humans), Gangestad makes a number of subtle points. For example, he explains why a mate's genetic quality is not reduced to triviality due to the importance of finding a mate who will effectively invest in offspring and the interaction of mate preferences for material benefit.

Having removed the theoretical obstacles to significant good genes selection, Gangestad goes on to discuss a number of empirical tests of the good genes hypotheses focusing on developmental imprecision, which arises from genetic and environmental perturbations in adaptive design, especially including mutations and pathogens. Developmental asymmetries measured are on the order of one to two millimeters, too small to be detected in social interaction, and so are not cues by which individuals assess others' developmental imprecision. However, they are posited to be markers of underlying developmental imprecision, which substantially affects overall individual fitness, and so Gangestad investigates a number of reasons why fluctuating asymmetry (FA) is associated with human and nonhuman male mating success. Developmental imprecision is highly correlated with variation in fitness traits and mating success in some species; we would be surprised if humans were an exception to that empirical rule, an expectation that generates a number of hypotheses about what the perceived basis of the mate preference for men with good genes might be.

Gangestad's tests show adaptive design due to good genes selection in humans. Symmetrical men enjoy mating advantages (which implies current adaptiveness but not evolutionary adaptation). Their mating advantages accrue in virtue of female design for good genes preferences (which show evolutionary adaptation via the specificity in female orgasm, female scent preferences across the menstrual cycle, and their preferences for symmetrical men as a function of mating context). In contrast to the tendency of some evolutionists to make big claims on the basis of little evidence, Gangestad cautiously concludes that, "To date, evidence that good genes sexual selection has played an important role in the evolution of human mating is strongly suggestive but not compelling." His empirical case—that although there may be alternative explanations for individual findings but not for the whole set of empirical findings—is the sort of argument that changes minds and inspires similar research. This argument is an explicit example of the scientific method of inference to the best explanation: The explanation to be preferred above others is the one that accounts for the totality of the evidence, and whether one should rationally believe that explanation depends on the nature of the evidence (its quantity, its quality, its diversity, its decisiveness, etc.).

Ketelaar's and Todd's chapter makes the case for a "satisficing" approach to how organisms represent the environment, in contrast to Gangestad's "optimization" approach based on how evolution by selection works. They address the need to determine, given that the human mind contains a number of such domain-specific tools for generating adaptive behaviors, how these tools can work effectively. Their project clearly articulates two related postulates of the Cosmides-Tooby research program. First, minds did not evolve to execute arbitrary tasks, but to solve problems related to survival and reproduction. Second, these problems are not solved by programming rules that yield logically optimal solutions ("constraints" are not computable by organisms, much less "optima"); they yield pragmatically effective solutions that enhanced fitness in ancestral environments of evolutionary adaptedness. How do psychological mechanisms lead to adaptive behaviors? That raises the recalcitrant "frame problem." How should the organism frame an adaptive problem, possibly a choice among an indefinitely large array of different behavior sequences, so that only the subset of adaptively relevant choices are evaluated by a biologically plausible computational device (the mind)? An original conceptual insight in the chapter by Ketelaar and Todd shows how intelligent agents overcome the frame problem to generate adaptive cognition by distinguishing two sorts of rationality. They argue that the computer metaphor of the mind should be elaborated by appeal to information-processing mechanisms that use heuristics, which supply satisfactory solutions to adaptive problems, rather than optimal rational (algorithmic) solutions in the sense of optimal design models (as found in Gangestad's chapter).

Ketelaar and Todd make their case in three parts. The first part sets up the frame problem: How can successful decision-making be achieved under conditions of limited time and knowledge? By the time an organism has sifted through all possible inferences from current circumstances to possible behavior sequences, it may use up too much of its time budget. For instance, it needs to eat everyday but the computations last more than a day, or the time wasted on their consideration has heavy opportunity costs in terms of it being prevented from doing other things, or it is eaten by a predator in the midst of its calculations. The presence of system-dependent and time-dependent constraints within which organisms generate adaptive behaviors (recall Gangestad's chapter) motivate us to move from optimally rational solutions (i.e., optimal in the absence of such constraints) to satisfactory rational solutions in order to find an ecologically valid description of our evolved computational architecture.

The second part asks: How much information do intelligent agents really need to solve important information processing problems? The authors answer this question using the example of mate search, i.e., how to best search for the mate that maximizes one's fitness. In a thought experiment, suppose you know that you will examine 100 prospective mates in succession, cannot see any twice, and aim to minimize the total number of prospective mates examined. An optimal, algorithmic rule requires you to examine at least 37 candidates and finds the sole person with the highest mate value only 37 percent of the time. In contrast, a person employing a satisficing (satisfactory, but not optimal) solution starts with an aspiration level of a mate that is not optimal but only "pretty good" (the highest quartile) and will sample only 7 percent of the population and succeed 92 percent of the time. This result from probability theory can be made more realistic by incorporating the greater number of constraints found in actual human mating situations. By using one's own mate value (itself a changing result of ongoing encounters with potential mates, proposals, and rejections by members of the opposite sex) as the aspiration level, less than 20 encounters are needed for the majority in the population to form mated pairs of individuals with similar mate value. So, we see how in principle a surprisingly small portion of the available information can effectively guide successful mate search decisions, even when both sexes are choosy.

The third part asks: To what sort of information should an intelligent agent bother to pay attention in attempting to solve important information processing problems? The novel answer here is: the emotions. In contrast to the traditional view that emotions interfere with rational decision-making, the numerous emotions operate as numerous domain-specific mechanisms that provide information about the future consequences of pursuing various strategy choices, by affectively representing the cost-benefit structure of various environmental contingencies. These affective representations are more salient and more accurate than objective "cold" calculations of costs and benefit, as discussed in relation to the Prisoner's Dilemma and chess playing computers. This view is part of Ketelaar's (1999) ongoing research postulate that, "An organism built to infer the future state of its environment from its affective or emotional reaction to events in its current environment may have an adaptive advantage over other organisms that must directly experience consequences in order to learn the contingencies between current and future environments (e.g., that a snake bite right now can cause death later)." Incorporating emotion into reason should be of special interest to feminists, who have for years attacked value-laden dichotomies of reason over emotion in connection to men over women as found in our patriarchal culture. The authors conclude with methodological proposals for how to study ecological rationality.

Turning to the evaluation of Ketelaar's and Todd's chapter, the key issue is how best to formulate and solve the frame problem. Using thought experiments and work on artificial intelligence, they tell us about how mental representations work effectively as tools, where ecological constraints on the time, energy, and information-processing budgets of organisms favor satisficing heuristics rather than optimally rational algorithms. They draw on research in artificial intelligence to pose the frame problem in biological terms and suggest that the problem can be solved by understanding ecological rationality. The frame problem is a central problem for cognitive psychology. Adopting the computer metaphor of the mind, suppose we try to simulate actual organismic behavior by building the minimum cognitive content into the organism as a computer program. No one has yet been able to simulate a complex organism that can survive and reproduce successfully. In simplest terms, think of an organism with a small number of possible behaviors and a value function that would correlate to biological fitness (think of which movements are more likely to obtain food, shelter, a mate, or else win a dominance contest). The permutations of possible behavior sequences are astronomically large. By the time an organism could run through the computations of all these possible behavior sequences, assess their effects on the fitness-correlate, and choose the optimal sequence, it would be too late—the time for performing the behavior would have passed and fitness would be compromised. We can simulate how organisms solve adaptive problems by building into the computer problem only the assumptions relevant to solving the problem, "framing" the adaptive problem so that the enigma of a required but computationally intractable number of decisions is ruled out of the organism's decision-making architecture at the outset. Where do these "assumptions "come from, as they govern information-processing in real organisms? How do real organisms know which possible logical possibilities or consequences can be ruled out as "irrelevant" to which tasks? Ketelaar and Todd propose to solve the frame problem in three ways. First, the decision-rules used by organisms are "satisficing" rather than "optimizing" rules. Second, the decision-rules utilize information in the organism's environment to devise relevant solutions. Third, emotions are intrinsic to reasoning; the proverbial dichotomy between "being emotional" and "being rational" is obsolete and misleading. Emotions function to focus attention on what information in the environment to attend to, anticipating the fitness benefits and costs of a small handful of adaptively relevant possible behaviors. Emotions such as feelings of attraction for potential mates, sexual jealousy about one's current mate, sadness over the loss of a mate, and so on, guide present behavior according to goals or values inherent in the projected future consequences of behavior, where the goals or values are determined by the criterion of lifetime reproductive success.

The research strategy of moving from optimizing rules to satisficing rules is an important one in evolutionary theorizing. In order to open the way for this move, one has to see the difference between how natural selection operates and how organisms that evolved by natural selection operate. It has long been thought that natural selection is an optimizing process, as it favors those organisms with maximal fitness, and so the fitness-maximizing strategy would have to be an optimizing strategy. That thought is wrong on all three counts. Natural selection optimizes adaptation only ceteris paribus (other things being equal, which usually turns out to be other factors being absent, ceteris absentibus). There has been a long dispute in ecology and evolutionary biology over the constraints imposed on natural selection that lead to outcomes not nearly as fitness maximizing as those envisaged in optimal design models. Even if no organism bears a trait that maximizes fitness, natural selection can operate on variant traits so long as some traits are relatively better or worse, promoting the better trait. In frequency-dependent selection, natural selection maintains two or more traits as a function of their relative frequency in the population. Finally, satisficing strategies, which solve adaptive problems in a "quick and dirty" way as opposed to a logico-mathematically optimal "pristine" way, can work better than optimizing strategies simply because they fit better into the time, energy, and information-processing budgets of organisms. An organism that uses an optimal strategy but dies before it arrives at a decision is out of the evolutionary game, as compared to one that uses a sub-optimal but satisfactory strategy to find food, shelter and mates, and to win contests. Some satisficing strategies confer higher fitness than do others, and those that do so will evolve by natural selection.

The move to narrow down the combinatorial explosion of behavioral possibilities by attending to information in the environment divides the issue of where the organism's "assumptions" come from into internal and external sources. Nonevolutionary approaches typically identify environmental influences on behavior, but ignore the issue that this presupposes ability on the part of the organism to identify and process that environmental information. So, something has to be built-in. For traits such as "looks for resources and mates," it is trivially true to view the trait as a species-typical characteristic. Even primate species differ drastically in their resource-subsistence and mating systems. So, whatever is built-in to the modern human is far more specific than "a drive for survival and reproduction;" no species has that drive per se, it only has behavioral mechanisms for obtaining those things that correlate positively to reproductive success. So, these built-in assumptions are going to have to be far more specific than traditional psychology and social science allow. The raw materials that our conscious minds, learning, and culture operate on are not "raw" at all; they have already been processed by natural selection and so are full of specialized goal-oriented qualities that traditional scientists have overlooked to their own peril. "Built-in" does mean "genetic" in the sense that an environment must be in place for the expression of genes. "Built-in" need not mean "genetic" in the sense of being caused by genes in the absence of pertinent environmental stimuli. Specific environmental cues may be needed during critical stages of development in order for a human to express one rather than another alternate strategy, where the potential to develop any one of the array of alternative strategies is a species-typical characteristic.

Holcomb's chapter addresses the issues of the unity of psychology and the relevance of evolutionary theory to psychological phenomena. I argue that the methods of evolutionary psychology should be married to the methods of traditional psychology. In general, while the former focus upon the discovery of ultimate causes or mechanisms—those involved in the evolution of a given capacity—the latter focus upon the discovery of current and proximate mechanisms—those involved in exercising the capacity today. Is it true that all traditional (nonevolutionary) psychology explains by reference to proximate mechanisms? Although evolutionists typically draw the proximate/ultimate distinction in ways that assume an affirmative answer, that is not the case. Using anorexia as a case study, I analyze the standard view of eating disorders, namely, as malfunctional eating behaviors that symbolize deeper underlying problems. The standard view relies on "symbolization" rather than "mechanisms" to tell us what psychological phenomena are all about. How, then, could evolutionary explanations in terms of mechanisms ever be relevant to that preconceived view of eating disorders?

I answer: To construct a relevant evolutionary explanation, I propose three generalizable strategies. First, we eliminate the kind of explanation inherent in the nonevolutionary explanation that constitutes a key source of irrelevance by extracting the potential causal explanatory factors (distress, low self-esteem, body image) from the symbolization model. Second, we produce relevance by constructing a different kind of explanation that puts the preceding causal factors into a model that connects proximate mechanisms (lack of desire to develop sexual traits, dislike of one's body size and shape, fear of becoming fat, seeing control as a virtue, desire to conform) with ultimate evolutionary processes and evolutionary history (sexual selection, past adaptations triggered by current thought, emotion, and behavior). As a criterion of adequacy for preventing explanatory loss, we must do justice to cognitive aspects of eating disorders associated with the prior nonevolutionary model (symbolization). To do so, we connect faulty thought patterns in these proximate mechanisms to both domain-specific cognition (self-assessment of one's mate value) and domain-general cognition (costs of inhibiting behaviors such as eating, ironic control of mental processes that produce system information-processing overload and lead to breakdown). The chapter justifies and executes these strategies.

The chapter advocates and exemplifies a new way to do philosophy of science. Typically, philosophers of science focus on the methodology of evolutionary psychology, but have no positive, substantive evolutionary hypothesis of their own to offer (e.g., Griffiths, Richardson, this volume). I propose a "decreasing mate value" hypothesis of how dieting leads to eating disorders and illustrate these proposed methods in terms of it. The chapter's length derives from its novel aim: to contribute both to psychology and philosophy of science by exhibiting the sort of reasoning we can use to construct and develop a relevant evolutionary hypothesis in detail with conceptual, methodological, and theoretical clarity.

The symbolization model of eating disorders has three basic postulates: (1) in anorexia, and in eating disorders generally, food itself is not the primary problem, but is, rather, a symbol of serious distress; (2) the serious distress always involves low self-esteem; and (3) eating behavior disorders are arbitrary symbols of negative psychological states, so that the psychological states are manifested or expressed in abnormal eating behavior. Thus, anorexia, despite appearances, is not about food. It is about how people feel about themselves, which can be played out in different arenas of behavior. People, typically young women or girls, are living for others, trying to attain some goal, or for some other reason are obsessed with their own body image, which never meets their expectations and standards of self-esteem. They perceive their weight to be unbearable. That leads to an eating disorder, with anorexia related to bulimia and depression. As the anorectic starves herself, she is dying for food, starving herself as a symbol of her low self-esteem. These psychological vulnerabilities are combined with cultural norms that "the thinner the better," "being fat is bad," and so on to explain why self esteem problems are played out in the arena of eating behavior rather than an arbitrary domain of behavior.

The reproductive suppression model of eating disorders has three quite different basic postulates from those of the symbolization model. (1*) Food is literally, not symbolically, connected to anorexia, as eating too little is causally and functionally related to differential reproductive success. (2*) Females are expected as a result of natural selection to control the timing of sexual behavior and reproduction to produce the greatest number of healthy offspring within the constraints imposed by existing and ancestral circumstances. (3*) Anorexia typically involves suppressed and delayed reproduction; if a girl or woman perceives herself at a present reproductive disadvantage, adaptive self-suppression will be selected for, i.e., in order to enhance her biological fitness she should suppress current reproduction and delay reproduction until conditions improve.

My "decreasing mate value hypothesis" essentially links the suppression and symbolization models conceptually. It links the concepts of "low self-esteem" and "distress" emphasized in the symbolization model to the evolutionary concepts of "low mate-value" and "distress due to a perceived mate value disadvantage," providing a novel version of the reproductive suppression model that can be generalized to cover eating disorders not only in females but also in males. I shall not attempt to evaluate my own chapter, leaving others to assess it unbiased by my own interpretation. I will emphasize that, although the usual explanations of eating disorders rightly emphasize stress, distress, and low self-esteem, they are not the kind of explanations much of psychological science is looking for. Many researchers are looking beyond nonevolutionary "symbolic explanations" for appropriate "causal-functional" explanations. One of evolutionary psychology's contributions to psychological science is to reestablish the importance of functional explanation; and to show that understanding how our psychological mechanisms (cause-effect chains that serve functions) work requires showing how they make evolutionary sense. I construct a novel "decreasing mate value hypothesis" as a refinement of the reproductive suppression model. The larger point is to illustrate a host of methodological decisions involved in hypothesis formation, i.e., forming hypotheses worthy of scientific investigation and testing. Hypothesis formation is just as much a part of scientific method as hypothesis testing, and deserves far more attention than it has gotten in discussions of evolutionary approaches to human behavior or of specific evolutionary hypotheses.

Jankowiak's and Woodman's chapter explores the relation between evolved psychological strategies and culture. How do biology and culture work together to produce human symbolic behavior? They hope to advance this larger issue by examining a simple, mundane, symbolic activity: the naming of children by parents, specifically, by mothers. Providing a woman's perspective that is not directly centered on males is what makes their ethnographic findings significant. They argue that multiple motives affect the formation of women's parenting strategies. This argument gains further significance in terms of the recent emphasis in studies of science on the social context of science and the rise of feminist critique of common attitudes toward gender, which all too often are smuggled tacitly into science. The relation between science and ideology contains a tension between two extremes. On one hand, science can simply perpetuate status quo gender roles by telling us what we think we already know and rationalizing it through theory. On the other hand, science can critically examine status quo beliefs about gender to the point of undermining them with new findings. If feminist scholarship succeeded in problematizing women's position in society, then early evolutionary research exploring W.D. Hamilton's well-known hypotheses about kin selection also succeeded in problematizing men's position within the family. For much of the nineties, both positions have sought to return a balance to their studies by focusing on the opposite sex. In evolutionary theory, this includes a focus on women's inclusive fitness (i.e., kin-related) strategies, which often involve efforts not directly organized around obtaining investment from the male sexual partner.

Jankowiak and Woodman make their case in three parts. First, they review life history theory as it pertains to American naming patterns. This sets up three hypotheses about naming to be tested against the ethnographic evidence concerning Mormons: a standard evolutionary paternal investment hypothesis, a purely cultural patriarchy hypothesis, and the authors' new evolutionary hypothesis in terms of maternal investment. Second, they describe the Mormon polygynous community and the ethnographic methods they use to study naming in that community. Third, in discussing their results, they argue that some but not all phenomena are consistent with each of the two male-oriented hypotheses (paternal investment or patriarchy), and argue that the anomalies to these hypotheses can be explained by the maternal investment hypothesis. Recall that Sulloway had found it necessary to postulate within-family niches specific to each sibling in order to understand personality development, which implies that behavioral genetics methods for doing twin studies should be modified to recognize that twins have different within-family niches. Similarly, Jankowiak and Woodman found it necessary to postulate within-family niches specific to each co-wife in the status hierarchy of women in a polygynous family, which implies that family studies recognize female hierarchies despite the feminist identification of social hierarchical thought and practice with male attitudes. As one Mormon woman said, "We compete over everything; why not names?"

Women's competition over names is importantly divorced from male concerns, both from official prescriptions of patriarchy and from men's need for paternity assurance. Variations in naming arise from multiple motives that reflect female-female competition for dignity, gain, and autonomy within the family, as shaped by the woman's personality and relative status within the family and within social dominance hierarchies. Individual decisions based in part on a person's standing within the local family hierarchy shape how they structure extra-familial alliances as well as seek to sustain some kind of autonomy for themselves within that system. So, these women are not blindly following partriachal ideology, as the "culture is everything" school would have it. Nor are they solely using paternal reassuring strategies, as the "biology is everything" school has treated women, on the basis of making a debatable assumption. It has been widely assumed that the posited ancestral conditions in which males rather than females controlled most of the resources have made present day human female strategies into strategies for ensuring continual provision of resources from the father. Instead, these women are following what is for them clearly a maternal centered strategy, a broad-based strategy to elicit support from both maternal and paternal kin, support that would enhance their inclusive fitness.

The evaluation of Jankowiak's and Woodman's chapter raises several interesting larger issues over the evolutionary approach and culture. This broad-based maternal strategy, in virtue of these particular fitness factors, is a response to the new social conditions typical of a polygynous family environment within newly formed Mormon theocratic communities, i.e., in response to a particular cultural change. So, this finding is quite consistent with cultural anthropology. Perhaps we should regard the cultural environment as a source of selective pressures, just as the biotic and ancestral environments are. Even though cultural change often modifies the selective pressures at hand, an evolved psychology mediates the way human organisms respond to current selective pressures. Contrary to some evolution-advocates, we cannot dismiss culture as "a mere proximate cause of behavior," because the evolutionary story about natural selection cannot be told without incorporating culture into it. Contrary to some anthropologists who resist the evolutionary approach, we cannot dismiss evolved strategies as "a mere potential to act in the ways cultural norms tell us to act," because culture cannot explain variations within a culture that violate those norms, and do so in ways that make evolutionary sense. In this case, the family and social dominance hierarchies exert crucial current selective pressures. Accordingly, we concluded in Part One with Cummins' chapter on how social dominance hierarchies found in primates generally have exerted strong selection pressures on the human mind, thereby making it our greatest tool for engaging in adaptive social behavior. Also, Mealey's emphasis on kinship as "the tie that binds disciplines" might extend to fields as different as biology and cultural anthropology. That role of kinship would be reflected in the effort of female Mormons to name their offspring in ways that elicit the greatest amount of resources and social support from the totality of kin, including those on the mother's side and those on the father's side of the child's ancestry.

Jankowiak's and Woodman's case can be recast to bring out another dimension of the basic issue over how evolutionary psychology bridges biology and culture in its view of families. Once we have settled on a mate, we typically bear children, engaging in parenting activities in which we invest in our offspring. Jankowiak and Woodman apply the evolution of parental investment to the choice of names parents give their children; they are assuming that naming constitutes an investment, leaving it to others to demonstrate that fact in order to get on with their project. Whereas previous cultural anthropologists typically disregarded biology, Jankowiak and Woodman endorse the relevance of evolutionary theory for ethnographic analysis. They search out the current limits of selectionist hypotheses in evolutionary psychology as exposed by cultural anthropology. In particular, names had been thought of as a form of parental input that enhances the sense of relatedness experienced by the parents and other relatives to the child, e.g., by naming the child after one or other of the (alleged) parents. Empirical studies reinforce the common sense observation that Americans more often than not name their offspring, or at least their first-born child, after the father. So it had been hypothesized that namesaking would arise whenever there is a high level of paternity uncertainty or anxiety about the father's willingness to invest in offspring. It would not enhance a man's inclusive fitness to invest in children that are not his own or his kin's own offspring. A mother, however, would benefit by giving her spouse's name to her child, even if he were not the father, as a manipulative tactic for ensuring continual provision of resources from her spouse. So, this paternity confidence hypothesis about naming was tested using an ethnographic analysis of an American Mormon community (noted for its polygynous mating system).

Jankowiak's and Woodman's methodology for seeking out the limits of evolutionary psychology imposed by culture is worth getting clear on. To do so, they deliberately choose a simple symbolic behavior, the naming of children, in order to highlight the essentials of symbolic behavior as such. They show that it would be unduly restrictive to reduce naming to the particular motivation of bolstering the father's paternity confidence, as they find both motivations of reproductive investment and seemingly non-reproductive motivations. Such multiple motivations make sense only if the male-biased reduction of mother's naming practices to reassuring men about their paternity were complemented with a female-biased view. A larger issue here is, of course, that of the gendered, ideological, cultural construction of science (or of human nature). Their novel finding also illustrates how hard it is to actually make precise the preconception that the limits of our evolutionary legacy are imposed by culture. The women studied partly obey and partly violate the male-biased reproductive strategies posited by evolutionary psychologists and also partly obey and partly violate the patriarchal cultural norms posited by anti-evolutionary psychologists and anthropologists.

It is well known that the severity of a test can be increased by moving from the test of one hypothesis against the data to a test of several rival hypotheses against the same data. Here, Jankowiak and Woodman use the Mormon ethnography to test between prediction generated by (1) a narrow paternal investment evolutionary psychological explanation (need for father's reassurance in the face of low paternity confidence), (2) a purely cultural explanation (in terms of the relationship between cosmology, gender, and spirituality), and (3) a broad maternal investment explanation (in terms of women's eliciting investment from maternal and paternal kin by asserting their sense of identity, especially when their autonomy is undermined).

As often happens, the data the authors collect is consistent in a rough way with all three predictions, but the motives for namesaking cannot be reduced to either the purely evolutionary explanation (1), or the purely cultural explanation (2). The authors favor the broad maternal-investment hypothesis and find that theory supported by the general scientific method of inference to the best explanation. They say, "The best explanation for the naming patterns and their variations appears to be a maternal investment hypothesis, with its emphasis on woman's decision to focus on procurement of resources through engaging in strategies that range from becoming a favorite wife, to establishing economic self-sufficiency, to dominating the tacit female-centered domestic hierarchy." Note that this is an explicit statement of the use of inference to the best explanation as a method of reasoning for establishing scientific conclusions; we found this method explicitly used in Gangestad's chapter, and earlier in Cummins' chapter (see the Introduction to Part One). Actual practice in evolutionary psychology continues to support the idea that evaluation of evolutionary psychology should be cast in terms of inference to the best explanation (Holcomb, 1996), and many of the methodological themes about the way it was used in sociobiology (Holcomb, 1993) apply to the literature of evolutionary psychology.

Considerations about inference to the best explanation have implications for theory testing. Jankowiak's and Woodman's combination of life history theory and parental investment theory can do the entire explanatory work that the nonevolutionary "family systems theory" can do and more. So, the authors do not include family systems theory in their test—both approaches explore how the interrelationship between family members affects perception of opportunities and thus shapes behaviors within the family. Co-wives of a Mormon husband have their own status hierarchy, and a wife's motives for naming may vary according to her place in the family hierarchy. Social position, a factor not included in the original evolutionary hypothesis of procuring paternal investment, may lead different wife-types to use children's names to retain their husband's attention to them and their children, to maintain affective bonds with natal kin, to demarcate themselves as autonomous units within the family, or to reaffirm religious commitment.

It would be unduly restrictive to limit all possible evolutionary psychological hypotheses just to those of a simple paternity confidence type, as social position and striving for social status have always been key concepts of sociobiology and evolutionary psychology. Some of these motives appear to be "cultural, not evolutionary" on grounds that they constitute nonreproductive motives. That would be wrong, as we already recognize that nonreproductive sex in bonobo chimps is adaptive (e.g., when one chimp greets another, it lets them know they are dealing with a friend rather than a foe). If control over one's behavior and especially reproductive behavior aids fitness, then we should reconceive some or all of these "nonreproductive" motives as indirectly reproductive motives. At the same time, evolutionary psychology in the past has not connected parental investment to natal affection, ego enhancement, and identity assertion to the extent needed to explain the study's data about why women name their babies as they do. Ethnographic analysis can suggest new types of evolutionary hypotheses.

The "culture enables us to transcend evolution" diehards should remember that the cultural environment is a source of selective pressures, just as the natural and biotic environments are. Even though cultural change often includes changes in the selective pressures at hand, an evolved psychology mediates the way human organisms respond to current selective pressures. Moreover, every species, not just Homo sapiens, has new traits in current generations that have not yet been shaped by new selective pressures, in addition to new selectively neutral variations, so the mere fact that many of our traits have no evolutionary past and are not adaptations no more enables us to "transcend biology and evolution" than it does other species. The availability of various possible options in choices and behaviors reflects more types of factors in humans than in other species; in addition to individual circumstances, perception, and cognition, it reflects culture.

On one hand, we cannot dismiss culture as "a mere proximate cause of behavior." The evolutionary story about natural selection and humans cannot be stated without incorporating culture into it. There has long been a consensus in philosophy of science that scientific theories apply to actual cases only when general statements about relationships between variables and singular statements about particular circumstances are combined. Generalizations of theory by themselves imply nothing about actual cases (this is a core of truth in the logical empiricist view that explanation and prediction work by subsuming particular events under "covering laws"). Particular circumstances that include individuality, mind, and culture form "initial conditions" or "antecedent conditions" that are necessary in order for generalizations in evolutionary theory to apply to humans at all, just as location and physical forces are necessary initial conditions for the applicability of Newton's theory to actual physical objects. The same generalizations, combined with different initial conditions, deliver different explanations or predictions; this is how science explains diversity or variation in nature using a few basic theoretical principles. In this case, the broad maternal strategy is a response to a cultural change. On the other hand, we cannot dismiss evolved strategies as "a mere potential to act in the ways cultural norms tell us to act." Culture cannot explain variations within a culture that violate those norms—and do so in ways that make evolutionary sense. In this case, women violate the patrilineal naming pattern, with an especially strong father-son bias. In this case study, women violate cultural norms, not randomly, but in a manner so that their struggle for power, status, autonomy, and love within the polygamous family increases their biological fitness.

In conclusion, Jankowiak's and Woodman's chapter marks one of the many ways evolutionary psychology and evolutionary anthropology, which have been mired in controversy surrounding how best to apply evolutionary theory to humans (Crawford, 1998), can work together with good results, even if many foundational issues involving evolution and culture remain mysterious.

REFERENCES

Cosmides, L. & Tooby, J. 1992. "Cognitive Adaptations for Social Exchange", in Barkow, J., Cosmides, L., & Tooby, J., eds., *The Adapted Mind: Evolutionary Psychology and the Generation of Culture.* New York: Oxford University Press.

Crawford, C.B. 1998. "The Theory of Evolution in the Study of Human Behavior: An Introduction and Overview", in Crawford, C. & Krebs, D., eds., *Handbook of Evolutionary Psychology: Ideas, Issues, and Applications.* Mahwah, New Jersey: Lawrence Erlbaum, pp. 3-41.

Holcomb, H. 1993. *Sociobiology, Sex, and Science.* Albany, NY: SUNY Press.

Holcomb, H. 1996. "Just-So Stories and Inference to the Best Explanation in Evolutionary Psychology", *Minds and Machines*, 6, 525-540.

Ketelaar, T. 1999. "Emotion and Motivation", paper presented at the Meeting of the Human Behavior and Evolution Society, Salt Lake City, Utah.

CHAPTER 4

SEXUAL SELECTION, GOOD GENES, AND HUMAN MATING

STEVEN W. GANGESTAD
University of New Mexico

Abstract. Sexual selection is selection due to differential access to quantity and quality of mates. One form of it, intersexual selection, is due to differential access to mates driven by the mate preferences of individuals of the other sex. The major question about sexual selection pursued in theoretical investigation during the past two decades is, what accounts for mate preferences that drive intersexual selection? Several plausible models have been developed, one or more of which may apply to any particular species. This chapter has two major aims. First, evolutionary psychologists are interested in the linked tasks of inferring historical selection pressures that shaped psychological adaptations governing behavior, and describing the nature of those adaptations. Williams (1966) proposed that the criterion for inferring selection pressures, and thereby identifying adaptations that resulted, is special design. The logic of Williams' approach is briefly described. Second, one particular process of interest is good genes sexual selection. This form of sexual selection occurs when members of one sex (I focus here on females) prefer as sexual mates individuals who possess markers of good genes, whose gametes benefit the chooser's offspring. The criterion of special design applied to good genes sexual selection is: Are there any features of female mate choice that possess special design for choosing mates for their gametes, such that it is unlikely that such features would have evolved were they not designed by selection for this function? After discussing relevant background theory, I present several lines of evidence that are suggestive of such special design.

H.R. Holcomb III (ed.), Conceptual Challenges in Evolutionary Psychology, 143–178.

INTRODUCTION

Sexual selection refers to selection on genes due to differential access to quantity and quality of mates. Although it received very little attention in the first 100 years after Darwin introduced it, sexual selection has been a hotspot of theoretical and empirical research in the past two decades. Darwin (1871) distinguished two forms of sexual selection. Intrasexual selection refers to selection due to differential access to mates due to competition between individuals of the same sex (e.g., physical intimidation). Intersexual selection refers to selection due to differential access to mates driven by the mate preferences of individuals of the other sex.

Probably the major question pursued in theoretical investigation over the past two decades is, what accounts for mate preferences that drive intersexual selection? At least four answers have been offered (see, e.g., Andersson, 1994; Cronin, 1991). The first is that the preferences have been selected because of genetic benefits passed on to offspring of the chooser. This view, the "good genes" sexual selection theory, states that choosers prefer mates who possess markers of genetic fitness (Trivers, 1972). The benefits that select for preferences in this instance do not directly affect the chooser's fitness but rather affect the fitness of her genes through effects on offspring quality and, hence, this form of sexual selection is referred to as indirect (e.g., Kirkpatrick & Ryan, 1991). A second is that the preferences have been selected because of material benefits passed on to either the offspring of the chooser or the chooser herself (Trivers, 1972). These benefits may come in many forms, and different theorists have emphasized different benefits (e.g., food, direct parental care, physical protection from predators or conspecifics, lack of disease that could be passed onto mates). These benefits directly affect the chooser's reproductive success and, hence, this form of sexual selection is referred to as direct (Kirkpatrick & Ryan, 1991). A third is that sensory biases of the mate chooser, which are presumed to have evolved for other reasons, are exploited by members of the opposite sex. Cross-assortative mating between individuals who have particularly strong forms of the sensory bias and those who have the trait that exploits the bias can result in selection for greater forms of the trait. This view, though elaborated and refined in recent years, is the one Darwin (1871) favored. Fisher (1930) noted that selection could amplify traits favored by sensory bias, a process he called "runaway selection". A fourth view introduced very recently is "chase-away selection" (Holland & Rice, 1998). This view is based on the same initial premise of runaway selection that individuals of the chosen sex will evolve to exploit sensory biases of choosers. Because selecting mates on the bases of these exploitive traits does not benefit choosers, however, selection is presumed to operate on choosers to lower preference for the exploitive trait. This selection, like runaway selection, can result in exaggeration of the exploitive trait to "trigger" the preference due to sensory bias. The conflict between the sexes hence results in what is perhaps a counterintuitive outcome: a coevolution of greater degrees of the exploiting trait in the chosen sex and lesser preferences for the trait on the part of choosers. This last form of selection may explain why some highly exaggerated, sexually selected male traits (such as extravagant plumage in some species of birds) may not be associated with male mating success (e.g., widow birds: Andersson, 1982; jungle fowl: Ligon & Zwartjes, 1995).

These selection processes are not mutually exclusive. All may operate in nature and, indeed, all may operate on preferences for the different traits within the same species. Hence, if females prefer mates who possess traits X, Y, and Z, it is possible that trait X is a marker of good genes, trait Y promises material benefits and trait Z is disfavored by sensory biases. A primary question with regard to the sexual selection operating on any specific species is, then, what forms account for the selection of which traits?

In this paper, I will focus on attempts to document and test the effects of one form of sexual selection, specifically good genes sexual selection, on human mating. In so doing, I do not assume that this form of sexual selection has historically been the most important form operating on humans (for other views of the evolution of human mating, see Buss & Schmitt, 1993; Campbell, Simpson, & Orina, in press; Hrdy, 1981, 1997; Symons, 1979). Rather, I focus on this form of selection as one illustration of how selectionist theories can be applied to a study of human mating and address questions of the form just posed. Good genes sexual selection is perhaps a good theory to examine from this standpoint because it has been controversial over the years. Recent theoretical as well as empirical developments, then, have been important to its application to humans. Because evolutionists must have some means of distinguishing the role that different selection processes have had on traits, before discussing good genes sexual selection specifically I briefly discuss criteria for evaluating adaptationist accounts, including sexual selection accounts.

A DIGRESSION INTO METATHEORY

Evolutionary biology is an historical science. Evolution consists of changes in gene frequencies in populations over time. Evolutionary explanations purport to explain why a genetic system (as well as the phenotypes of individuals in the system) evolved from some prior state S to a later (e.g., current) state S'. These explanations invoke laws or phenomena that are not dependent on the precise system under consideration or the time period under consideration. Thus, for instance, only four phenomena are recognized as forces of evolutionary change and, hence, all evolutionary explanations must invoke at least one of these phenomena: mutation bias, migration, random drift, and selection (e.g., Crow, 1986). In addition to these time- and system-independent phenomena, evolution depends on events specific to time and system. These events can often be thought of as constraining the nature of evolutionary change that can take place. Thus, for instance, some phenotypes may be less subject to variation due to mutational changes, which limits their evolution. Similarly, developmental constraints may limit what array of traits can simultaneously evolve (Gould & Lewontin, 1979). An evolutionary account of the evolution of a trait in a population often cannot specify *a priori* time- and system-dependent constraints or even specify them after the fact. Hence, evolutionary explanations are rarely complete in the sense that they specify all conditions relevant to the evolution of a trait. Yet evolutionary explanations invoking time-independent phenomena (in particular, explanations of adaptations due to natural selection) can be compelling. An evolutionary account is often rendered persuasive (despite incompleteness) because: 1) it is based on a model that predicts a specific outcome that is observed, even without taking into consideration extraneous constraints; and/or 2) a compelling case is made that facts are coherent when viewed through the lens of the evolutionary account, even when constraints have not been specified. In both instances, the epistemological appeal of the account is due to the fact that, given what we know in terms of general phenomena responsible for evolution, no other evolutionary account can do as good a job at fitting the facts. (See Holcomb, 1993, 1996, for excellent treatments of the epistemological issues underlying evolutionary explanation of human behavior.)

Adaptationism is an attempt to understand the evolutionary outcomes of natural selection. Within the adaptationist framework, there are two major forms of explanatory argument, each following these general lines: tests of optimization models and arguments from design.

Tests of optimization models

An optimality approach attempts to specify the strategy or set of strategies that would result from natural selection in the absence of genetic or developmental constraints (see Parker & Maynard Smith, 1991). An optimization model must first identify a set of all possible strategies in a given domain. For instance, a model of the optimal sex ratio of offspring should identify all possible sex ratios of offspring, a model of optimal allocation of effort to investment in an offspring should identify all possible amounts of investment, and a model of optimal preference for a trait should specify all possible preferences. Next, the benefits and costs of each strategy associated with each strategy must be specified. They must be expressed in terms of fitness gains or losses to include those associated with giving up the opportunity to allocate the effort to other endeavors). In some cases, the benefit and cost functions may be empirically estimated through experimentation. In other cases, they must be specified on an intuitive or theoretical basis. The optimal strategy is that which maximizes net benefits. Complexities can be taken into account, such that optimal strategies may be contingent on the relative frequencies of each strategy in the population (frequency-dependent optima) or the condition or phenotype of the individual (conditional or phenotype-limited optima) (See Parker & Maynard Smith, 1991, for a fuller discussion.)

A test of an optimization model entails asking whether the optimal strategy is found in nature. If so, the data are taken as evidence that natural selection has acted on the organism under investigation in ways specified within the model. If not, the model is not supported. A model may fail for a variety of reasons: The entire range of strategies may not have been specified, the cost-benefit functions may be incorrectly specified (perhaps due to a missing component of the selection process), or the model may be right in so far as specifying what natural selection would produce in absence of genetic or developmental constraints, but constraints prohibit the evolution of what is optimal. Model failure gives reason to explore these different possibilities (see Mayr, 1983).

Arguments from design

Dynamic sufficiency of a model is the extent to which the model can state fully the parameters and laws by which evolutionary change takes place (Lewontin, 1974). In a dynamically sufficient optimization model, all parameters relevant to the evolution of the trait under consideration would have to be specified (within tolerance limits; naturally, parameters with very weak effects may not have to be specified). These parameters include all major benefits and costs of the evolution of the trait, as well as the function forms of the effects of the trait under consideration on benefits and costs. While this may be possible in some cases, in many cases current knowledge does not permit development of such models. A powerful form of argument for the operation of specific selective forces can be made even in the absence of a dynamically sufficient model. This form of argument—the argument from special design—was explicated by George Williams in his 1966 classic, *Adaptation and Natural Selection* (see also Thornhill, 1991).

Although natural selection is not the only force responsible for evolutionary change, it is the only force that can account for the evolution of species-universal, complex traits. A trait exhibits "special design" for facilitating a particular beneficial effect if it shows specificity, efficiency, or economy in producing the effect (Williams, 1966). Natural selection is the only evolutionary force known to account for special design. Migration (which does not affect allele frequencies on species-universal traits), mutation, and drift (which are chance events) cannot generate functional design. Thus, when one has shown that a trait exhibits special design for a given effect, one has shown that natural selection was by far the most likely force responsible for the trait's evolution. One has also demonstrated that the effect for which the trait exhibits special design was by far the most likely cause of the trait's evolution (i.e., the effect is the function of the trait). In the words of Williams (1966), "The designation of something as the *means* or *mechanism* for a certain *goal* or *function* or *purpose* will imply that the machinery was fashioned by natural selection for the goal attributed to it" (p. 9).

An argument for special design does not require a demonstration that the trait under consideration is an *optimal* response to natural selection. It need only establish that the trait in question appears to be designed for a special purpose, whether or not it is the truly optimal design. (Obviously, the criteria of efficiency and economy for establishing special design imply that the trait is reasonably *good* at achieving a purpose, but need not imply optimality.) As a result, evidence about design may be particularly useful when theory does not allow precise identification of an optimal strategy.

Sometimes, however, optimization models and arguments from design supplement each other in an argument for some specific effect of natural selection. Because of incomplete specification of all parameters or function forms (e.g., the precise levels of benefits and costs associated with a particular trait), an optimization model may not tell us precisely the level of a trait that should have evolved under the specified selection forces with no additional constraints. A formal model may nonetheless inform us that *some* level of the trait should have evolved under the specified selection forces and hence can be used in an argument for special design. For example, parent-offspring conflict theory (Trivers, 1974) involves a game-theoretic model based on the notion that the genetic interests of parents and offspring differ (e.g., Parker & McNair, 1978); offspring will be benefited by an amount of parental investment greater than that which is optimal for parents. While parent-offspring conflict theory cannot specify what adaptations will develop as a result, it suggests that the resulting conflict can escalate, with each side developing costly tactics that offset each other. Haig (1993) has used the existence of costly traits used to regulate maternal levels of hormones in both human fetuses and their mothers to argue that the selection pressures specified in parent-offspring conflict theory have led to adaptations that importantly affect human pregnancy.

Good genes sexual selection has been controversial over the years. Only recently have models been articulated suggesting that, indeed, it can operate. In the following discussion, I will review theory and evidence relevant to the operation of this form of sexual selection, as well as suggest some forms of its outcomes specified by optimization models. I will then review evidence pertinent to its operation in humans, with particular attention given to special design.

GOOD GENES SEXUAL SELECTION: CONTROVERSY AND CURRENT STATUS

In their influential article on the evolution of human mating, Buss and Schmitt (1993) stated:

> Men may gain reproductively . . . in the currency of . . . better genes that are passed on to his sons and daughters. *The potential genetic benefit remains theoretically controversial in evolutionary biology*, appears difficult to test at the present (Trivers, 1985; Williams, 1975), but represents a viable theoretical possibility. [p. 215]

Two specific concerns are particularly pertinent. First, there's the question of whether natural populations contain sufficient genetic variation in fitness to fuel the process. Second, there's the question of whether genetic quality could possibly be important relative to material benefits in species characterized by substantial biparental care, such as humans.

The Existence of Genetic Variation in Fitness

The view that fitness has little genetic variance is a long-standing one. Fisher's fundamental theorem of natural selection states that directional selection on a trait reduces its genetic variation. Fitness is by definition under directional selection, and thus should have its genetic variation depleted:

> It is a well-known result of population genetics theory that *natural selection tends to exhaust the additive genetic variance in fitness* (Fisher 1930, 1941). . . . This creates a serious difficulty for the good genes hypothesis (Maynard Smith 1978; Partridge 1983; Taylor and Williams 1982). [Charlesworth, 1987, p. 22]

Naturally, if there's no genetic fitness variation in a pool of potential mates, there is no choosing for good genes; nobody's are better than anybody else's. Here, Mark Kirkpatrick, the most ardent opponent of the good genes process through the 1980s, cites himself as having used the fundamental theorem to *dismiss* good genes selection —*but to have done so in error.* As he notes, there are reasons *not* to expect zero variation in genetic fitness.

> An unfortunate legacy of Fisher's (1958) "fundamental theorem of natural selection" is that many evolutionists have been led to believe that there should be little heritable variation for total fitness in natural populations (Price and Schluter 1991). *This view is the basis for the argument made by a number of workers that the good genes process does not work (e.g., Kirkpatrick 1986).* In fact, mutation, gene flow, and fluctuating selection create heritable variation for total fitness in natural populations. [Kirkpatrick, 1996, p. 2134, emphasis added]

Perhaps the most evolutionarily meaningful measure of the genetic variation in a trait is the additive genetic coefficient of variation (CV_a), the square root of the trait's genetic variance standardized by the phenotypic mean times 100 (to remove decimal places). This measure is evolutionarily meaningful because Fisher's fundamental theorem states that the expected proportional change in a trait in a single generation due to selection (and hence the rate of evolution due to selection) is a function of the square of this value (without multiplication times 100). Hence, the larger the CV_a, the greater the potential rate of evolution of the trait. Again, the argument based on Fisher's fundamental theorem is that fitness and traits that importantly affect fitness would have their genetic variation exhausted by past selection and hence should have little genetic variation (e.g., a small CV_a) and little potential for evolution now. To address the issue empirically, Houle (1992) and Pomiankowski and Møller (1995) recently compared CV_as across different sorts of traits based on the extensive literature on many organisms: ordinary morphological traits, a subset of these known to be under stabilizing selection, and fitness components such as longevity and fecundity. As it turns out, not only do fitness traits *not* have smaller genetic variance, they actually have substantially *greater* CV_as. Specifically, life history traits such as fertility, fecundity, and survivorship have CV_as in the range of 10-20. The amount of genetic variance in fitness itself (which has not been directly measured) has been indirectly estimated using a variety of methods to be between 10 and 30 (Burt, 1995). By contrast, ordinary morphological traits and those known to be under stabilizing selection have CV_as that average about 5. Life history traits, then, have about 2-4 times the CV_as that ordinary traits have.

Another way to think about these results is in terms of the ratio of the trait values of individuals at the extreme ends of the trait distribution—say, 2 standard deviations above and below the mean. CVs nicely translate into these ratios. For CV_as of ordinary morphological traits, such as height, this ratio is about 1.2. The tallest couple percent are about 20% taller than the shortest couple percent. For fitness traits, it is over 2. (Specifically, if the CV_a is 20, it is [1+.2x2]/[1-.2x2] » 2.3.) Individuals at the top have over twice the longevity or fecundity of those at the bottom, *due to genetic factors alone*. This amount of genetic variation would exist for height only if a couple percent of the population were genetically predisposed to be 8 feet tall or more, and a couple percent genetically predisposed to achieve an adult height of only 4 feet. This amount of variation is astonishingly high.

Given Fisher's fundamental theorem, how could this be? Fisher

(1930) himself provided the framework for the answer. Populations, he said, are characterized by a shifting equilibrium, rather than one that remains constant often characterizes populations. Each generation, those who successfully reproduce have mean fitness greater than that of the whole population. So too, however, within each generation various processes degrade the fitness of progeny relative to their parents, including harmful changes in the environment, such as changes in pathogens' ability to parasitize their hosts, or harmful changes in the organism itself due to mutation. At a stable equilibrium natural selection and degrading processes balance out so that the population fitness remains unchanged. The fundamental theorem applies only to natural selection, not processes that degrade fitness. Degrading processes not only ensure that populations are continually characterized by some level of maladaptation, but they also prevent populations from becoming uniform; some individuals will inevitably be less adapted than others (Burt, 1995).

One question that arises concerns the extent to which the variation in fitness is due to mutation alone and to what extent it is due to degradation of the environment (e.g., host-pathogen coevolution; Hamilton, 1982; Hamilton & Zuk, 1982). Based on empirical parameter estimates such as the mutation rate, selection against mutations, and so on, the CV_a of fitness due to *mutation alone* can be theoretically estimated. Recently, Charlesworth and Hughes (1998) estimated that figure to be about 8 for *Drosophila*. (Charlesworth [1990] had earlier estimated a value of about 17.) The figure for other larger organisms is likely to be similar. Perhaps half of the genetic variance in fitness, then, is due to mutation alone, the other half to other factors. More precise estimates must await further research. (In this model, incidentally, the vast majority of deleterious mutations that exist in the population at any given time have been in the pool for multiple generations. Whereas the per genome mutation rate is assumed to be about 1, an average of at least 20 mutations per individual are assumed to reside in the population because, given weak selection against them, they have not yet been eliminated by selection [Charlesworth, 1990].)

It might be noted that alleles that are good genes at one point in time need not be intrinsically better than alternate alleles. Host-parasite coevolution imposes changing selection pressures on host genes, maintaining heritable fitness in individuals. Thus, an allele that is a good gene today might be selected against in future generations, and an allele that is selected against now could become a good gene in the future. Because no gene is inherently better than alternate genes, the population does not become more fit through selection. (Hence, the existence of genetic variance in fitness does not mean that the long-term genetic fitness of the population can be improved through eugenics.)

The important point here is that within natural populations there is abundant genetic variance in fitness, which might well lead to preference for mates possessing good genes. Kirkpatrick has recently modeled the process as follows: Assume a trait taps genetic fitness (for now, simply imagine ability to hold a central territory, as in a lekking species such as black grouse). There is no benefit to be gained by mating with a particular individual aside from the genetic benefits to offspring fitness. That is, material benefits are zero. Mate preference is a parameter that produces an exponential favoritism for mates possessing the good genes trait. The CV_a of fitness is 17. And, there is stabilizing selection against extreme mate preferences due to the costs they entail (in terms of assessment, search, etc.) According to the model, good genes selection will *always* produce preferences for individuals who possess the good genes trait, resulting in up to a few standard deviations of change in the trait. This is a story much different from that Kirkpatrick (1986) told 10 years ago. As he notes, his model may account for several good, empirical examples of good genes selection. Specifically, it has been demonstrated in several species that male traits preferred by females are correlated with offspring survivorship, independent of direct material benefits (see Kirkpatrick, 1996). With or without an explicit model, these examples provide evidence for special design of mate preferences for identifying mates with good genes.

If, indeed, the good genes process underlies the evolution of many sexually selected traits (in particular, those preferred by females), these traits should tend to reflect the overall condition of males and thereby "capture" much of the genetic variance in male fitness (Rowe & Houle, 1996). In conjunction with the finding that fitness traits tend to have much higher genetic variance than ordinary morphological traits under stabilizing selection, a prediction follows from the good genes hypothesis: Sexually selected morphological traits should have CV_as closer to those typical of life history traits such as fecundity and mortality (10-20) than to those typical of non-sexually selected morphological characters. Pomiankowski and Møller (1995) surveyed the extant literature and found precisely this result. On average, sexually selected traits had a CV_a of 17, over three times larger than that typical of non-sexually selected traits and highly similar to those of fitness traits. Though other models of sexual selection may be compatible with this finding as well, it is a finding that can be predicted *a priori* by the good genes hypothesis.

Good Genes Selection with Biparental Care

Humans obviously biparentally invest in offspring, with intensity and duration unparalleled in the biological world. Given the importance of finding a mate who will effectively invest in offspring, can the mate's genetic quality play anything more than a trivial role?

Alterations in the Kirkpatrick model

Let us first consider this question with Kirkpatrick's model. Specifically, imagine that the trait covarying with genetic quality—let us now say more generally competitiveness for resources—also covaries with investment capabilities. In that instance, selection favors the trait preference and the trait itself, independent of the good genes process. What, then, happens to the selection due to the good genes process? If the other parameter values in the model are unchanged, good genes selection remains *unaffected*; it affects preferences and preferred traits the same as when there is no other directional selection on preferences.

In fact, parameter values *will* change and, in some instances, reduce the potency of good genes sexual selection (indeed, *any* sort of sexual selection). Some changes, however, should actually *enhance* selection for good genes. One very important parameter in Kirkpatrick's model is the strength of selection against any preference due to the costs of mate assessment. But if an individual is investing in offspring considerably, it can pay to incur greater costs to assess a mate's genetic quality, and particularly when, as in humans, errors in mate choice compound due to multiple offspring with the same mate.

To illustrate this point, consider an analogous effect outside the arena of sexual selection. I just noted that increases in parental investment could result in increased willingness to pay for mate assessment. The same applies to assessment of and "search" for offspring. Figure 2 shows a model of disinvestment in offspring—adaptive abortion—by Haig (1990). Offspring are assumed to vary in fitness, which is represented here as a two-fold difference. A mother can invest in all offspring and achieve a specified level of fitness payoff per unit of investment, represented by this slope of the line tangent to the mean fitness returns per offspring. If she can detect differences in fitness among offspring, however, she might be able to increase fitness payoffs by selective investment in less than all, but only if the cost of aborting some is not too heavy. In this instance, if she invests in only the top half, the cost of aborting one can only be an amount that allows her to achieve at least the same payoffs to investment as investing in all (see Figure 1). If her costs are greater, her payoffs per unit of cost will be less than what she could get by investing in all offspring. The cost itself must be thought of in relative terms; it must be small *relative* to the amount of total investment per offspring. If the investment is very high, then adaptive abortion and selective investment in only a subset of conceptuses is more likely.

Human parental investment is enormous. If abortion pays in any organism it should pay in humans. In fact, data on human conceptions suggests that embryonic mortality *is* very high, with as many as 70% of human conceptuses not making it, most due to transplantation failure. As some have noted, this high level of embryonic mortality may be adaptive quality control (Forbes, 1997). It appears the time window of opportunity for proper implantation is a fairly narrow one, such that conceptuses that don't make it are often developmentally delayed and hence probably *are* less fit genetically.

The economics of pregnancy loss:
Haig's (1990) model

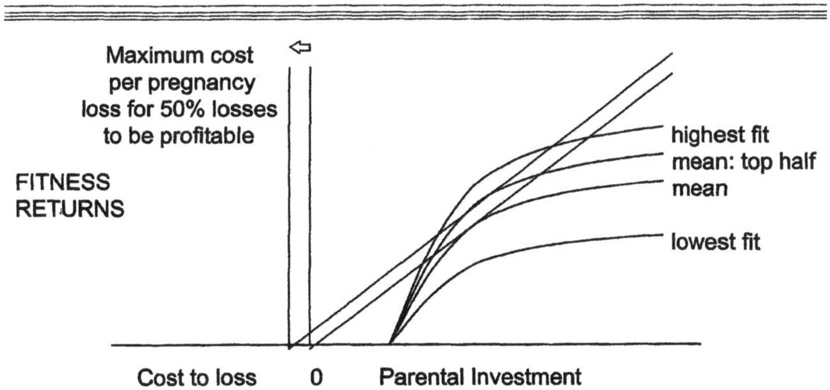

Note: The smaller the cost of loss relative to "full" parental investment (e.g., the greater the parental investment) the greater the benefit of loss.

Figure 1. A model of pregnancy loss (Haig, 1990). See text.

Again, just as high costs to rearing offspring can mean that greater costs associated with "finding" the right quality offspring can pay off, they may mean that greater costs to detecting genetic quality in mates can pay off, with correspondingly greater shifts toward good genes preferences. Though good genes sexual selection might be less potent when there's substantial biparental investment due to other reasons, it is not clear that the net reduction need be sizable.

Interaction between biparental investment and good genes sexual selection

Perhaps the most important effect of substantial biparental investment on the good genes process is not its potency, but rather the fact that good genes selection and good provider selection may interact. I've assumed that investment capabilities—such as competitiveness for resources—positively covary with good genes, such that the individual who offers greater genetic benefits also offers greater material benefits. This assumption has intuitive appeal in that individuals who have higher genetic fitness do so because of their phenotypes; they are presumably more competitive. There's an important difference, however, between *being able* to provide material benefits and *actually providing* them to any individual mate.

When individuals with intrasexual competitive advantages can also offer genetic benefits, what will they do? One way of thinking about this question says they will actually invest *less*. The argument goes as such: Let us assume effective polygyny, so that males can pursue multiple mates. Assume males differ in resource competitiveness, which covaries with genetic fitness. Because, independent of material benefits, a male with better genetic quality will be preferred by a female, males who have better genes pay a smaller cost per mating independent of investment in the offspring. That is, their mating effort is more efficient. How much will these males invest in individual offspring? If each maximizes his fitness payoffs, each will invest *less*. (See Figure 2.) Stated intuitively, because their mating effort is more efficient (they're more successful per unit time), these males will do better by allocating less energy, time, etc. to parental investment. This prediction is not novel; Nancy Burley set a version of it forth back in the 1980s (Burley, 1986).

The preceding argument is not the only way to think about this issue; it has its assumptions. One main assumption is that mating effort and provisioning of material investments are significantly exclu-

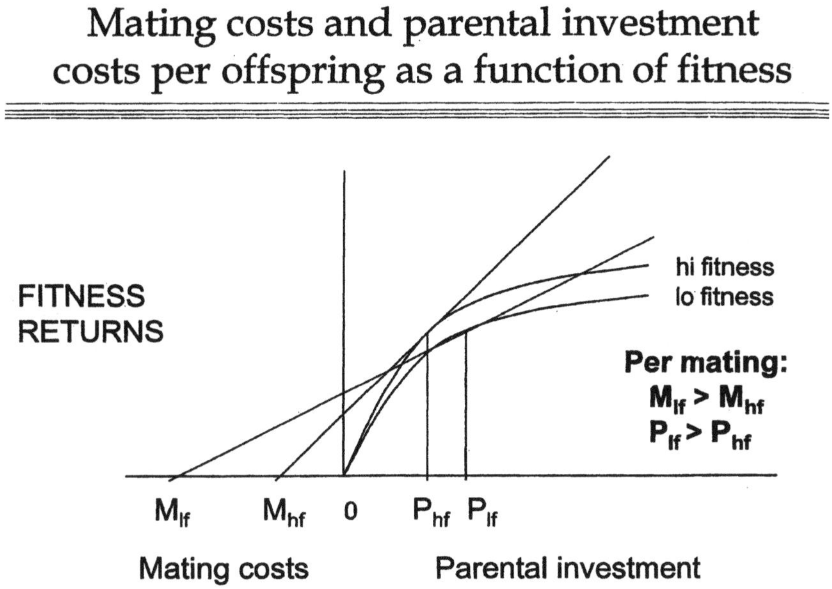

Figure 2. A model of optimal parental investment per offspring by males differing in fitness and, as a result, mating costs per offspring. See text.

sive activities. In some instances, that simply cannot be assumed. If demonstrations of physical dominance attract mates and also promise that offspring will be well-protected against conspecifics, one and the same expenditure has mating benefits as well as parenting benefits. Nonetheless, in many cases it should hold: For instance, the time and energy an individual puts into demonstrating physical dominance may well take away from time and energy devoted to feeding offspring.

Another main assumption is that multiple matings by males are assumed to be possible. What may be particularly compatible with this model is female multiple mating—mating with a main in-pair partner as well as one or more extra-pair partners. In such a case, males may have offspring with very little parental investment, despite a large typical amount of investment in offspring. (In the model, these males are of course overrepresentatively those who exhibit markers of good genes.) Yet this mating system itself must be considered a result of selection. The evolution of extra-pair sex on the part of females has received a fair amount of attention as of late, but remains incompletely understood (Petrie & Kempenaers, 1998). Hence, the precise conditions that would render the model in Figure 2 plausible are not known. Dispersion of food sources (and breeding density), amount of genetic variation between males (e.g., due to differences in the mutation rate), importance of male investment, and ecological constraints on male guarding of female mates may be important determinants of the extra-pair paternity rate in biparentally investing species (Petrie & Kempenaers, 1998).

If males with markers of good genes actually provide fewer net material benefits to females, what is the effect on selection for preferences? Independent of the material benefits, we should expect that females would prefer good genes males. Independent of genetic benefits, however, we expect that females would actually prefer males without good genes markers. Taken together, the net benefits might produce little or weak preference for good genes males. According to this thinking, then, good genes males should not be highly preferred as co-investing mates. Rather, their mating advantage should be particular to when they are being chosen as a non-investing mate (i.e., for their genes only), as in extra-pair mating. That males with good genes should be particularly preferred in mating contexts in which genes alone are considered may seem intuitively obvious. But note that if these males also provide material benefits, the opposite prediction must be made: Genetic plus material benefits (as in investing mate) would produce stronger preference for good genes males than genetic benefits alone. Hence, preference for good genes males particularly in extra-pair mating contexts is a prediction made by one model of the covariation between genetic benefits and material benefits provided by males to offspring; it is not a general prediction.

Møller and Thornhill (1998) surveyed the literature on bird species in which there is biparental care and found that some species, males who possess sexually selected traits (such as colorful displays or long ornamental tails) do indeed exert less parenting effort (as measured by, for instance, the proportion of feeding visits a male makes) and others in which they do more. There is little doubt that male traits can evolve to advertise good parenting, and so the latter cases are not surprising. Perhaps the most interesting finding in Møller and Thornhill's analysis is that a strong predictor of whether attractive males invest more or less is the extra-pair paternity rate. When there is a lot of extra-pair paternity (in excess of 20% of all conceptions), attractive males tend to invest *less*. When the extra-pair paternity rate is very low (less than 5% of all conceptions), attractive males tend to invest more. As just noted, this finding can be predicted from the above model. When females engage in extra-pair copulations (EPCs) and thereby provide males with multiple mating opportunities, males attractive by virtue of possessing markers of good genes might be expected to allocate a lot of effort to pursuing those opportunities and hence invest less in parenting.

Two conclusions about good genes selection within biparentally investing species, then, can be offered. First, the process is not eliminated. Genetic variation in fitness between potential mates remains substantial, and it can pay to monitor information relevant to that variation. Second, the good genes process may interact with material benefits processes. Under certain conditions, (specifically, when females provide males a pool of multiple mating opportunities), males with good genes may provide fewer material benefits, even if they have potential to provide more.

TESTS OF THE GOOD GENES MODEL

A Criterion: Fluctuating Asymmetry and Developmental Imprecision

Theoretical considerations provide a reason to suspect that good genes sexual selection can occur, even in biparentally investing species such as humans. I now turn to consider empirical tests of good genes hypotheses.

To test whether the good genes process has occurred in any particular species, we would optimally like to measure differences in genetic fitness directly. Unfortunately, no such measure exists. That being the case, a measure that taps maladaptation *partly* due to genetic factors is the next best thing. In the last decade, individual differences in *developmental imprecision* (or developmental instability) have become a focal point of research on sexual selection (see Møller & Swaddle, 1997; Møller & Thornhill, 1997b). Developmental imprecision is the imprecise expression of developmental design due to genetic and environmental perturbations, importantly including mutations and pathogens. Hence, those factors that contribute to genetic variation in fitness also contribute to developmental imprecision. Indeed, it is not inconceivable that mildly deleterious mutations (which, again, may account for about half or more of the genetic variance in fitness) affect fitness largely through "random" deleterious effects on the expression of adaptive design.

The primary measure of developmental imprecision used in biology is *fluctuating asymmetry*, absolute asymmetry in bilateral traits due to random errors in the development of the two sides. We find that the greater the developmental imprecision, the greater the number of random errors in development, and, if the errors on the two sides are independent, the greater the absolute asymmetry.

In research conducted at the University of New Mexico, we have measured a number of asymmetries in humans—of the ears, elbows, wrists, ankles, feet, and fingers (see, for instance, Furlow et al., 1997, 1998; Gangestad & Thornhill, 1997a,b, c; Gangestad, Thornhill, & Yeo, 1994; Thoma et al., 1998; Thornhill & Gangestad, 1994, 1998; Thornhill, Gangestad, & Comer, 1995). The asymmetry that exists in these traits is very small. The mean asymmetry is 1-2 mm., so small that you cannot detect it reliably through normal social interaction. Hence, the asymmetries we measure cannot and do not serve as cues by which individuals assess others' developmental imprecision. The reason we measure it, then, is because *these asymmetries purportedly are markers of underlying developmental imprecision, which substantially affects the overall phenotypic fitness of individuals.* Some people find it astonishing, if not unbelievable, that tiny departures from symmetry can tell anything about the phenotypic, let alone genotypic, fitness of an individual. Recently, Randy Thornhill and I developed a simple quantitative model for analyzing the validity of the "signal" in FA measurements and looked at the literature (mostly on nonhuman organisms) in light of this model. The results, we think, suggest that the signal in asymmetry measures *can* tell us about the phenotypic and genotypic fitnesses of individuals. At the same time, there is a caveat, which I discuss later.

I will briefly describe this model (see Gangestad & Thornhill, 1999, for more details see also Gangestad & Thornhill, in press). First, imagine a population in which all individuals experience exactly the same number and intensity of developmental errors. That is, there's no variation in developmental imprecision. That being the case, do all individuals have identical asymmetry of a given character? No. Asymmetry will vary simply because, by random processes, errors in some individuals accumulate in one direction or the other, and in many individuals they "cancel out." A population of individuals with *identical* developmental imprecision will have normally distributed asymmetry.

Let us now add individual differences. That is, assume some individuals have more mutations or pathogen effects than others and hence more developmental errors. These differences affect the variance in asymmetry produced by errors: more imprecision, more errors, more variance. In Figure 3, I've depicted two distributions in asymmetry that differ in variance. The one with greater variance contains individuals who have greater developmental imprecision than those within the distribution with lesser variance. The former distribution has a greater mean absolute asymmetry (ignoring signs) than the latter distribution, which is why absolute asymmetry (FA) partly taps these individual differences. Of course, individual differences are more realistically represented as continua. Hence, we really ought to think of a population as mixture of many overlapping, normally distributed distributions of right-left differences that differ in mean. Again, the more the imprecision there is, the more variance in asymmetry there is.

When we look at a distribution of asymmetry, however, we naturally cannot see subdistributions corresponding to different levels of imprecision. We see a single distribution. How can we tell, then, whether asymmetry reflects *individual differences* in imprecision or just the random effects of error? That is, can we know that individuals actually fall into different distributions of individuals with differing propensities toward developmental imprecision and, if so, how? As it turns out, it seems we can, *by looking at the kurtosis of the distribution*. Compounds of normal distributions with the same mean but different variances are not normal. They are leptokurtic (peaked

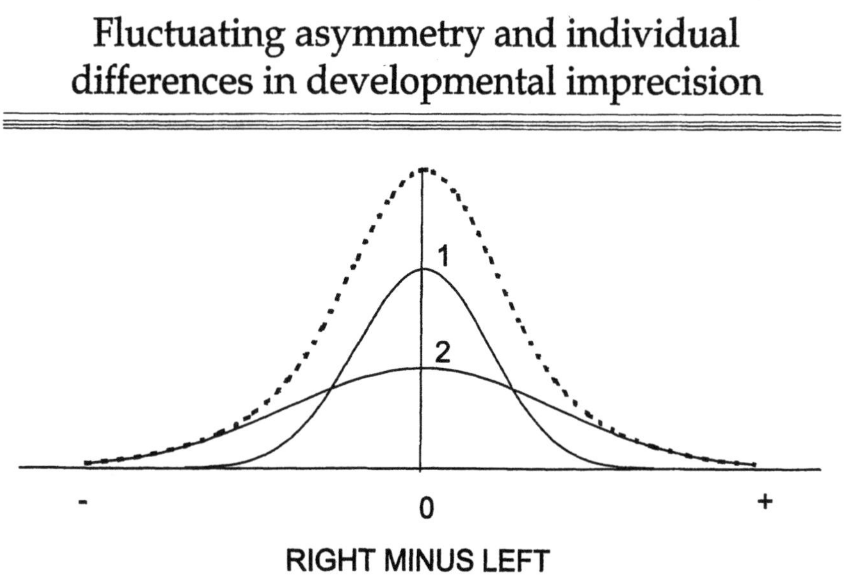

Figure 3. The distribution of left-right differences (asymmetry) due to developmental errors for two subpopulations differing in proneness to developmental error (solid lines labeled 1 and 2) and the total population (dotted line). See text.

with long tails). Moreover, the greater the differences are in underlying predispositions to developmental error, the greater the leptokurtosis is.

> Using some simulation modeling, we estimated the way that kurtosis of asymmetry changes as a function of variation in developmental imprecision (Gangestad & Thornhill, in press). As it turns out, there appears to be a systematic relationship between the CV of the individual differences and kurtosis of asymmetries, one highly insensitive to the exact form of the distribution of the individual differences. The greater the CV is, the greater the leptokurtosis is and in a highly specific way. (For instance, a CV of individual differences in developmental imprecision yields a kurtosis [g_2] of about .4-.5 under a variety of different models of the distribution of individual differences.) This is useful because it permits estimation of the CV of the underlying individual differences in developmental imprecision using the kurtosis of asymmetry (see also Van Dongen, 1998; Whitlock, 1999).

Using this logic, we estimated the CV of developmental imprecision in a number of different species and found something rather striking: No matter what the species, the estimated CV of underlying individual differences in developmental imprecision is 1) very similar, and 2) very high—over 20. (As noted earlier, a CV of 20 implies a two+-fold difference in extreme values; such variation is not typical of ordinary morphological traits.) Since we performed our survey of the literature, Hallgrimsson (1998) published kurtosis data on over a dozen skeletal asymmetries in nine different species, thereby adding a substantial amount of data to the literature. For all species, the kurtosis was significantly leptokurtic. The estimated CVs for individual species ranged from 17 to 25 and averaged 21.

Though the CV for developmental imprecision is estimated by these procedures to be very high, it is not in fact high relative to the theoretical expectation that developmental imprecision is strongly associated with fitness. Fitness traits *have* CVs on the order of 20+. Based on heritability studies of FA (which, incidentally, show the asymmetry of a single character to have a heritability of just a few percent; Gangestad & Thornhill, in press; see also Møller & Thornhill, 1997a), the heritability of underlying developmental imprecision can be estimated to be about .35-.55. These figures, used in conjunction with the estimated CV of phenotypic developmental imprecision, yield an estimated additive genetic CV (CV_a) of developmental imprecision of 14 to 18. Earlier, I discussed the fact that the estimated CV_as of sexually selected characters average about 17, impressively close to what would be expected if these traits tended to "capture" overall condition and hence the genetic variance underlying fitness (Rowe & Houle, 1996). Something similar can be said for the CV of developmental imprecision. It is impressively close to what would be expected if developmental imprecision "captured" much if not most of the genetic variance in fitness.

On this basis, then, developmental imprecision looks like a fitness trait and, thereby, FA appears to convey important information about the condition of an organism. But the relationship between FA and fitness can be examined more directly. Lots of studies have examined associations of FA with fitness traits and mating success in a wide variety of nonhuman organisms. Recent meta-analyses and reviews indicate that reliable nonzero associations clearly exist (Leung & Forbes, 1996; Møller, 1999; Møller & Thornhill, 1997b; Thornhill & Møller, 1997). Admittedly, the correlations reported in these reviews are small to moderate: an average of -.2-.25, figures that may not seem very impressive on the face of it.

But here is the caveat I mentioned: Despite the fact that large variations in developmental imprecision appear to exist (as evidenced by the estimated CVs), the individual differences in developmental imprecision account for a small amount of variance in a single trait's asymmetry. The simulation models (Gangestad & Thornhill, in press) suggest (based on kurtosis) that they account for only about 7-8%, the remaining 90+% due to the random effects of developmental error. What this means is that, even when developmental imprecision correlates very strongly with a fitness trait, you can't expect a single trait's FA to correlate highly with it; it is too noisy a measure. To estimate the correlation between underlying developmental imprecision and a trait, then, you must adjust it for the noisiness of FA.

Let us consider the specific values reported in meta-analyses. The mean correlation between a single nonsexually selected trait's asymmetry and mating success in the literature is about -.13, reliable though small (Møller & Thornhill, 1997b; see also Gangestad & Thornhill, in press). But the correlation adjusted for the noisiness of FA is -.13/Ö.07, about -.4 to -.5, a value that is moderately high. The same is true of fitness traits more generally. Leung and Forbes (1996) estimated the mean correlation between FA and these traits to be -.26. Adjusted for the noisiness of FA this value becomes close to 1. Of course, fine points can get lost in a meta-analysis; in some species, correlations may be close to 1, in others near zero. Moreover, these values may be overestimated due to publication bias and the file drawer problem. Even if these values are overestimated by a factor of 2, however, the analyses make a very important point at a gross level: Developmental imprecision must be at least moderately correlated with variation in fitness traits and mating success in at least some species.

Fluctuating Asymmetry in Humans

Let us now turn to humans. A number of researchers have explored the effects of human developmental imprecision, including my colleagues Randy Thornhill, Ron Yeo, Bob Thoma, Bryant Furlow, and Tara Armijo-Prewitt, as well as John Manning and his colleagues, Bob Trivers, Linda Mealey, Joanna Scheib, and more. In our studies, we have measured 6-10 asymmetries of the body (again, including the ears, fingers, wrists, etc.) and summed them into a single index. (Using multiple traits increases the signal in the measure so that developmental imprecision accounts for 25% rather than less than 10% [Gangestad & Thornhill, 1999].) Much of this work has investigated sexual selection through associations of asymmetry. Other work has addressed associations between purportedly important phenotypic traits and asymmetry. For instance, Thoma et al. (in press) measured brain characteristics on MRI images of 28 normal individuals, the major ones of interest being atypical anatomic asymmetries. For instance, humans typically have a larger planum temporale on the left hemisphere, presumably due to language specialization. But some have atypical asymmetries, either a somewhat larger right than left planum, or an extremely large left relative to right planum. The correlation between a composite of these atypical brain asymmetries and the index of minor body asymmetries we use was examined and, lo and behold, a strong relationship was found. Another feature associated with more developmental imprecision was a smaller corpus callosum, the structure that connects the two hemispheres. Interestingly, individuals with a variety of neurodevelopmental disorders also have smaller callosa as well as atypical asymmetries (see Yeo & Gangestad, 1998, for a review). These findings are consistent with the broad claim just made, that body asymmetry reveals something about developmental integrity of the whole phenotype.

Some of the research we have done has been aimed at the question of whether FA is associated with human male "mating success," as it is in many other species. One way in which this issue has been examined has been simply to correlate FA with a variety of self-reported sexual behaviors such as total number of sex partners in a lifetime, partners outside of a committed relationship, and being chosen as an extra-pair partner. An assumption made here is that men preferred by potential sex partners will, on average, have more of them. Table 1 combines results from all studies. With all measures, correlations are highly significant, though not large. As I've noted, however, FA measures are noisy. Correlations disattenuated for noise average about .4, fairly substantial (see also Gangestad, Bennett & Thornhill, in press). In a noncontracepting group of Mayans in Belize, Waynforth (1998) found that more symmetrical men tended to have more partners (the effect size was similar to that reported in Table 1, though not statistically significant given small sample size) and, moreover, had significantly greater numbers of offspring.

Two questions arise. The first is: what traits mediate these effects? The asymmetries we measure are undetectable in normal social interaction and hence cannot directly affect mate choice. Table 1 also shows a few variables associated with asymmetry, which females may prefer. I would not argue "fighting" per se is a trait females

Table 1. Estimated Correlations Between Developmental Imprecision and Male Traits in a College Population

Trait	Estimated Correlations With Developmental Imprecision	N
# of sex partners lifetime	-.37	531
# of EPC partners	-.32	170
# times an EPC partner	-.26	90
"physicality" (vigor, robustness)	-.59	99
informal "status"	-.59	99
# of intrasexual fights	-.44	95
escalation of intrasexual fights	-.86	42
intelligence	-.41	95
investment in relationship partner	.41	199

Note. Estimated from results in Thornhill & Gangestad (1994), Gangestad & Thornhill (1997, 1998a, 1998b, 1998c), Furlow et al. (1997, 1998), Simpson et al. (in press). See Gangestad & Thornhill (in press) for estimation of correlations with developmental imprecision from correlations with FA.

prefer, but at this age fights are often over minor status issues. More symmetrical men, we think, are more likely to stand up for themselves and not take a back seat to other men. That is essentially what this variable of social "status" presented also taps. More generally, then, low FA men may tend to join the fray of male intrasexual competition, as it occurs in informal social situations.

The second question is: what sexual selection forces account for the mating advantages of symmetrical men? We might be tempted to conclude that, given the traits associated with asymmetry (male intrasexual competitive traits and intelligence) it is probably sexual selection due to preferences for material benefits. In fact, after presenting these asymmetry-trait associations at a Human Behavior and Evolution Society meeting a few years ago, a couple of colleagues suggested that something that had been very interesting was now not so exciting. The idea, I think, was that we already know that women should prefer male competitive advantages because ancestral men who possessed them had greater access to resources. If symmetry taps this dimension of male competitiveness, what is the need for good genes sexual selection theory? Given what I've said to this point, the reply should be predictable: For a number of reasons, we shouldn't *assume* that men who enjoy intrasexual competitive advantages actually provide more material benefits—not, at least, when these advantages are indicators of a two-fold difference in genetic fitness. As I have argued, one model of selection suggests that males preferred for their genetic benefits will actually provide fewer material benefits to the females who choose them. This is a prediction that can be made by a model of good genes selection but not by alternative models.

One critical issue, then, is whether symmetrical men provide more or fewer benefits to mates. To explore this issue, we had men and women in long-term romantic relationships rate both themselves and their partners on a variety of relationship investment scales (developed by Ellis [1998]). Correlations were entered into a causal model with total male investment as the outcome variable and male expected income and female attractiveness as additional predictors. In line with some of the nonhuman data, males with greater developmental precision invested *less* (Gangestad & Thornhill, 1997a).

On specific measures, men with greater precision were significantly less honest and more sexualizing of other women, precisely what you might expect if they were engaging in mating efforts outside the relationship. There was one (not quite significant) exception to the overall trend: Men with greater precision were seen as more physically protective. Analysis of individual items indicates that this effect is due to symmetrical men being seen as more *able* to protect rather than more willing to take the time to protect. The way that we interpret this result relates to what I mentioned earlier: Some mating efforts are not exclusively mating efforts. Intrasexual competition and extra-pair mating efforts may not compromise ability to physically protect a mate in the same way they do, say, spending time with a mate. Of course, this interpretation needs further exploration (for additional discussion of relevant theory and data, see Gangestad & Simpson, 2000).

Female Design for Good Genes Preferences?

It appears, then, that more symmetrical men tend to have more sex partners (at least in populations of college students) and possess traits related to intrasexual competition that may be responsible for these effects. Moreover, they appear not to provide their mates with greater material benefits. These results, while consistent with good genes sexual selection, are not compelling evidence for such selection. Even if intrasexual competitive traits mediate symmetrical men's success at finding sex partners, we do not know that the effects involve female choice rather than direct male competition (e.g., intimidation). Furthermore, the fact that symmetrical men do not invest as much in their dating relationships as their asymmetrical counterparts is not nearly so convincing evidence that females have evolved preferences for them, despite receiving fewer material benefits, as finding that they invest less in offspring in circumstances more similar to those of our evolutionary past in which female preferences might have evolved (i.e., in hunter-gathering societies).

To further test the good genes hypothesis, we have sought evidence of another kind. Specifically, we have asked whether this selection process has left telltale signs for its operation in design features of female adaptations that account for the preference. Earlier, I argued that, if females have evolved to engage in extra-pair copulations (under specific circumstances) and hence in evolutionary history have provided males with a pool of multiple mating opportunities, the good genes model suggests that females should have evolved to particularly prefer males possessing markers of good genes as extra-pair partners. The argument is based on the notion that these males will provide fewer material benefits to their offspring and hence will be less preferred as long-term, investing mates. Again, alternative views do not lead to this prediction. A question we might ask, then, is whether there exist design features of female preferences that specifically advantage men possessing markers of good genes for their genetic benefits? Three lines of evidence suggest yes.

Female orgasm

The first concerns female orgasm. Baker and Bellis (1995) have proposed that female orgasm functions to retain sperm. Fertilization itself does not require a sperm retention device. Yet such an adaptation would have been useful in the context of multiple mating: a mechanism that could bias fertilization in favor of one mate's sperm over another's. In multiple mating contexts, at least one male is probably a noninvesting male. And, Baker and Bellis (1995) found that, in multiple mating women, patterns of female orgasm would have favored the sperm of their extra-pair partners over that of their in-pair partner in absence of contraception. This result makes sense if female adaptations underlying extra-pair sex are designed to garner genetic benefits. If so, then one might also expect that female orgasms favor the sperm of men who possess putative markers of good genes, such as symmetry. Consistent with this expectation, we found that the partners of more symmetrical men orgasmed more frequently than the partners of less symmetrical men (Thornhill et al., 1995). Our interpretation of this result depends on Baker and Bellis being right about female orgasm having evolved to retain sperm. More evidence bolstering that claim is needed.

Scent preferences across the cycle

The second line of evidence addressing special design concerns changing female preferences across the menstrual cycle. Bellis and Baker (1990) have shown that, whereas normally ovulating (non-pill-using, pre-menopausal) women's sex with in-pair partners is fairly evenly distributed over the menstrual cycle, sex with extra-pair partners occurs most commonly during the fertile, follicular phase (see also Baker & Bellis, 1995). In surveys of British women, the ratio of extra-pair copulations to in-pair copulations at women's modal fertility peak (approximately day 12 of the cycle) was about 2.5 times what it was in the late luteal phase. These data suggest that women's preferences slightly change over the course of their cycle, with the shift resulting in more extra-pair sex partners during the fertile period. This shift makes sense if, ancestrally, women could receive genetic benefits through extra-pair sex at some potential cost, a cost which should not be paid if there's no chance of receiving the genetic benefits.

Women claim that male scent is an important determinant of their sexual desire (Herz & Cahill, 1997), and some evidence indicates that their evaluation of scents changes over the course of their menstrual cycle (Grammer, 1993). We thought that, if women prefer symmetrical men particularly for genetic quality, they might selectively prefer their scent during their fertile period. We had 41 men wear t-shirts for two nights, during which time they used no fragrances, ate a bland diet, and refrained from sex. We then had normally ovulating women sniff the shirts and rate their odor's pleasantness and sexiness. Obviously, women cannot literally "smell symmetry," but there could be odors associated with symmetry. For each woman, we calculated a preference for symmetry versus asymmetry (the regression slope of their preferences on men's symmetry). As predicted, women did favor the scent of symmetrical men—but only during the fertile, follicular phase. During the luteal phase, there was absolutely no preference for symmetry or asymmetry (Gangestad & Thornhill, 1998b).

To test with greater sensitivity the prediction that women's preference for "the smell of symmetry" increases with their probability of conception, we estimated each woman's probability of conception at the time of smelling from tabled values based on day of cycle in the medical literature. These actuarial probabilities of conception strongly predicted preference for symmetry ($r = .54$). This effect replicated in a second, slightly larger sample ($r = .42$; Thornhill & Gangestad, 1999), as well as in a study conducted in Austria by Rikowski and Grammer (1999). What sort of chemical signal discriminates symmetrical and asymmetrical men remains unknown at this time. Such a cue must exist, however, and it appears that women use it such that, in a multiple mating context, they would be more likely to sire a symmetrical man's offspring.

More recently, Penton-Voak, Perrett, Castles, Burt, Koyabashi, and Murray (1999) reported that women's preferences for men's facial features change across the menstrual cycle as well. Whereas women show no preference for a hyper-masculine face over average or feminized male faces during the luteal phase, they prefer hyper-masculine features during the fertile follicular phase. This effect has been demonstrated in four studies—two by Pentan-Voak et al. (1999) and studies by Penton-Voak and Perrett (2000) and Johnston, Hagel, Franklin, Grammer and Fink (in press). It appears that men's facial masculinity is positively associated with symmetry and hence the preferences for symmetry and masculinity at mid-cycle appear to be related (Gangestad & Thornhill, 2001; see also Penton-Voak, Little, Baker, Tiddeman, Burt, & Perrett, in press; Scheib, Gangestad, & Thornhill, 1999). Whether hyper-masculine features are associated with symmetry or other markers of good genes is unknown at this point.

Preferences for symmetrical men as a function of mating context

Finally, Jeff Simpson, Niels Christensen, Kira Leck, and I videotaped men being interviewed by an attractive woman for a potential "lunch date" (Simpson, Gangestad, Christensen, & Leck, 1999). In addition, they were videotaped while telling another guy—a potential competitor—why they should be chosen for the date. In fact, both the woman and the other guy were not "live" but rather were on videotape. We coded what the men said about themselves both to the woman and the other guy. Factor analysis revealed two main dimensions: An "I'm a nice guy and I'll treat you well" factor and an "I'm just plain better" factor. We examined correlations between these dimensions and men's symmetry. As expected, symmetrical men were more likely to say, "I'm just plain better." There was no significant association between men's symmetry and their saying, "I'll treat you well." (We speculated that the reason for the latter is that there is no process that keeps this verbal signal honest.)

More to the point of what female preferences look like, we showed segments of these tapes to undergraduate women and asked them to rate the attractiveness of these men as both: a) long-term mates, and b) short-term partners (either sex partners or extra-pair partners) (Gangestad, Simpson, Cousins, & Christensen, 2000). Two sets of women's ratings were examined: Those who score above the mean on the Sociosexual Orientation Inventory (women of unrestricted sociosexual orientation) and those who score below the median on the measure (women of restricted sociosexual orientation). This measure, which taps willingness to engage in sex in absence of commitment and emotional closeness (Simpson & Gangestad, 1991), is highly associated with women's tendency to engage in extra-pair sex. Of all the extra-pair sexual relationships women in our sample reported, nearly 90% were reported by unrestricted women.

We performed contrasts of correlations between women's ratings of men's attractiveness and men's symmetry in the different mating contexts. Two major effects were found. First, those women most likely to engage in extra-pair sex (unrestricted women) tended to prefer the symmetrical men more than women least likely to engage in extra-pair sex. Second, mating context interacted with these individual differences such that women most likely to engage in extra-pair sex preferred symmetrical men, particularly in short-term mating contexts. Women unlikely to engage in extra-pair sex did not prefer symmetrical men in any context. This result may be due to the fact that these women do not have a good notion of what they'd look for in a short-term mate, given that they claim to be uninterested in short-term sexual relationships. In sum, then, the overall pattern of results is consistent with women particularly preferring men who possess markers of good genes in short-term mating contexts (e.g., as extra-pair partners).

Each of three lines of evidence, then, at least *suggest* that female adaptations are designed to favor certain men for their genetic benefits. Naturally, there are probably alternative explanations for each of these individual findings. I am unaware of one that accounts for all three.

SUMMARY

Research on non-human species has shown that, in many, males who exhibit evidence of greater developmental precision and perhaps underlying genetic fitness enjoy greater mating success (Møller & Thornhill, 1997b). In some cases, evidence strongly points to the operation of female choice selected by indirect genetic benefits accounting for these effects. In humans, evidence also suggests that males who exhibit evidence of greater developmental precision and perhaps underlying genetic fitness may be preferred by females as mates in at least some mating contexts. A critical question from an evolutionary perspective is what sexual selection process accounts for these findings.

Models of the good genes selection process suggest that: 1) males who exhibit markers of good genes may provide fewer material benefits than males who do not exhibit these markers, so long as females provide a pool of multiple mating opportunities; 2) males who exhibit markers of good genes may therefore be selectively advantaged as extra-pair mates (e.g., chosen for their genetic benefits only) and less advantaged (or not advantaged) as long-term, investing mates. Tests of the good genes hypotheses have attempted to examine these predictions. With regard to the latter hypothesis, evidence has been sought showing that female choice has been specially designed to favor certain males for their gametes (e.g., during the fertile phase of their cycle as extra-pair mates).

How strong is the evidence for the good genes model? The evidentiary strength is partly a function of the specificity of the theoretical derivations that follow from a good genes model. One limitation of the theoretical derivations, already noted, is that they depend on females providing a pool of multiple mating opportunities in the form of extra-pair sex. Yet we do not fully understand the conditions that lead to extra-pair sex in general or account for extra-pair mating in humans specifically (e.g., Smith, 1984). Indeed, the prevalence of extra-pair mating in ancestral human populations is unknown. Additional theoretical and empirical work on the evolution of polyandry in biparentally investing species such as humans is needed (see Jennions & Petrie, 2000; Petrie & Kempanaers, 1998).

Although evidence is consistent with female choice being designed to favor certain males for their genetic benefits, more evidence supporting special design is needed. First, evidence that bolsters our interpretation of the effects found thus far would be helpful. For instance, females must use some signal in male scent to discriminate males during their fertile phase. What signal is associated with male symmetry? What selection process ensures the honesty of this signal (Grafen, 1990)? Answers to these questions would increase confidence that female preference is indeed for good genes.

Second, evidence for additional special design would be useful. Female olfactory preferences appear to change across the menstrual cycle. If this is an instance of special design created by selection for good genes preferences, it would be surprising if selection did not create any additional design features involving changing preferences across the menstrual cycle. What visual or olfactory preferences do females selectively have during the fertile phase? Do females particularly prefer men who exhibit certain behaviors (e.g., successful intra-sexual encounters) during this phase?

To date, evidence that good genes sexual selection has played an important role in the evolution of human mating is strongly suggestive but not compelling. Further research is needed and future avenues for this research are clearly laid out.

Address correspondence to Department of Psychology, University of New Mexico, Albuquerque, N.M.

REFERENCES

Andersson, M.B. 1982. "Female Choice Selects for Extreme Tail Length in a Widowbird", *Nature, 299*, 818-820.

Andersson, M.B. 1994. *Sexual Selection*. Princeton, NJ: Princeton University Press.

Baker, R.R. & Bellis, M.A. 1995. *Human Sperm Competition: Copulation, Masturbation, and Infidelity*. London: Chapman & Hall.

Bellis, M.A. & Baker, R.R. 1990. "Do Females Promote Sperm Competition? Data for Humans", *Animal Behaviour, 40*, 997-999.

Burley, N. 1986. "Sexual Selection for Aesthetic Traits in Species with Biparental Care", American *Naturalist, 127*, 415-445.

Burt, A. 1995. "Perspective: The Evolution of Fitness", *Evolution, 49*, 1-8.

Buss, D.M. & Schmitt, D.P. 1993. "Sexual Strategies Theory: A Contextual Evolutionary Analysis of Human Mating", *Psychological Review, 100*, 204-232.

Campbell, L., Simpson, J.A., & Orina, M. (in press). "Sex and Mating: Sexual Strategies, Trade-offs, and Strategic Pluralism", in Rosen, D. & Luebbert, M., eds., *Evolution of the Psyche*. Westport, CT: Greenwood.

Charlesworth, B. 1987. "The Heritability of Fitness", in Bradbury, J.W. & Andersson, M.B., eds., *Sexual Selection: Testing the Alternatives*. New York: Wiley, pp. 21-40.

Charlesworth, B. 1990. "Mutation-selection Balance and the Evolutionary Advantage of Sex and Recombination", *Genetical Research, 55*, 199-221.

Charlesworth, B. & Hughes, K.A. 1998. "The Maintenance of Genetic Variation in Life History Traits", in Singh, R.S. & Krimbas, C.B., eds., *Evolutionary Genetics from Molecules to Morphology*. Cambridge, England: Cambridge University Press.

Cronin, H. 1991. *The Ant and the Peacock*. Cambridge: Cambridge University Press.

Crow, J.F. 1986. *Basic Concepts in Population, Quantitative, and Evolutionary Genetics*. New York: Freeman.

Darwin, C. 1871. *The Descent of Man and Selection in Relation to Sex*. London: Murray.

Ellis, B.J. 1998. "The Partner-specific Investment Inventory: An Evolutionary Approach to Individual Differences in Investment", *Journal of Personality*, 66, 383-442.

Fisher, R.A. 1930. *The Genetical Theory of Natural Selection*. Oxford: Clarendon. (Also, 1958, New York: Dover)

Forbes, L.S. 1997. "The Evolutionary Biology of Spontaneous Abortion in Humans", *Trends in Ecology and Evolution, 12*, 446-450.

Furlow, B., Armijo-Prewitt, T., Gangestad, S.W., & Thornhill, R. 1997. "Fluctuating Asymmetry and Psychometric Intelligence", *Proceedings of the Royal Society of London* B, *264*, 823-829.

Furlow, B., Gangestad, S.W., & Armijo-Prewitt, T. 1998. "Developmental Sta-

bility and Human Violence", *Proceedings of the Royal Society of London B*, *266*, 1-6.
Gangestad, S.W., Bennett, K., & Thornhill, R. (in press). "A Latent Variable Model of the Developmental Instability in Relation to Men's Sexual Behavior", *Proceedings of the Royal Society of London B*.
Gangestad, S.W., Simpson, J.A., Cousins, A.J., & Christensen, N.P. 1998. "Fluctuating Asymmetry, Sociosexuality, and Women's Context-specific Mate Preferences", manuscript submitted for publication.
Gangestad, S.W. & Thornhill, R. 1997a. "Human Sexual Selection and Developmental Stability", in Simpson, J.A. & Kenrick, D.T., eds., *Evolutionary Personality and Social Psychology*. Hillsdale, NJ: Erlbaum, pp. 169-195.
Gangestad, S.W. & Thornhill, R. 1997b. "The Evolutionary Psychology of Extra-pair Sex: The Role of Fluctuating Asymmetry", *Evolution and Human Behavior*, *18*, 69-88.
Gangestad, S.W. & Thornhill, R. 1998a. "An Evolutionary Psychological Analysis of Human Sexual Selection: Developmental Stability, Male Sexual Behavior, and Mediating Features", manuscript submitted for publication.
Gangestad, S.W. & Thornhill, R. 1998b. "Menstrual Cycle Variation in Women's Preference for the Scent of Symmetrical Men", *Proceedings of the Royal Society of London* B, *265*, 727-733.
Gangestad, S.W, & Thornhill, R. 1998c. "Sexual Selection and Relationship Dynamics: Trade-offs Between Partner Investment and Fluctuating Asymmetry", manuscript submitted for publication.
Gangestad, S.W. & Thornhill, R. 2001. "Facial Masculinity and Fluctuating Asymmetry", manuscript under review.
Gangestad, S.W. & Thornhill, R. (in press). "Fluctuating Asymmetry, Developmental Instability, and Fitness: Toward Model-based Interpretation", in Polak, M., ed., *Developmental Instability*. Cambridge, UK: Cambridge University Press.
Gangestad, S.W. & Thornhill, R. (in press). "Individual Differences in Developmental Precision and Fluctuating Asymmetry: A Model and Its Implications", *Journal of Evolutionary Biology*.
Gangestad, S.W., Thornhill, R., & Yeo, R.A. 1994. "Facial Attractiveness, Developmental Stability, and Fluctuating Asymmetry", *Ethology and Sociobiology*, *15*, 73-85.
Gangestad, S.W. & Simpson, J.A. 2000. "The Evolution of Human Mating: The Role of Trade-offs and Strategic Pluralism", *Behavioral and Brain Sciences*, *23*, 575-587.
Gould, S.J. & Lewontin, R.C. 1979. "The Spandrels of San Marco and the Panglossian Paradigm: A Critique of the Adaptationist Programme", *Proceedings of the Royal Society of London* B, *205*, 581-598.
Grafen, A. 1990. "Biological Signals as Handicaps", *Journal of Theoretical Biology*, *144*, 517-546.
Grammer, K. 1993. "5-a-androst-16en-3a-on: A Male Pheromone? A Brief Report", *Ethology and Sociobiology*, *14*, 201-214.

Haig, D. 1990. "Brood Reduction and Optimal Parental Investment When Offspring Differ in Quality", American Naturalist, 136, 550-556.
Haig, D. 1993. "Genetic Conflicts in Human Pregnancy", *Quarterly Review of Biology*, *68*, 495-532.
Hallgrimsson, B. 1998. "Fluctuating Asymmetry in the Mammalian Skeleton: Evolutionary and Developmental Implications", in Hecht, M.K., ed., *Evolutionary Biology*, v. 30. New York: Plenum, pp. 187-251.
Hamilton, W.D. 1982. "Pathogens as Causes of Genetic Diversity in Their Host Populations", in Anderson, R.M. & May, R.M., eds., *Population Biology of Infectious Diseases*. New York: Springer-Verlag.
Hamilton, W.D. & Zuk, M. 1982. "Heritable True Fitness and Bright Birds: A Role for Parasites", *Science*, *218*, 384-87.
Herz, R.S. & Cahill, E.D. 1997. "Differential Use of Sensory Information in Sexual Behavior as a Function of Gender", *Human Nature*, *8*, 275-286.
Holcomb, H.R. 1993. *Sociobiology, Sex, and Science*. Albany, NY: State University of New York Press.
Holcomb, H.R. 1996. "Just So Stories and Inference to the Best Explanation in Evolutionary Psychology", *Minds and Machines*, 6, 525-540.
Holland, B. & Rice, W.R. 1998. "Chase-away Sexual Selection: Antagonistic Seduction Versus Resistance", *Evolution*, *52*, 1-7.
Houle, D. 1992. "Comparing Evolvability and Variability of Traits", *Genetics*, *130*, 195-204.
Hrdy, S.B. 1981. *The Woman That Never Evolved*. Cambridge, MA: Harvard University Press.
Hrdy, S.B. 1997. "Raising Darwin's Consciousness: Female Sexuality and the Prehominid Origins of Patriarchy", *Human Nature*, *8*, 1-49.
Jennions, M.D. & Petrie, M. 2000. "Why Do Females Mate Multiply? A Review of the Genetic Benefits", *Biological Review*, *75*, 21-64.
Johnston, V.S., Hagel, R., Franklin, M., Fink, B., & Grammer, K. (in press). "Male Facial Attractiveness: Evidence for Hormone Medicated Adaptive Design", *Evolution and Human Behavior*.
Kirkpatrick, M. 1986. "The Handicap Mechanism of Sexual Selection Does Not Work", *American Naturalist*, *127*, 222-40.
Kirkpatrick, M. 1996. "Good Genes and Direct Selection in the Evolution of Mating Preferences", *Evolution*, *50*, 2125-2140.
Kirkpatrick, M. & Ryan, M.J. 1991. "The Evolution of Mating Preferences and the Paradox of the Lek", *Nature*, *350*, 33-38.
Lewontin, R.C. 1974. *The Genetic Basis of Evolutionary Change*. New York: Columbia University Press.
Leung, B. & Forbes, M.R. 1996. "Fluctuating Asymmetry in Relation to Stress and Fitness: Effects of Trait Type as Revealed by Meta-analysis", *Ecoscience*, *3*, 400-413.
Ligon, J.D. & Zwartjes, P.W. 1995. "Ornate Plumage of Male Red Junglefowl Does Not Influence Mate Choice by Females", *Animal Behaviour*, *49*, 117-125.

Maynard Smith, J. 1978. *The Evolution of Sex*. Cambridge, UK: Cambridge University Press.

Mayr, E. 1983. "How to Carry Out the Adaptationist Programme?", *American Naturalist*, *121*, 324-334.

Møller, A.P. 1994. *Sexual Selection in the Barn Swallow*. Cambridge: Cambridge University Press.

Møller, A.P. 1997. "Developmental Stability and Fitness: A Review", *American Naturalist*, *149*, 916-932.

Møller, A.P. 1999. "Asymmetry as a Predictor of Growth, Fecundity and Survival", *Ecology Letters*, *2*, 149-156.

Møller, A.P. & Swaddle, J.P. 1997. *Developmental Stability and Evolution*. Oxford: Oxford University Press.

Møller, A.P. & Thornhill, R. 1997a. "A Meta-analysis of the Heritability of Developmental Stability", *Journal of Evolutionary Biology*, *10*, 1-16.

Møller, A.P. & Thornhill, R. 1997b. "Bilateral Symmetry and Sexual Selection: A Meta-analysis", *American Naturalist*, *151*, 174-192.

Møller, A.P. & Thornhill, R. 1998. "Male Parental Care, Differential Parental Investment by Females and Sexual Selection", *Animal Behaviour*, *55*, 1507-1515.

Parker, G.A. & Maynard Smith, J. 1991. "Optimality Theory in Evolutionary Biology", *Nature*, *348*, 27-33.

Parker, G.A. & McNair, M.R. 1978. "Models of Parent-offspring Conflict: I. Monogamy", *Animal Behaviour*, *26*, 97-110.

Partridge, L. 1983. "Non-random Mating and Offspring Fitness", in Bateson, P., ed., *Mate Choice*. New York: Cambridge University Press, pp. 227-256.

Penton-Voak, I.S., Little, A.C., Baker, S., Tiddeman, B., Burt, D.M., & Perrett, D.I. (in press). "Male Facial Attractiveness: No Evidence for a Trait Size/Symmetry Correlation", *Proceedings of the Royal Society of London* B.

Penton-Voak, I.S. & Perrett, D.I. 1998, June. "Effects of Exaggerated Femininity and Masculinity on Facial Attractiveness", paper presented at the annual meetings of the Human Behavior and Evolution Society, Davis, CA.

Penton-Voak, I.S. & Perrett, D.I. 2000. "Female Preference for Male Faces Changes Cyclically—Further Evidence", *Evolution and Human Behavior*, *21*, 39-48.

Penton-Voak, I.S., Perrett, D.I., Castles, D., Burt, M., Koyabashi, T., & Murray, L.K. 1999. "Female Preference for Male Faces Changes Cyclically", *Nature*, *399*, 741-742.

Petrie, M. & Kempanaers, B. 1998. "Extra-pair Paternity in Birds: Explaining Variation Between Species and Populations", *Trends in Ecology and Evolution*, *13*, 53-58.

Pomiankowski, A. & Møller, A.P. 1995. "A Resolution of the Lek Paradox", *Proceedings of the Royal Society of London* B, *260*, 21-29.

Price, T. & Schuler, D. 1991. "On the Low Heritability of Life History Traits", *Evolution*, *45*, 853-861.

Rikowski, A. & Grammer, K. 1999. "Human Body Odour, Symmetry and Attractiveness", *Proceedings of the Royal Society of London* B, *266*, 869-874.

Rowe, L. & Houle, D. 1996. "The Lek Paradox and the Capture of Genetic Variance by Condition Dependent Traits", *Proceedings of the Royal Society of London* B, *263*, 1415-1421.

Scheib, J.E., Gangestad, S.W., & Thornhill, R. 1999. "Facial Attractiveness, Symmetry, and Cues of Good Genes", *Proceedings of the Royal Society of London* B, *266*, 1318-1321.

Simpson, J.A. & Gangestad, S.W. 1991. "Individual Differences in Sociosexuality: Evidence for Convergent and Discriminant Validity", *Journal of Personality and Social Psychology*, *60*, 870-83.

Simpson, J.A., Gangestad, S.W., Christensen, P.N., & Leck, K. (in press). "Fluctuating Asymmetry, Sociosexuality, and Intrasexual Competitive Tactics", *Journal of Personality and Social Psychology.*

Smith, R.L. 1984. "Human Sperm Competition", in Smith, R.L., ed., *Sperm Competition and the Evolution of Human Mating Systems*. London: Academic, pp. 601-660.

Symons, D. 1979. *The Evolution of Human Sexuality*. Oxford: Oxford University Press.

Taylor, P.D. & Williams, G.C. 1982. "The Lek Paradox is Not Resolved", *Theoretical Population Biology*, *22*, 392-409.

Thoma, R.J., Yeo, R.A., Gangestad, S.W., Lewine. J., & Davis, J. (in press). "Fluctuating Asymmetry and the Human Brain", *Laterality.*

Thornhill, R. 1991. "The Study of Adaptation", in Bekoff, M. & Jamieson, D., eds., *Interpretation and Explanation in the Study of Behavior*. Boulder, CO: Westview.

Thornhill, R. & Gangestad, S.W. 1994. "Fluctuating Asymmetry and Human Sexual Behavior", Psychological *Science*, *5*, 297-302.

Thornhill, R. & Gangestad, S.W. 1999b. "The Scent of Symmetry: A Human Pheromone That Signals Fitness?" *Evolution and Human Behavior, 20*, 175-201.

Thornhill, R., Gangestad, S.W., & Comer, R. 1995. "Human Female Orgasm and Mate Fluctuating Asymmetry", *Animal Behaviour*, *50*, 1601-1615.

Thornhill, R. & Moller, A.P. 1997. "Developmental Stability, Disease and Medicine", *Biological Reviews*, *72*, 497-548.

Trivers, R. 1972. "Parental Investment and Sexual Selection", in Campbell, B., ed., *Sexual Selection and the Descent of Man, 1871-1971*. Chicago: Aldine.

Trivers, R. 1974. "Parent-offspring Conflict", *American Zoologist*, *14*, 249-264.

Trivers, R. 1985. *Social Evolution*. Chicago: Aldine.

Waynforth, D. 1998. "Fluctuating Asymmetry and Human Male Life History Traits in Rural Belize", *Proceedings of the Royal Society of London* B, *255*, 1497-1501.

Williams, G.C. 1966. *Adaptation and Natural Selection*. Princeton, NJ: Princeton University Press.

Williams. G.C. 1975. *Sex and Evolution.* Princeton, NJ: Princeton University Press.

Yeo, R. & Gangestad, S.W. 1998. "Developmental Instability and Phenotypic Variation in Neural Organization", in Raz, N., ed., *The Other Side of the Error Term: Aging and Development as Models in Cognitive Neuroscience*. Amsterdam: Elsevier, pp. 1-51.

CHAPTER 5

FRAMING OUR THOUGHTS: ECOLOGICAL RATIONALITY AS EVOLUTIONARY PSYCHOLOGY'S ANSWER TO THE FRAME PROBLEM

TIMOTHY KETELAAR
University of California at Los Angeles

PETER M. TODD
Max Planck Institute for Human Development

Abstract. Decision makers challenged by real-world adaptive problems must often select a course of action quickly, despite two computational obstacles: There is usually a vast number of possible courses of action that can be explored, and a similarly vast number of possible consequences that must be considered when evaluating these options. Thus, decision makers routinely face the *frame problem*: how to focus attention on adaptively relevant information and how to keep this information set small enough that the mind can actually perform the computations necessary to generate adaptive behavior. By identifying simple, effective heuristics that work within the computational and informational constraints of an organism and still result in adaptively adequate behavior, the new ecological rationality perspective within evolutionary psychology suggests how intelligent agents can overcome this frame problem. In this chapter we show how ecologically rational psychological mechanisms can address two challenges derived from the frame problem: 1) Using the example of sequential mate choice, we demonstrate that a satisficing information search strategy can lead to good decisions using only a limited amount of information; and 2) We argue that emotions can guide information search, by focusing the computational mind on precisely that information which is most useful for making decisions that lead to good outcomes over the long run.

H.R. Holcomb III (ed.), Conceptual Challenges in Evolutionary Psychology, 179–211.

We further describe how a bottom-up approach to studying ecological rationality, which first identifies simple heuristic decision-making components and then builds more sophisticated cognitive modules out of these components, can generate fruitful new directions for psychological scientists attempting to illuminate the structure of our evolved computational minds.

I. THE WISE BRAHMIN AND THE CHESSBOARD: HOW SEEMINGLY SIMPLE CALCULATIONS CAN REQUIRE MORE COMPUTATION THAN OUR MINDS CAN HANDLE

There is a story of an ancient rajah of India who wished to reward his royal adviser, a wise Brahmin, for inventing the game of chess (see Kasparov, 1997). When the rajah asked what he would desire for compensation, the Brahmin motioned to the grid of the 8x8 chessboard and requested that a single grain of wheat be placed on the first square of the board. He asked that the rajah's servants then double the amount of wheat on the second square and continue doubling the amount of wheat (i.e., 4 grains on the third square, 8 on the next, etc.) until all 64 squares on the chessboard had been covered. The rajah was perplexed by the Brahmin's seemingly simple request and asked him whether he indeed desired only grains of wheat—albeit in continuing doubled amounts on consecutive squares of the chessboard—as his compensation? The Brahmin replied, "Yes," and the deal was set.

A day later the rajah's senior mathematician informed him that they were having trouble granting the Brahmin's demands. By the time his servants had arrived at the 21st square of the chessboard they required over one million (1,048,576 to be exact) grains of wheat, and by the 30th square (still not quite half finished) they needed more than 10 billion grains of wheat to satisfy the rajah's promise to the Brahmin. According to the senior mathematician's calculations, to continuously double the amounts of wheat across all 64 squares of the chessboard would require more wheat than the kingdom, indeed the entire world, could produce in countless millennia of bountiful harvests.

The point of this story is not that one should distrust their advisers, but rather that problems that appear quite simple and straightforward at first glance can lead to computations on humanly unmanageable scales. Many of the human capacities we most take for granted—

vision, language, finding a mate, or deciding which strategy to play in a competitive bargaining game—involve problems that can potentially involve a vast number of possible inferences. In this chapter we explore what evolutionary psychology has to offer in explaining how the finite human mind can cut through this computational thicket and arrive at reasonable solutions to a variety of information processing problems that appear to involve considerably more resources than our computational minds have to offer.

The computational mind's dilemma

Some cognitive scientists (e.g., Fodor, 1995; Haugeland, 1995; also Baumgartner & Payr, 1995) have argued that the traditional computer metaphor of the mind, instantiated in artificial intelligence (AI) in particular, runs smack into the following dilemma: How can successful decision-making be achieved under conditions of limited time and knowledge? This dilemma has come to be called the *frame problem* (McCarthy & Hayes, 1969; Pylyshyn, 1987; Ford & Pylyshyn, 1996). This dilemma manifests itself in many domains of human computation where the number of potentially relevant inferences that can be drawn far outpaces the human mind's capacity to evaluate them. From the perspective of evolutionary psychology, a central information-processing dilemma concerns how to "frame" a problem with a limited set of inferences that a biologically plausible computational device (i.e., a mind) can handle. Even in cases where the mind could, in principle, generate and evaluate all possible inferences there are often severe ecological costs to attempting such computational omnipotence (see below).

Far from being a strictly conceptual puzzle for philosophers of mind, the frame problem emerges as a legitimate concern for computational modelers in a variety of real-world psychological domains. These domains include: (1) language comprehension: How do individuals actually draw appropriate inferences from the vast sea of possible interpretations of even the simplest speech acts? (see Sperber & Wilson, 1996; Chiappe & Kukla, 1996), (2) economic decision making: How do individuals actually select an appropriate strategy in a competitive bargaining game like the repeated Prisoner's Dilemma without mapping the incalculably vast decision tree of possible strategies[1] and the infinity of possible consequences? (see Rasmussen, 1994; Samuelson, 1997), and even (3) perception: How does an individual actually determine from a single retinal image which of the incomputably vast number of possible arrangements of objects in the world gave rise to this particular pattern of retinal activation? (see Pinker, 1984, 1997; Poggio, 1984). In all of these domains, drawing appropriate inferences could involve sifting through an astronomical number of relevant and irrelevant inferences. Any agent that actually attempts to locate the "optimal" conclusion in the vast haystack of alternative inferences may find itself bogged down in a sea of computation that puts it at an adaptive disadvantage compared to less omnipotent, but more computationally frugal organisms. Moreover, as one philosopher notes:

> unless some drastic preselection can be effected among the alternatives their evaluation could never be completed. This gives rise to what is known among cognitive scientists as the 'Frame Problem': in deciding among any range of possible actions, most of the consequences of each must be eliminated from consideration a priori, i.e., *without any time being wasted on their consideration....* (deSousa, 1994, p. 276, emphasis in the original)

An ecologically valid description of our evolved computational architecture must account for how intelligent agents overcome this frame problem to generate adaptive cognition.

Explosive Consequences of the Frame Problem

Philosopher and cognitive scientist Daniel Dennett (1987) describes the essence of the frame problem in a thought experiment involving a robot whose main mission in life is to survive. Suppose that its most vexing challenge at the moment is that its backup battery is in a room next door and that a time bomb is also in that room. The robot enters the room and deduces that the battery is on a wagon. It further deduces that pulling the wagon out of the room will remove the battery from the room. Although it deduces that the time bomb is also on the wagon, its limited deductive powers unfortunately prevent it from considering the further implications of pulling the wagon—along with the time bomb—out of the room and our mechanical friend is promptly blown to smithereens. Now imagine that the robot's design engineers develop what they believe will be an obvious remedy to the robot's limited deductive capacity: simply equip the robot with greater computational might such that it can deduce *all possible implications* of its alternative courses of action. Yet, this "obvious" solution, like the Brahmin's seemingly simple request of the rajah, quickly proves to be computationally intractable. Dennett (1998) notes that the imaginary robot began its mission:

> as designed, to consider the implications of such a course of action. It had just finished deducing that pulling the wagon out of the room would not change the color of the room's walls, and was embarking on a proof of the further implications that pulling the wagon out would cause its wheels to turn more revolutions than there were wheels on the wagon—when the bomb exploded (p. 42).

Even when the robot is given the additional computational power to calculate still more possible implications, it is obvious that this strategy is doomed. What gives rise to the frame problem is that:

> we need to know whether a consequence will turn out to be relevant *before drawing it.* If it is relevant and we have not retrieved it, we may act irrationally. But, if it is irrelevant and we have already drawn it, we have already wasted time... (deSousa, 1987, p. 194, emphasis in the original).

How can the human mind escape this quandary?

Two aspects of the frame problem

In Dennett's (1987) thought experiment, the cost of the robot's long-winded process of deducing implications far outweighed the benefits it could receive from considering those outcomes. There are simply too many implications to be considered, too much information to be computed. We identify this problem of *too much information,* as one aspect of the frame problem that confronts any computational device, be it a robot or a rabbit. So the first step that the human mind can take in overcoming the frame problem is to limit the amount of information used to make inferences—but will it still be possible to make good decisions with this limitation? We address this question in Section II of this paper.

A second important aspect of the frame problem emerges as Dennett's robot story continues: Faced with too many possible implications for their robot to consider (the problem of too much information), the robot engineers resolved to "... teach it the difference between relevant implications and irrelevant implications ... and teach it to ignore the irrelevant implications" (Dennett, 1987, p. 42). As straightforward as this solution appears, it too results in a combinatorial explosion of computations, followed in the story by a physical explosion of robot parts, as the robot starts churning through the virtually infinite sea of irrelevant implications trying to locate the few relevant implications it could employ to solve its task. Thus, the second challenge stemming from the frame problem is that of determining *what information to attend to.* In those situations where there is too much information to make use of, it is of course similarly unreasonable to decide which set of information to focus on by looking at all that is available and determining which data are relevant and which are not, i.e., this would just add another, possibly even longer, step to the already infeasible decision process. Fortunately, the human mind possesses a few built-in mechanisms that can provide shortcuts to highlighting the relevant information for different problems, as we will see in Section III.

Simple (but effective) is sometimes better

When AI researchers design computational algorithms to tackle specific tasks, they evaluate the performance of these mechanisms by considering not just whether they can accomplish the task, but also whether they require more or fewer resources than competing algorithms. The number of inputs required and the "worst case" and "average case" running time are often used to evaluate the performance of a computational mechanism (Cormen, Leiserson, & Rivest, 1997). Although the robot in Dennett's thought experiment clearly failed to accomplish its mission, perhaps its greatest weakness lies not in whether it can, in principle, solve the battery/bomb problem, but in the unreasonable number of inputs and amount of running time its algorithm requires to achieve a successful solution.

The use of performance criteria such as number of inputs and running time is even more appropriate for evaluating the cognitive abilities of biologically intelligent agents. This is the case because any organism that can quickly solve an important adaptive problem, particularly by using less and only the most appropriate information, has an adaptive advantage over competing organisms employing cognitive mechanisms that require more time to reach an adequate solution. The rabbit that has successfully avoided becoming the dinner of the fox by employing several crude but fast heuristics can begin working on other important adaptive problems (e.g., the task of searching for its own dinner). By contrast, the rabbit that attempts the more mathematically elegant but time consuming strategy of calculating the single optimal escape path by considering all possible escape trajectories (e.g., left, right, zigzagged, vertical?) may well end up doing most of its computations inside the stomach of a less elegantly calculating predator. In the grand Darwinian steeplechase of natural selection, finishing ahead of one's competitors is more important than the mathematical elegance of one's solution.

Outline of this chapter

These considerations of what kind of cognition is actually most functional and adaptively beneficial to an organism lead to the idea of *ecological rationality* (Gigerenzer, Todd, & the ABC Research Group, 1999; Todd, Fiddick, & Krauss, in press). The ecological rationality perspective focuses on identifying simple, fast, and effective heuristics that work within the computational constraints of an organism and the informational constraints of the environment and still result in adaptively adequate behavior—a perspective that meshes tightly with evolutionary psychology (Todd, 2000; Todd, Fiddick, & Krauss, in press). In the rest of this chapter we show how evolutionary psychology's focus on ecologically rational descriptions of psychological mechanisms can provide insights into two challenges derived from the frame problem:

1. How much information do intelligent agents really need to solve important information processing problems?
2. What sort of information should an intelligent agent bother attending to, in attempting to solve important information processing problems?

We address the first question by showing that only a limited amount of information is needed to solve important adaptive questions, focusing on the example of mate search. Once we have determined that only a limited amount of information is necessary to guide some aspects of adaptive behavior in a particular context, we must of course next consider what specific information to seek out or attend to. We present an approach to this problem by describing how a simple emotion mechanism might help us surmount the frame problem inherent in strategy choice for risky choice decisions in much the same way that computer chess programs "think before they leap."

II. HOW MUCH INFORMATION IS REALLY NEEDED TO SOLVE ADAPTIVE PROBLEMS? THE EXAMPLE OF MATE SEARCH

The ordinary chessboard referred to in our introductory story has provided a challenging set of problems for chess players and psychologists alike: how to choose the best move in a chess game, from a particular arrangement of pieces on the board. The typical solution suggested for this problem is to consider as many options as possible in the available time, focusing on the most promising ones, and then selecting the best option discovered when decision-time finally runs out (see Section III for more detail on approaches in computer chess). But many real-world decisions are rather different from this artificial setting: Some options may take a long time to assess, and some may no longer be available after they are first considered and not immediately chosen. What approach can a decision maker take in such situations outside of the chess parlor?

Consider the situation that presents itself to men and women searching for a mate in many modern Western cultures. This type of choice usually consists of a sequential search through successive potential mates, in which each one is evaluated and decided on in turn in a process that can take minutes, hours, days, or years. (Here the decision can be thought of as whether or not to settle down and have children with a particular person, though other definitions are possible.) A significant cost associated with this search is that it is often difficult, and sometimes impossible, to return to a potential mate that has been previously discarded (because they remain in the "mating pool" and will often pair up with someone else while one's own search continues, as countless romantic tragedies attest). To further complicate matters, one does not know ahead of time what the range of potential mates may be: how can we know, during our first love, whether someone else might be able to incite still deeper passions, if only we just kept searching long enough to find them? We cannot even tell how many more potential mates we may encounter during our lifetime.

Given these constraints on the search process, how can we go about finding a mate? Certainly there is a great deal of information available about the mate values of all the possible potential mates out there—but how much of it must we gather before we can make reasonable mate choice decisions? Are we doomed to an extensive search and evaluation of all possible mates, or is there a better, faster way?

In this section, we develop a view of mate choice that we feel will be a fruitful area for future research in evolutionary psychology: exploring the role of satisficing heuristics in mating decisions.

Formalizing mate search: How to solve the dowry problem with little information

We present the mate search problem in a more precise form that we can explore in detail, a model inspired by the *dowry problem*, a well-known puzzle from probability theory (Mosteller, 1987; Gilbert & Mosteller, 1966). Imagine that the Indian rajah we introduced earlier, still stinging from the enormous (and yet unmet) cost of his Brahmin's chessboard wheat ploy, wishes to regain some measure of pride by putting this adviser to a new and deadly test. The rajah arranges to have 100 women from his territory brought before his adviser in succession, and all the Brahmin has to do to retain his post is to choose the woman with the highest dowry. If he chooses correctly, the Brahmin keeps his post, and if not, he loses his head. He can see one woman at a time and ask her dowry; then he must decide immediately if she is the one with the highest dowry out of all 100 women, or else let her pass by and go on to the next woman. He cannot return to any woman he has seen before—once he lets them pass, they are gone forever. Moreover, the adviser has no idea of the range of dowries before he starts seeing the women. What strategy can he possibly use to have the highest chance of picking the woman with the highest dowry?

The "optimal" solution: The 37 Percent rule

It turns out that the algorithm the adviser should use, to guarantee the highest chance of making the right choice, is to look at the first 37 women, letting them all pass but remembering the highest dowry from that set, and then, starting with the 38th woman, select the first candidate with a dowry larger than the maximum from the first 37. (For derivations of this procedure, see Mosteller, 1987; Gilbert & Mosteller, 1966; Ferguson, 1989). More generally, if the adviser knows that he will be presented with a succession of women from a total set of a given size, he should check the first 37% of those women, remembering the highest dowry, and then select the first one after those 37% who exceeds the previous highest value. This *37 Percent rule* finds the highest value more often than any other algorithm—but it only succeeds in finding that highest value in 37% of the times it is applied. The rest of the time, the rajah has his revenge: The adviser loses his head and the rajah gets his wheat back. Not only is this rule's success rate limited, but it also requires a large amount of search: A minimum of 37% of the population must be checked (by definition), but then on average the adviser will end up going through 74% of the women before coming across one with a dowry that exceeds the previously-set threshold that stops his search (Todd & Miller, 1999). Clearly this is a rather exhaustive (and exhausting) approach to mate search, gathering information about three quarters of the population of potential mates before making a decision. But as we will see, it is possible to make a successful mating decision by employing a much more frugal approach, using much less information.

Satisficing solutions using less information

Any animal searching for a mate is probably not restricted to only settling for the single "best" member of the opposite sex that can be found—other "pretty good" potential mates will often be selected instead, to save search time or energy. It may suffice for instance, in terms of having an adaptive advantage over other competing mate-seekers, to quickly locate and pair up with an individual possessing a mate value in the top 10% of the population. It turns out that to maximize the chance of finding a mate under these criteria, an individual searching through a set of 100 potential partners would only have to check 14% of them before using the highest value seen in that initial sample as the aspiration level to use for the rest of the search (see Todd & Miller, 1999). This kind of aspiration-level-guided search is an instance of what Simon (1990) calls a *satisficing* heuristic for guiding behavior in domains where complete information—here of the distribution of mate values—is not immediately available. This search strategy would succeed in finding a mate in the top 10% of the population over three-quarters of the time (83%, compared to the 37 Percent rule's success rate of only 37%). If the standards are relaxed a bit further, so that a mate in the highest quartile will do, then only 7% of the population of 100 need be checked to form an aspiration level (and this will succeed 92% of the time). Finally, rather than searching for a mate in the top ranks of the population, an animal may be loss-averse, preferring only to minimize its chances of landing a mate with a value in the bottom quartile of the population. The way to achieve this goal is to use a much shorter search time, checking only 3% of the population before setting the aspiration level. Such a loss-avoiding search strategy would select a mate from the bottom (quarter) of the barrel only 1% of the time. In contrast, following the 37 Percent rule would pick these poor mates more than 9% of the time—much worse performance by loss-averse standards.

Thus, successful mate search—defined as meeting or exceeding these biologically plausible aspiration levels—can be achieved using much less information than is required by the optimal 37 Percent rule. We do not in general know what criteria animals, including humans, may have evolved to satisfy in their mate choice, but our results show that for a broad class of possible criteria, relatively little search is necessary to gain enough information for setting an appropriate aspiration level in this one-sided search situation. But we must still ask how appropriate such a one-sided search model is—it may reasonably reflect some search situations such as buying a house (where the house does not usually search for a buyer in turn), but does it capture the realities of human mate search?

The more realistic problem of mutual mate search

In chess, the challenge and interest of the game come from playing with someone who makes moves in response to one's own. The same is true for the mating game as well. One-sided mate search, in which the members of one sex do all the searching and make all the decisions, is not a realistic model of the problem actually faced by many species, including humans. The problem for these species is that at the same time one sex is evaluating members of the other sex as prospective mates, they are themselves being evaluated in turn. If a particular male does not meet a particular female's standards, for instance, then no amount of proposing on the male's part is going to win the female (in the restricted scenario we are considering, at least). Furthermore, if everyone in the population has been one-sidedly setting their aspiration levels based on the highest mate value seen in the first few potential mates they encounter, then everyone will have a rather high aspiration for whom they will agree to mate with. The trouble is then that if an individual does not possess a high mate value (relative to the aspiration levels of others in the population), they will not be selected by anyone else as a potential mate, and will end up alone—and thus for most individuals, a one-sided search strategy in a mutual mating market will be unsuccessful.

We can explore how different mate search rules work in a mutual search situation by constructing a more biologically plausible model. Consider a population of 100 males and 100 females, each with a distinct mate value between 0.0 and 100.0, and each with the ability to accurately assess the mate values of members of the opposite sex, but lacking knowledge of their own mate value. We give each of the 200 individuals the same search strategy, which first involves assessing some specific number of members of the opposite sex during an "adolescence period." During this time, individuals can adjust an aspiration level, if their search rule uses one, up or down from an initial value of 50. After this adolescence period, pairs of males and females meet at random, and they can either make a proposal (an offer to mate) to their partner, or decline to do so. If both individuals in a pair make an offer to each other, then this pair is deemed mated, and the two individuals are removed from the population. Otherwise, both individuals remain in the mating pool to try again. This pairing-offering-mating cycle is repeated until every individual is mated, or until every individual has had the opportunity to assess and propose to every member of the opposite sex. We are interested in how many individuals in the population get paired up in this setting using different search rules, and how well matched the pairs end up being.

As we indicated above, trying to use a one-sided mate search rule in the two-sided setting has rather disastrous results for most of the population. For instance, if everyone checks a dozen members of the opposite sex and sets an aspiration level equal to the highest mate value seen, then only 7% of the population will end up in mutually-agreeable pairs. Furthermore, it will only be the very highest-valued individuals that end up mated. This is certainly counter to human experience, as well as that of other mutually selecting (e.g., some monogamous) species, where the majority of individuals, across a wide range of relative mate values, are able to find mates. Clearly, a different kind of search rule must be used in these cases.

A little search of the right kind is still enough

A much more successful two-sided mate search can be achieved by simply using *one's own mate value* as the aspiration level for deciding which members of the opposite sex to propose to. With this strategy, most of the population can succeed in finding and pairing up with mates of a similar value to their own. But there is a problem: Accurate knowledge of one's own mate value is not necessarily an easy thing to come by. We cannot be born with it, because it is both context-sensitive (it depends on the others around us) and changes over time as we develop. Because we initially lack this knowledge, we must somehow estimate our own mate value if we are to use it to form our aspiration level. Can this estimation occur rapidly, or does the two-sided mate search situation doom us to pursue the long-winded information-gathering process that could be avoided in the one-sided search case?

Once again, we have found that mate search decisions can be guided by limited information, even when both sexes are choosy. By using a learning rule that adjusts one's estimate of one's own mate value, and thus one's aspiration level, upwards with every proposal received from a higher-value member of the opposite sex and lowers the level every time a lower-value individual does not propose, individuals can rapidly estimate their own mate value and use this estimate to pair up with similarly-valued mates. With such a rule, less than twenty encounters with members of the opposite sex are necessary for the majority of the population to form mated pairs of individuals with similar mate value (Todd & Miller, 1999). An extensive information-gathering process is not necessary—in fact, basing one's mate choice on a more ambitious information gathering process almost always leads to fewer (rather than more) individuals finding agreeable mates.

What these models leave out

These simple models certainly do not capture the full complexity and subtlety of real mate search in nature. Humans for instance can often return to previously-encountered partners for another try, and can learn about their own mate value (and those of others) through indirect observation in addition to direct (dating) experience. But such additional considerations will often serve to further shorten the search process, and thereby further strengthen our point, that mate search is a prime example of an important adaptive domain in which making good decisions does not require amassing vast amounts of information. Simple satisficing heuristics for one-sided and mutual search could be used by decision makers to learn appropriate aspiration levels after checking only a few candidates. As such, these heuristics show how considerations of ecological rationality can provide the computational mind with a way around the frame problem: the mate choice heuristics considered here use only a small portion of the available information and still yield satisfactory performance.

Adopting this ecological rationality perspective opens up a variety of other research topics. In the domain of mate choice specifically, in addition to enhancing the realism of the search model as just described, more work is needed to understand the mechanisms that guide perception and display of appropriate (and deceptively inappropriate) traits and the process by which such perceptions are used in the assessment of mate quality (see Miller & Todd, 1998, for other research questions in mate choice). And this domain is only one among many: A variety of simple decision heuristics are being found for other decision applications, ranging from categorization and estimation to parental investment and predator intention judgment, where similar frugality of information-use nonetheless leads to reasonable decisions (Gigerenzer, Todd, & the ABC Research Group, 1999; Todd, Gigerenzer, & the ABC Research Group, in press). In all of these domains, the main questions stemming from the ecological rationality viewpoint are: What simple evolved heuristics can solve important decision-making problems in a fast and frugal manner, and how is information structured in the environment to allow these heuristics to accomplish their tasks with minimal calculation? In answering these questions, we seek to fill in the often-missing component of specific mechanisms in evolutionary psychology, while remaining faithful to the actual environments in which those mechanisms function.

At least one question, though, remains in our consideration of mate search. Any but the most dispassionate of readers is likely to wonder where love has disappeared to amidst all this rather mechanical description of mate search heuristics. Have we eliminated emotion from the process of finding a mate? Not at all—we seek to find a central place for emotions in our view of ecological rationality. This is currently another important area for further work, but at present we do have some ideas about the role that emotions can play in adaptive decision-making. In particular, emotions might help address the second aspect of the frame problem: determining *which* information to attend to in the first place. We turn now to the question of how emotions could guide information gathering in risky choice situations.

III. WHAT SORT OF INFORMATION SHOULD AN INTELLIGENT AGENT BOTHER ATTENDING TO? EMOTIONS AND THE FRAME PROBLEM IN RISKY CHOICE DECISION-MAKING

The idea that emotions might serve an adaptive information-processing function in risky choice contexts is not a mainstream view. Indeed, emotions have traditionally been seen as impairments to decision-making. This view is commonplace whether one considers pop culture icons of good judgment (such as Mr. Spock, the usually hyper-rational and emotionless Vulcan from *Star Trek*) or the scientific literature on decision-making (including social cognition research on emotional biases in information processing, see Ketelaar & Clore, 1997 for a recent review). One economist notes that:

> The standard view of the relation between rationality and emotions is, of course, that emotions *interfere* with rationality. They are, as it were, sand in the machinery of action. Nobody would deny that this is often true. (Elster, 1995, p. 1394, emphasis added).

Most theorists consider emotions to function primarily in producing behavior, not cognition. According to this view, emotions promote adaptive behavior through the activation of so-called "action tendencies" (see Ekman & Davidson, 1994; Scherer & Ekman, 1984 for good reviews). Researchers who adopt this behaviorally oriented perspective are generally quite suspicious of any claims that emotions might function primarily to promote adaptive cognitive functioning (for a rare exception, see Demasio, 1995). However, some evolutionary psychologists have argued that such a strict focus on behavior and its adaptive consequences leap-frogs from descriptions of behavior (e.g., fear makes us run from bears) to adaptive explanations (e.g., it is fitness-promoting to avoid danger) without specifying the actual mechanisms that provide the "missing link" between these behaviors and their adaptive implications (Buss, 1995; Cosmides & Tooby, 1987). From the perspective of evolutionary psychology, decision-making behavior (i.e., enacting one's preferences) is *contingent upon* the existence of cognitive (information processing) mechanisms that select information from the environment and use this information to draw appropriate inferences, which in turn, guide behavior toward the appropriate "adaptive" strategies. Might emotions play an important role in the cognitive processes underlying decision-making behavior?

In this section, we present a view of emotion that we feel will be a fruitful area for future research in evolutionary psychology: namely, the view that emotions operate, in part, as specialized cognitive (information processing) mechanisms that provide potent information about the future consequences of pursuing various strategy choices. In particular, we explore the view that emotions function to help decision-makers overcome the frame problem inherent in risky choice decisions. (Here we define risky choice decisions as situations where the individual must select one of several options—e.g., potential courses of action—under conditions of uncertainty regarding the future consequences of pursuing each option). According to this view, affective representations of the prospective utility of different strategic options —in the form of good and bad feelings generated while considering these options—may serve to focus attention on those strategies with adaptively relevant outcomes, ignoring the rest. The existence of numerous distinct emotions—each with a corresponding positive or negative affective state—suggests that we possess numerous domain- specific mechanisms that enable us to affectively represent the cost-benefit structure of a variety of distinct environmental contingencies (see also Ketelaar & Clore, 1997). In this sense, specific emotions might help solve the problem of what information to attend to in specific environmental circumstances by providing an "affective" guide to important features of the adaptive landscape.

Thus, when the future outcome of various courses of action cannot be objectively "calculated," often because there are simply too many plausible consequences to consider (i.e., the frame problem), it may pay to "let your emotions be your guide" in selecting which course of action to pursue. This view of "affective information processing" is consistent with a more general view of emotion endorsed by some economists (Frank, 1988; Hirshleifer, 1987) and evolutionary psychologists (Nesse, 1990; Nesse & Williams, 1994; Tooby & Cosmides, 1990) who contend that emotions function, in part, to focus attention on information about the future cost-benefit structure of one's environment. In this sense, emotions are seen as evolved software packages designed to alert an intelligent agent to the most propitious courses of action among one's current options (Nesse, 1990).

Affective representations of value: A guide to strategic decision-making?

If evolutionary psychologists focus attention on "real world" asymmetries in the prospective utility of gains and losses, emotional representations of value might provide a better guide to strategy choice than more "objective" valuations, even though emotional decisions often override lots of seemingly "rational" information. One promising view of emotion would be to explore whether affective representations of the cost-benefit structure of the environment are: 1) *more salient* than objective, "cold" calculations of costs and benefits (thereby provoking individuals to attend *more* to their emotional perceptions when making decisions), and 2) a *more accurate* guide to the long-term consequences of pursuing various strategy choices (thereby provoking individuals who make "emotional" decisions to perform better over the long-run than their "emotion-less" counterparts).

Affective feelings are an immediately salient source of information

Emotions are typically seen as having both a cognitive structure and an accompanying affective (e.g., good-bad) feeling component (Ortony, Clore, & Collins, 1988). Experiencing a pleasant or unpleasant feeling state is, by definition, a large part of what it means to have an emotional experience; as such, these affective feelings can be seen as a necessary, but not sufficient component of emotion (Clore, 1992). Perhaps emotions so often intrude in our most important decisions precisely because the affective feeling component accompanying an emotion is so immediately salient and unable to be ignored. One can't help but attend to one's affective feelings while one is experiencing them. This is the case because feeling states, as opposed to other types of cognitive states, are by definition, states of awareness. Thus, it makes little sense to say that "I feel really bad right now, but I'm not aware of it," or that "I feel really good right now, but I am not aware that I do."[2] Indeed, there is ample evidence that affective feeling states, including those that accompany specific emotions, intrude on a wide variety of seemingly "purely cognitive" judgments and decisions (see Clore, Schwarz, & Conway 1994; Ketelaar & Clore, 1997 for reviews). This robust finding has been labeled the "Affect-as-information" model because individuals appear to consult their affective reactions as one source of information when evaluating objects and events (see Schwarz & Clore, 1983, 1988).

But is affective feeling an accurate indicator of long-term consequences?

The question remains, is the information conveyed by affective feeling states actually useful in decision-making? In the rest of this section we focus on the idea that specific emotions may help solve the problem of what information to attend to in specific environmental circumstances, by virtue of their accompanying affective feeling states. According to this view, the affective feeling state that accompanies a specific emotion can function as a strategic information state—in the form of an "affective" decision weight—that accurately represents the prospective[3] utilities corresponding to the specific set of circumstances in which the emotion arose. We now explore how the affective feeling component of emotion might help us surmount the frame problem inherent in risky choice decisions in much the same way that computer chess programs operate, by focusing attention on a limited number of strategic options and making the future consequences of those options accurately and powerfully salient.

The Frame Problem in risky choice decision-making: A computer chess analogy

Although forecasting the future consequences of alternative courses of action (to help select the best strategy) is an important adaptive problem, it also presents considerable computational obstacles. Strategy choice in a risky choice context entails a version of the frame problem, resulting from the exponential explosion of potential outcomes that one must consider when calculating the future consequences of pursuing each of the possible alternative strategies at one's disposal. Considering an analogy between risky choice decisions and chess can readily highlight this point.

The frame problem occurs whenever constructing the decision tree of possible alternative courses of actions and inferring their likely future consequences becomes computationally intractable (remember the robot in Dennett's thought experiment). The game of chess also involves considering a large number of strategies and their future consequences. For example, performing at a reasonable level of skill in chess requires more than simply selecting the one move that will generate the best *immediate* outcome (say, capturing your opponent's queen). Instead, a better strategy is to look ahead several consecutive moves to consider the likely outcomes of various alternative sequences of moves (known in chess parlance as continuations). If one could simply map out the entire decision tree of all possible continuations and their final consequences, chess would become equivalent to tic-tac-toe where the sure-fire winning (or perpetually tying) strategies are well documented. Yet, any computational device that attempts to select its next move in the game of chess by considering all possible legal moves and all possible future consequences of such moves will soon find itself tangled in a decision tree of immeasurable complexity. The number of possible sequences of moves that can be played from the start of a game to the end is astronomically large. It is literally impossible to generate the entire decision tree of possible sequences of consecutive chess moves (see Newborn, 1996 for an accessible overview of computer chess). Therein lies the rub, and the frame problem.[4] Could AI's solution to the problem of selecting the "best" chess move suggest a workable analogy for how emotions might surmount the frame problem inherent in risky choice contexts?

A fast and frugal answer to the frame problem in chess: Limited search with an evaluation function

In light of the computational difficulties inherent in selecting the "best" chess move, AI researchers had to find an alternative to mapping out the entire decision tree of chess continuations. Today the very best computer chess programs rely on an alternative strategy outlined by the grandfather of computer chess, Claude Shannon (see Newborn, 1996). Shannon (1950) proposed that the goal of mapping the entire decision tree of chess moves be abandoned in favor of a simpler strategy in which a move "evaluation function" played a central role. Successful chess playing computers, such as Deep Blue, employ this relatively simple strategy and map only a minuscule portion of the decision tree when selecting their next move. In one variant of this approach, known as the "fixed depth" search, a move is selected by:

> exploring all lines of play to some fixed depth and then assigning a score to the position *at the end* of each continuation. The score assigned to the position by a "scoring function" is a measure of how good the position is for the side on the move (Newborn, 1996, p. 9, emphasis added).

This "scoring function" or "evaluation function" considers a number of factors including relative material (piece) advantage, the mobility of one's pieces, et cetera. This simple strategy can be summarized as follows: 1) look ahead a *limited* number of moves, 2) assign a score to the terminal position at the end of each continuation, and then 3) select a move that guarantees progress toward the terminal position with the highest score.

Although hardly omnipotent, this strategy has performed quite well in competition. In the most recent encounter with arguably the best human chess player ever, Grand Master Gary Kasparov, Deep Blue's successor (Deeper Blue) employed this method to achieve the first-ever computer victory over a human world champion (see King, 1997). Thus, Shannon's (1950) strategy of searching a limited portion of the decision tree with an appropriate evaluation function not only appears to overcome the frame problem (as it is manifest in this chess example), but equally important, is able to perform at (or above) par with the best human chess players. How similar is this to how human reasoners use emotions to help select future courses of action?

Emotional Gambits: Affect, risky choice, and the computer chess evaluation function

We can cross the bridge from computer chess programs to more day-to-day risky choice decision making through the example of a strategy game known as the repeated Prisoner's Dilemma. This is a popular if somewhat overused example of the common problem of having to select one of multiple courses of action (e.g., some pattern of cooperating or defecting in social encounters) with incomplete knowledge of the long-term consequences of one's decision. In the single-shot, no-future version of this game, each player encounters the other only once, and the immediate consequences of either cooperating or defecting are obvious: Defection yields the higher expected utility. However, in the repeated version of this game where players encounter each other multiple times, this strategy choice (i.e., always defecting) must contend with several alternative strategies (e.g., tit-for-tat, generous tit-for-tat, GRIM, mutual cooperation, etc.) that can lead to a higher *long-term* payoff (see Axelrod, 1984; Samuelson, 1997). If the number of iterations to be played is large or unknown (as it often is in real life), trying to search the whole decision tree of strategies and counter-strategies is no longer computationally feasible for determining one's best next move. Just as in the game of chess, the number of possible strategies and their ultimate consequences that can be considered soon becomes mind-boggling. How else then can one determine which strategy to play in a repeated Prisoner's Dilemma game where the number of iterations to be played is large or unknown? One view is that the affective component of emotion can play a functional role in evaluating alternative courses of action in situations such as these (see also Frank, 1988).

This is not a particularly popular notion. As in other domains of judgment and decision-making, emotions are more commonly viewed as impairing, rather than improving, rational judgment. Kahneman (1999), for example, has found that when individuals evaluate the quality of an experience extended over time, their affective reactions lead them to assign a disproportionate weight to how the experience ends. Kahneman (1999) argues that the normatively "correct" strategy for evaluating the quality of episodes extended over time is to calculate the temporal integral across all momentary affective (good-bad) sensations experienced during these episodes. Rather than integrating the hedonic values (affective reactions to the task) across all momentary instants, individuals instead appear to judge the utility of the entire episode in terms of its peak and how it ends. Accordingly, this judgment strategy has been labeled the "peak-end" rule (Kahneman, 1999). In several experiments participants have provided real-time recordings of their hedonic experiences during an episode, after which they were asked to provide global evaluations of the entire episode. Time and time again, it is revealed that individuals judge experiences as being worse if the peak amount of negative affect occurs later, rather than earlier in the episode (see Kahneman, 1999 for an overview). This leads to the paradoxical finding that individuals sometimes prefer episodes with more (rather than less) overall summed pain. In one study, individuals undergoing an uncomfortable medical procedure (colonoscopy) were shown to evaluate the entire episode as being less painful overall if the intensity of the pain was made to drop off slowly rather than quickly (a modification that does not involve changing the peak height, but does move its location from the end of the procedure to the middle) even though the former strategy results in more overall pain than the latter strategy (Redelmeier & Kahneman, 1996). If one assumes that one should utilize all available information in judging the overall quality of an outcome stretched over time, then this consistent focus on the affective representation of only "the peak and the end" of the episode has the appearance of a crude and irrational heuristic (see Kahneman, 1999). But is the "peak-end" rule really so irrational?

Consider our chess-playing computer again. It is interesting to note that both computers and humans appear to assign a disproportionate weight to the endings of events when considering possible courses of action (see Kahneman, 1999; Newborn, 1996). In chess, it is the score at the final or "end" positions of the decision tree of possible continuations that matters most. If an initially attractive line of play concludes with a high negative score at the end of the limited series of moves being considered, it will be appropriately seen as a dubious long-term strategy, despite its initial attraction. By contrast, if a particular line of play exhibits a high negative score early on in that particular sequence of moves, this does not necessarily portend a significant blunder. It may be the case that the course of action that exhibits its "peak" negative value early on in the sequence of moves merely reflects that one has employed a strategic gambit. A gambit is a strategic sacrifice, often made very early in the game, where one incurs a significant immediate loss in exchange for a long-term strategic advantage. The economist Robert Frank (1988) proposes that emotions operate in this same way, as devices for provoking strategic gambits in the game of life.

Affective gambits

The default information state of most organisms appears to be one where the environment is represented in terms of its immediate consequences (e.g., that piece of cake looks good *right* now, see Frank, 1988, p. 76-80). However, just as in the game of chess, relying on immediate outcomes for strategy choice can have dubious consequences for the long-term strategist. What if an initially attractive strategy (such as defection in the Prisoner's Dilemma) actually leads to a large negative score (e.g., a very low payoff) at the end of a limited series of moves? Because decision-makers are not clairvoyant, the final "negative" consequences of an initially attractive strategy cannot be directly perceived. Moreover, even if one could imagine all possible future scenarios in order to "simulate" their outcomes, by calculating the ultimate consequences of adopting a particular strategy and comparing it to the ultimate consequences of all possible alternative strategies (in order to select the best strategy), we run headlong into the frame problem. Just like chess-playing computers, human decision-makers need some means of focusing attention on a limited set of adaptively relevant strategic options.

Frank (1988) argues that one's affective reactions while considering various options can "inform" the individual of the future consequences of pursuing each option. Analogous to the "evaluation function" of a chess-playing computer, the emotions aroused during the process of making important decisions might aid individuals in strategy choice by shifting attention away from certain dubious long-range strategies that initially appear quite attractive, and toward other strategies that fare better over the long haul. For example, an individual may experience the emotion of guilt while considering the initially attractive strategy of defection in the repeated Prisoner's Dilemma. Because the affective feeling state that accompanies guilt is aversive, the individual might be provoked to select an alternative, less aversive, course of action. One interpretation of Frank's (1988) model is that emotions such as guilt function to move the *future* costs of actions such as defection into the *current* situation, in the form of a powerful and salient feeling state such as the negative affect associated with guilt (see Ketelaar & Clore, 1997). In a similar fashion, the consideration of alternative strategies (mutual cooperation, tit-for-tat, etc.) might evoke positive affective reactions, suggesting that their prospective future utilities are considerably higher, even if their immediate consequences (lower payoffs) are less attractive. These affective states can thus serve as psychological "stand-ins" for the imperceptible future consequences of pursuing particular courses of action. In this way, emotions are analogous to the "evaluation function" of chess-playing computers in that they assign disproportionate weight to the final outcome of the strategy choice that one is considering.

From one point of view, this emotional focus on "how things end" seems irrational (Kahneman, 1999). By contrast, an evolutionary perspective argues that, far from being maladaptive, emotions such as guilt can actually provoke an individual to pursue emotional gambits that, while incurring an immediate cost, lead to higher payoffs over the long haul (Frank, 1988). These emotional gambits provoke the individual to experience the peak amount of pain (a lower initial payoff) in the earlier stages of the decision-making episode in order to avoid selecting a strategy where the peak amount of pain is incurred at the end of the series of moves. An evolutionary psychological interpretation of this model suggests that the affective component of emotion functions as a sort of evolved decision weight where the perceived positivity or negativity of one's mental state while considering a particular strategic option reflects the future "adaptive" utility of pursuing that option. In circumstances where the future consequences of different strategy choices cannot be directly perceived, it may make sense to "let your emotions be your guide" in strategy selection. This can be the case if affective representations of value reflect the long-term evolutionary history of the particular cost-benefit calculations one is considering. While a mortal individual cannot look into the future to determine the ultimate consequences of pursuing a particular strategy, by using the "affective weights" attached to that option the individual may be able to "look back into the past" evolutionary history of similar scenarios to assign appropriate weights according to the likely fitness consequences of pursuing that strategy (see Tooby & Cosmides, 1990 for a similar view). We now describe preliminary research aimed at testing these ideas, which is producing promising support (Ketelaar & Au, 2001).

To examine whether this functional information-processing model of emotion is valid, one could examine the effects of so-called "emotional gambits" on strategic behavior in repeated social bargaining games such as the Prisoner's Dilemma. One could examine whether individuals who experience an aversive affective state (e.g., guilt) while considering "defection" are more likely to incur an immediate cost (i.e., lower payoff) by forgoing this initially attractive strategy in favor of adopting an alternative, "more cooperative," strategy. Specifically, one could compare individuals who experience guilt to those who do not, to determine who is more likely to display higher levels of cooperation indicative of adopting the better long-run strategies (see Axelrod, 1984; Frank, 1988). This is exactly the direction being pursued in an ongoing set of studies.

In one experiment college students were randomly assigned to guilty mood and neutral mood inductions and asked to play an indefinite number of iterations of the Prisoner's Dilemma on a computer (see Ketelaar & Au, 2000). Players in the guilt condition displayed higher levels of cooperation (53 per cent cooperative responses) compared to players in the control condition (39 per cent cooperation). Similar findings were observed in a repeated ultimatum game where participants had to negotiate how to divide a sum of money with a partner twice over a two-week period. Of primary interest was whether individuals who felt *guilty* after proposing a *selfish* division of the money at Time 1 would be more likely to propose a more *generous* split of the money when they repeated this negotiation (with the same partner) one week later (at Time 2). The idea here is that experiencing guilty feelings after proposing a selfish offer may serve to alert the individual that continued selfishness is a dubious long-term strategy in a repeated social bargaining game. Thus, one predicts that only those selfish individuals who felt guilty about their behavior at Time 1 will reverse their offers (in a "tit-for-tat" fashion consistent with reciprocal altruism), and propose a more generous split of the money one week later (at Time 2). The differences in responses at Time 2 as a function of guilt at Time 1 were quite large: 91% of the individuals who felt guilty after making a selfish offer at Time 1 proposed a generous offer at Time 2, compared to only 22% of the individuals who felt no guilt after a selfish offer. These preliminary findings are consistent with Frank's (1988) view that emotions (here we considered guilt) can provoke strategic gambits in repeated social bargaining games, leading individuals to incur an immediate cost (e.g., a lower payoff) by forgoing an initially attractive strategy option in order to reap a long-term strategic advantage.

Emotions, the "peak-end" rule, and the frame problem

The philosopher Ronald deSousa (1987, p. 195) has argued that emotions limit "the range of information that the organism will take into account, the inferences actually drawn from a potential infinity, and the set of live options among which it will choose." This view is consistent with our claim that emotions function, in part, to overcome the frame problem by both: 1) focusing attention on a limited number of strategic options and 2) making the future consequences of those options immediately salient. It would appear that under ecological conditions where future outcomes are uncertain and the frame problem runs rampant (as illustrated by the artificial example of chess-playing computers), the attention-limiting focus provoked by emotions that adhere to something like the "peak-end" rule makes good adaptive sense as an alternative to evaluating all possible strategy options (remember the vast decision tree of chess continuations). Under ecological conditions where strategic gambits need to be taken into consideration (as in the game of chess), endings really do matter more than earlier moments. Rather than viewing emotions as impairments to decision-making, this alternative model describes emotions as adaptively designed information processing mechanisms that provide information—via affective decision weights—about the future consequences of pursuing various strategy choices. In some circumstances, an affective focus on peaks and ends may not be so irrational after all.

In this manner, emotions may thus be crucial in generating intelligent behavior, one hallmark of which is the capacity to "look before you leap," by considering the consequences of actions *before* embarking upon them:

> Sophisticated—or just risk-averse—creatures avail themselves of systems of temporally *distal* anticipation. They use their brains to produce more far-ranging versions of the relevant future. So far as I can see, this is really just an extension of letting the

world warn you. It attempts to build less direct, more devious sorts of warnings out of the masses of individually unthreatening tidbits of information gleaned from the passing show, which it fabricates into internal alarm systems. (Dennett, 1996, pp. 2-3, emphasis in the original)

Emotions, we suggest, can perform the role of these internal alarm systems, alerting us to adaptive dangers and benefits by inferring the distal future from one's current circumstances and strategy choices. In this way, emotions can help the computationally limited human mind to circumvent the pitfalls of the frame problem by determining *which* information to attend to in the first place.

IV. CONCLUSIONS—A FAST AND FRUGAL APPROACH TO STUDYING ECOLOGICAL RATIONALITY

For any given problem that faces us in the real world, the number of ways we can think about solving the problem, and the number of consequences that follow from any particular course of action, are limitless. Considering all possibilities is clearly impossible. To function effectively and behave adaptively, we must have some way of cutting the possibilities down, putting a very narrow frame around any given problem we face and staying within it. As we have shown in this chapter, an ecological view of how an organism can behave rationally makes it clear that even a narrow frame—even taking only a small amount of information into account—can still allow for reasonable performance in important adaptive domains. Furthermore, we have argued that emotions can guide where this narrow frame is placed, by focusing the computational mind on just that information that will be most useful in making adaptive decisions with appropriate long-term outcomes.

But there is also a frame problem lurking for researchers trying to understand the nature of human cognition. There are too many possible mechanisms that *could* produce human behavior, so how do we search for appropriate ones? Even a progressive meta-theory like evolutionary psychology vastly underdetermines the scope of psychological mechanisms that could have arisen, evolutionarily speaking, to solve a particular information-processing problem (Ketelaar & Ellis, 2000). Ecological rationality gives us a way forward within this larger frame problem, by suggesting a bottom-up research approach of first identifying and exploring simple heuristic decision-making components, and then building more sophisticated cognitive modules out of these simpler components. We now briefly sketch this research methodology and how it contrasts with that inspired by the more traditional view of rationality.

The view of ecological rationality we have presented stems from illuminating simple adaptive heuristics that can solve the information-processing problems recurrently faced by our ancestors in their natural environment. As we have indicated, this stands in sharp contrast to traditional views of rationality that often presuppose a complex, content-free apparatus of logical or probabilistic reasoning instantiated in people's heads. This contrast does more than just draw argumentative lines in the sand between two camps—it has deep implications for how one believes the reasoning mechanisms of people and animals should be explored. If reasoning is to be logically and mathematically sound, then it makes sense to start looking for psychological reasoning mechanisms by finding the mathematically optimal solution to some particular behavioral challenge or problem, and then whittling that down through successive simplifications—all mathematically derivable and testable—until finally a psychologically plausible mechanism is found. Note that this "top-down" approach to deriving human psychology from lofty optimal heights presents several challenges, the foremost being that one must first construct an optimal solution to the problem being considered. In many cases, this will simply be impossible; no optima exist for a wide class of realistic behavioral challenges (Simon, 1990). In those cases where it is still possible, deciding on the appropriate steps of successive derivation and simplification to bring the optimal solution down to the psychological level of human computation presents a further challenge, again one of deciding which of many possible approaches to a problem is the best to take. In the end, if this approach succeeds, we will have gained a model of human behavior that is mathematically guaranteed and logically sound (without undesired intransitivities, etc.). This strategy also promises a clear paper trail of derivations leading back to the optimal solution, so that direct comparisons can be made in terms of performance on various criteria, range of applicability, mechanisms of operation, etc. But the path down the mountain can be arduous, and even finding the peak in the first place, let alone an appropriate descending path out of the multitude of possible derivations, presents a search problem all too familiar from our earlier discussions of the frame problem.

But there is an ecologically rational way around this problem as well. We start instead with the view that reasoning heuristics evolved because they performed well for people living and behaving in certain types of environments. With this perspective, it is reasonable to begin our search for mental mechanisms not empty-handed, but rather with some adaptive building blocks that people actually use—like emotions, or aspiration levels—and see how these can be combined into strategies that achieve sufficient levels of performance. This bottom-up approach to the construction of fast and frugal heuristics starts with psychological plausibility and ends with adaptive behavior in a particular problem domain. Determining the initial set of building blocks to work with should be simpler than finding an optimal solution in the top-down approach. Finding suitable combinations of fast and frugal building blocks for a particular problem may often prove challenging—we do not yet know how to specify this (re)search process—but trial-and-error methods (including heuristic-guided search) and exploratory shuffling will work here in a way that is impossible with the mathematical derivations required by the top-down method.

We hasten to point out that much work still has to be done once a fast and frugal heuristic is put forth for testing. The ecological rationality approach demands a demonstration of how a putative adaptive heuristic performs in different environments and an explanation for why it acts the way it does. Without the mathematical derivation pedigree of heuristics found by the top-down whittling away of optimal solutions, these questions must be answered instead by empirical observation, computer simulation, or even new mathematical analysis. Both the top-down derivation and bottom-up exploration approaches can yield the same results on a particular problem, identifying the same heuristic in the end. However, we expect that the bottom-up approach will illuminate plausible mechanisms faster, and with fewer false starts (e.g. those caused by the impossibility of specifying an optimal solution in the first approach). More important, because the bottom-up approach can discover mental mechanisms that do not lie within traditional mathematical or logical bounds—for instance, heuristics that violate transitivity—we expect this research plan to identify a broader range of psychologically plausible but potentially a-rational (by traditional standards) fast and frugal heuristics.

By looking for specific heuristic solutions to particular environmental challenges, an ecological rationality perspective on evolutionary psychology aims to reveal a domain-specific collection of well-adapted tools. In this chapter we illustrated how an ecological rationality perspective on decision-making, even in domains that have traditionally been viewed as non-cognitive—such as finding a mate and thinking while under the influence of emotion—can generate fruitful new directions for psychological research by shining the spotlight of the Darwinian approach inside the framework of the computational mind.

FOOTNOTES

[1] In approaches to decision-making such as game theory, equilibrium solutions to social bargaining games are sometimes determined by introducing assumptions such as mutual knowledge or perfect information states (e.g., "I know the rules of the game, you know the rules of the game, I know that you know the rules of the game, you know that I know that you know the rules of the game, I know that you know that I know that you know the rules of the game," ad infinitum). An equilibrium solution is the strategy that—in the case of a two-person game, for example—yields the mutually "best" strategy in the sense that each agent's strategy is optimal given the strategy of the other agent (see Rasmussen, 1994; Samuelson, 1997). Although such methods of equilibrium selection work in the abstract, they typically cannot be considered biologically plausible explanations of human decision-making because, in many cases, obtaining a "perfect information state" and "mutual knowledge" would entail versions of the frame problem: The mind is incapable of spinning through an *infinity* of recursive information states to locate an equilibrium solution.

[2] One could claim, of course, that they were experiencing an emotion, but were unaware that they were experiencing it at the time. This perception is quite common. The point here is that there is little empirical evidence to support anecdotal claims that individuals can experience full-blown affective "feelings" without awareness of these feelings. There is some evidence, however, that individuals can have valenced, affective reactions to stimuli without awareness (see Winkielman, Zajonc, & Schwarz, 1997 and Bargh, 1997 for one view of unconscious attitudes).

[3] These decision weights are referred to as "prospective" utilities to distinguish them from evaluations of the current worth of an option (e.g., strictly in terms of its immediate benefits). From an evolutionary standpoint, what is important is not just the current value of an option, but rather its future or "prospective" value. Drinking salt water might be just as effective in the short-run for decreasing current feelings of thirst as drinking fresh water, but the long-term consequences of these two options may differ. The use of the term "prospective" is meant to convey the idea that organisms can benefit from using decision weights that somehow represent the future (as opposed to the immediate) value of an option (see Frank, 1988 for a similar view).

[4] Although some definitions of the frame problem exclude domains such as chess (see Baumgartner & Payr, 1995, p. 314) because they are "well de-

fined," we include the game here as a clear example of a problem where the optimal solution involves an incomputable search and reasonable solutions require intelligent compromises to frame the search appropriately. Selecting the best possible move in a game of chess entails specifying the decision-tree of all possible continuations from the current position. The task of constructing and searching through this tree is beyond the scope of any computational device yet known. For example, the average midgame scenario has around 30-40 possible legal moves that can be played, over 1000 possible two-move continuations to ponder, over 1 million three-move continuations to consider, over 1 billion possible four-move continuations to think over, and the number grows exponentially as one looks further ahead (see Newborn, 1996). Not even computers such as Deep Blue (which can consider some 250 million moves per second) can make a dent in the complete search. Thus, selecting a move in chess requires addressing a version of the frame problem, deciding which options and consequences to explore and which possibilities to leave out.

REFERENCES

Axelrod, R. 1984. *The Evolution of Cooperation*. New York, NY: Basic Books.

Bargh, J.A. 1997. "The Automaticity of Everyday Life", in Wyer, R.S., ed., *The Automaticity of Everyday Life: Advances in Social Cognition*, vol. 10. Mahwah, NJ: Lawrence Erlbaum Associates, pp. 1-61.

Baumgartner, P. & Payr, S., eds., 1995. *Speaking Minds: Interviews with Twenty Eminent Cognitive Scientists*. Princeton, NJ: Princeton University Press.

Buss, D.M. 1995. "Evolutionary Psychology: A New Paradigm for Psychological Science", *Psychological Inquiry,* 6, 1-30.

Chiappe, D.L. & Kukla, A. 1996. "Context Selection and the Frame Problem", *Behavioral and Brain Sciences,* 19, 529-560.

Clore, G.L. 1992. "Cognitive Phenomenology: The Role of Feelings in the Construction of Social Judgment", in Tesser, A. & Martin, L.L., eds., *The Construction of Social Judgments.* Hillsdale, NJ: Erlbaum, pp. 133-164.

Clore, G.L., Schwarz, N., & Conway, M. 1994. "Affective Causes and Consequences of Social Information Processing", in Wyer, R.S. & Srull, T., eds., *The Handbook of Social Cognition*, 2nd Ed. Hillsdale, NJ: Erlbaum, pp. 323-417.

Cormen, T.H., Leiserson, C.E., & Rivest, R.L. 1997. *Introduction to Algorithms*. Cambridge, MA: The MIT Press.

Cosmides, L. & Tooby, J. 1987. "From Evolution to Behavior: Evolutionary Psychology as the Missing Link", in Dupre, J., ed., *The Latest on the Best: Essays on Evolution and Optimality.* Cambridge, MA: MIT Press, pp. 277-306.

Damasio, A.R. 1994. *Descartes' Error: Emotion, Reason and the Human Brain.* New York: Grosset/Putnam.

Dennett, D.C. 1987. "Cognitive Wheels: The Frame Problem in AI", in Pylyshyn, Z.W., ed., *The Robot's Dilemma: The Frame Problem in Artificial Intelligence*. Norwood, NJ: Ablex, pp. 41-64.

Dennett, D.C. 1996. "Producing Future by Telling Stories", in Ford, K.M. & Pylyshyn, Z.W., eds., *The Robot's Dilemma Revisited: The Frame Problem in Artificial Intelligence*. Norwood, NJ: Ablex Publishing Corporation, pp. 1-7.

deSousa, R. 1987. *The Rationality of Emotion*. Cambridge, MA: The MIT Press.

deSousa, R. 1994. "Emotion", in Guttenplan, S., ed., *A Companion to the Philosophy of Mind*. Cambridge, MA: Blackwell Publishers, pp. 270-276.

Ekman, P. & Davidson, R.J. 1994. *The Nature of Emotion: Fundamental Questions*. New York: Oxford University Press.

Elster, J. 1995. "Rationality and the Emotions", *The Economic Journal, 116*, 1386-1397.

Ferguson, T.S. 1989. "Who solved the Secretary Problem?", *Statistical Science, 4*, 282-296.

Fodor, J.A. 1995. "The Folly of Simulation", in Baumgartner, P. & Payr, S., eds., *Speaking Minds: Interviews with Twenty Eminent Cognitive Scientists*. Princeton, NJ: Princeton University Press, pp. 84-100.

Ford, K.M. & Pylyshyn, Z.W. 1996. *The Robot's Dilemma Revisited: The Frame Problem in Artificial Intelligence*. Norwood, NJ: Ablex Publishing Corporation.

Frank, R.H. 1988. *Passions within Reason: The Strategic Role of the Emotions*. New York: W.W. Norton.

Gigerenzer, G., Todd, P.M., & The ABC Research Group. 1999. *Simple Heuristics that Make Us Smart*. New York: Oxford University Press.

Gilbert, J.P & Mosteller, F. 1966. "Recognizing the Maximum of a Sequence", *American Statistical Association Journal, 61*, 35-73.

Haugeland, J. 1995. "Farewell to GOFAI?", in Baumgartner, P. & Payr, S., eds., *Speaking Minds: Interviews with Twenty Eminent Cognitive Scientists*. Princeton, NJ: Princeton University Press, pp. 101-114.

Hirshleifer, J. 1987. "On the Emotions as Guarantors of Threats and Promises", in Dupré, J., ed., *The Latest on the Best: Essays on Evolution and Optimality*. Boston: Bradford Books-MIT Press, pp. 307-326.

Kahneman, D. 1999. "Objective Happiness", in Kahneman, D., Diener, E., & Schwarz, N., eds., *Well-Being: The Foundations of Hedonic Psychology*. New York: Cambridge University Press. pp. 3-25.

Kasparov, G. 1997. *Lessons in Chess*. London: Cadogan Books.

Ketelaar, T. & Au, W.T. 2001. "The Role of Guilty Feelings in Cooperative Behavior in the Prisoner's Dilemma: An Emotion-as-Information Interpretation of Frank's (1988) Commitment Model", unpublished manuscript, UCLA.

Ketelaar, T. & Clore, G.L. 1997. "Emotions and Reason: The Proximate Effects and Ultimate Functions of Emotions", in Matthews, G., ed., *Personality,*

Emotion, and Cognitive Science. Advances in Psychology Series, Amsterdam: Elsevier Science Publishers (North-Holland), pp. 355-396.

Ketelaar, T. & Ellis, B.J. 2000. "Are Evolutionary Explanations Unfalsifiable? Evolutionary Psychology and the Lakatosian Philosophy of Science", *Psychological Inquiry*, *11*, 1-21.

King, D. 1997. *Kasparov v. Deeper Blue: The Ultimate Man v. Machine Challenge*. London: B.T. Batsford, Ltd.

McCarthy, J. & Hayes, P.J. 1969. "Some Philosophical Problems from the Standpoint of Artificial Intelligence", in Meltzer, B. & Michie, D., eds., *Machine Intelligence* (Vol. 4). Edinburgh: Edinburgh University Press, pp. 463-502.

Miller, G.F. & Todd, P.M. 1998. "Mate Choice turns Cognitive", *Trends in Cognitive Sciences*, *2*, 190-198.

Mosteller, F. 1987. *Fifty Challenging Problems in Probability with Solutions*. New York: Dover. (Reprint of the 1965 Addison-Wesley edition).

Nesse, R. 1990. "Evolutionary Explanations of Emotions", *Human Nature*, *1*, 261-289.

Nesse, R. & Williams, G. 1994. *Why We Get Sick: The New Science of Darwinian Medicine*. New York: Times Books.

Newborn, M. 1996. *Kasparov versus Deep Blue: Computer Chess Comes of Age*. New York: Springer-Verlag, Inc.

Ortony, A., Clore, G.L., & Collins, A. 1988. *The Cognitive Structure of Emotions*. New York: Cambridge University Press.

Pinker, S. 1984. *Visual Cognition: An Introduction*. Cambridge, MA: MIT Press.

Pinker, S. 1997. *How the Mind Works*. New York: W.W. Norton & Company.

Poggio, T. 1984. "Vision by Man and Machine", *Scientific American*, April, pp. 106-116.

Pylyshyn, Z.W., ed., 1987. *The Robot's Dilemma: The Frame Problem in Artificial Intelligence*. Norwood, NJ: Ablex.

Rasmussen, E. 1994. *Games and Information: An Introduction to Game Theory* (2nd Edition), Cambridge, MA: Blackwell.

Redelmeier, D. & Kahneman, D. 1996. "Patients' Memories of Painful Medical Treatments: Real-time and Retrospective Evaluations of Two Minimally Invasive Procedures", *Pain*, 116, 3-8.

Samuelson, L. 1997. *Evolutionary Games and Equilibrium Selection*. Cambridge, MA: The MIT Press.

Scherer, K.R. & Ekman, P., eds., 1984. *Approaches to Emotion*. Hillsdale, NJ: Lawrence Erlbaum Associates.

Schwarz, N. & Clore, G.L. 1983. "Mood, Misattribution, and Judgments of Well-being: Informative and Directive Functions of Affective States", *Journal of Personality and Social Psychology, 45*, 513-523.

Schwarz, N. & Clore, G.L. 1988. "How Do I Feel About It? Informative Functions of Affective States", in Fiedler, K. & Forgas, J., eds., *Affect, Cognition, and Social Behavior*. Toronto: Hogrefe International, pp. 44-62.

Shannon, C.E. 1950. "Programming a Computer to Play Chess", *Philosophical Magazine, 41*, 256-275.
Simon, H.A. 1990. "Invariants of Human Behavior", *Annual Review of Psychology, 41,* 1-19.
Sperber, D. & Wilson, D. 1995. *Relevance: Communication and Cognition.* Cambridge, MA: Harvard University Press.
Todd, P.M. 2000. "The Ecological Rationality of Mechanisms Evolved to Make Up Minds", *American Behavioral Scientist (Special issue on evolutionary psychology), 43*(6), 940-956.
Todd, P.M., Fiddick, L., and Krauss, S. (in press). "Ecological Rationality and Its Contents", *Thinking and Reasoning.*
Todd, P.M., Gigerenzer, G., & The ABC Research Group (in press). "How Can We Open Up the Adaptive Toolbox?" (Reply to commentaries), *Behavioral and Brain Sciences.*
Todd, P.M. & Miller, G.F. 1999. "From Pride and Prejudice to Persuasion: Realistic Heuristics for Mate Search", in Gigerenzer, G., Todd, P.M., & The ABC Research Group, *Simple Heuristics That Make Us Smart.* New York: Oxford University Press, pp. 287-308.
Tooby, J. & Cosmides, L. 1990. "The Past Explains the Present: Emotional Adaptations and the Structure of Ancestral Environments", *Ethology and Sociobiology, 11*, 375-424.
Winkielman, P., Zajonc, R. B., & Schwarz, N. 1997. "Subliminal Affective Priming Resists Attributional Interventions", *Cognition & Emotion, 11*, 433-465.

CHAPTER 6

FOOD AND FITNESS: ARE EATING DISORDERS ABOUT SELF-ESTEEM, MATING STRATEGIES OR BOTH?

HARMON R. HOLCOMB III
University of Kentucky

Abstract. Psychologists, academics, and the public often find evolutionary studies irrelevant to what they take for granted, e.g., eating disorders are about self-esteem, not mating strategies. This (nonevolutionary) explanation derives from the "symbolization model": eating disorders are symbols of underlying distressful mind-sets involving low self-esteem. Evolutionists counter that evolutionary theory can unify all of psychology, e.g., eating disorders are about mating, not mere symbols of ego-identity problems. That evolutionary explanation derives from the "reproductive suppression model": eating disorders are extreme forms of evolved adaptations for postponing reproduction to increase lifetime reproductive success. The two models yield different kinds of explanations. Because relevance is relative to the kind of explanation, evolutionary theory cannot unify all psychology theory (symbolic explanation has a place) but it is relevant to and can unify most psychological phenomena. Yet evolutionary explanations that posit functions without considering causes of behavior lose explanatory power and persuasive power. Causes of eating disorders include: lack of desire to develop sexual traits, dislike of one's body size and shape, striving to meet demanding physical standards, fear of becoming fat, desire to conform, and seeing control as a virtue. Psychology should be informed by evolution, e.g., I suggest research methods for how to explain these proximate mechanisms upon reconceiving "self-esteem" partly as "mate value." Conversely, evolutionary studies should be informed by causal findings in psychology, e.g., I connect the evolutionary model's domain-specific mating strategies to domain-general "nonevolutionary" mechanisms of control of cognition and inhibition.

H.R. Holcomb III (ed.), Conceptual Challenges in Evolutionary Psychology, 213–269.

1. THE RELEVANCE OF EVOLUTIONARY PSYCHOLOGICAL EXPLANATIONS

What are eating disorders all about? In this chapter I shall draw implications of fact that the standard nonevolutionary answer is based on, what I shall call, "the symbolization model," which purposely bypasses all proximate mechanisms of eating disorders:

> Individuals may develop eating disorders for a variety of reasons. Experts do not completely agree on the exact causes of eating disorders. It is theorized that some individuals may fear becoming fat, some may dislike the shape and size of their body; some may not want to develop sexual characteristics; some may have a need to meet standards for dance, modeling, wrestling or gymnastics; some may see control as a virtue; and some may want to do what others do. In all cases, food itself is not the primary problem, but rather a symbol of serious distress (National Live Stock and Meat Board, 1992, p.13).

Most people think that eating disorders are not "about food" but symbolize a different underlying stressful condition involving low self-esteem, which that individual typically tries to handle by dieting; call this the "symbolization model." Evolutionary psychology needs research strategies for constructing and evaluating explanations in ways that aim to prevent a common outcome: People reject evolutionary psychological explanations because they do not fit people's preconceived ideas. The standard evolutionary model is that eating disorders are literally about food, which is causally and functionally related to mating success via evolved strategies of suppressing current reproduction; call this the "reproductive suppression model." Evolutionary psychology often confronts the situation that use of symbols rather than causal or functional relationships are preconceived kinds of explanatory factors for understanding human behavior. Eating disorders provide a test case for going to the deeper part of the question of what sorts of reasons can persuade people that evolutionary psychological explanations are, contrary to first impressions, relevant.

This chapter addresses issues of relevance. We agree that relevance is necessary but insufficient for truth or practical application. Why believe and act on an irrelevant explanation? Yet standard images of science treat scientific method in terms of testing, with no guidelines for establishing explanatory relevance. The importance and difficulty of devising adequate tests of evolutionary hypotheses make prior relevance arguments all the more significant. Economy of research directs us to test only those hypotheses we have good reasons to deem relevant.

We all know the types of facts to which any explanation of eating disorders must be relevant. Eating disorders have reached "epidemic" proportions in the U.S. population. In a group of 25 adolescents, odds are that one will have an eating disorder and others are at risk of developing one. These odds are derived from an array of statistics about mutually relevant facts, such as: 5 percent of college women are bulimic in the United States; many people alternate between anorexia and bulimia, with 60 percent of people who have dieted extensively or starved themselves resorting to bingeing and purging to keep the weight off; 1 percent to 1 in 250 teen-age girls develop anorexia, 80-90 percent of women dislike their bodies; 70 percent of women are preoccupied with their weight; 40 percent of women continually gain and lose weight; 50 percent of adolescents are dieting; 90 percent of people who suffer from eating disorders are female, and the diet industry is currently a 35 billion-dollar industry annually (source: *http://www.aabainc.org*).

How should we explain these statistics? A nonevolutionary answer enjoys consensus, involving culture plus the symbolization model. Our culture puts such a premium on appearance in women that it is hard for adolescents to develop a healthy lifestyle of eating habits and a positive body image, and to avoid obsession for developing a "perfect" body weight or shape. The psychological part is that an eating disorder is symptomatic of social, family and/or individual problems. Females try to handle their problems through dieting. Females with eating disorders think that if they get thinner, they can deal with the stresses and challenges of life and feel better about themselves (National Live Stock and Meat Board, 1992).

This nonevolutionary answer can be reconceived in evolutionary terms, involving culture plus an additional fact that motivates the reproductive suppression model. The preceding answer ignores the reproductive significance of eating disorders. By definition, anorectics starve themselves so much that they have not menstruated for at least three consecutive months and bulimics have binged and purged so much that their menstrual cycle is often interrupted. Culture plus the symbolic transference of problems in one domain (e.g., family life) to another domain (eating behavior) cannot possibly explain this cognitive-behavioral-physiological connection. Because sexual and natural selection operate through differential reproduction, the loss of ovulation is probably relevant to the origin of eating disorders in the human species. The cultural part is that human males have evolved to choose female mates for indicators of ability to bear healthy babies, which include youth, health, and attractiveness. Our culture's emphasis on attractive appearance in sexually maturing females creates evolutionarily novel and intense sexual selection pressures for females to compete for the most perfect body weight and shape possible. Cultural variation is taken into account by the fact that people in different cultures may have different motivations that lead to eating disorders. The psychological part is that an eating disorder is an unconscious evolved tactic, which can be currently adaptive or maladaptive, to execute a strategy for solving an adaptive problem: delaying reproduction in females when conditions for reproducing are unfavorable but expected to improve. Present problems put at risk a woman's present ability to find a high quality mate, reproduce, and be a good parent to a healthy child. Her thought that she can solve her problems by eating less to get thinner inadvertently triggers an evolved psychological mechanism, to have insufficient body fat for ovarian function until her problems are resolved. This enhances her lifetime reproductive success. The idea of postponing reproduction is very familiar to women, and puts this model within the realm of possibility.

So, which answer to our question, " What are eating disorders all about?" should we buy into? We expect any satisfactory scientific answer to explain the phenomena associated with eating disorders through factors that are relevant to the phenomena, using appropriate research methods. So, any adequate answer should be both substantive, telling us what observed phenomena are most relevant to eating disorders, and methodological, telling us how to establish relevance between our explanatory hypotheses and the facts to be explained. The following analysis reveals that the two explanations differ on a crucial issue involving categories of thought that has rarely been noticed by people who think along the lines of the symbolization model. Are the explanatory factors in the categories *symbols* or else *causes-and-effects* and *functions*?

In the symbolization model, the relation between self-esteem problems and eating disorders is symbolic and hence arbitrary, i.e., the underlying problems could be transferred in principle to any behavioral domain, as their symbols. The kind of explanation inherent in the model implies a special linguistic interpretation of the common definition of "symptoms" as "signs of something", i.e., items associated with and serving to identify something else. In normal medical practice, physicians are trained to treat symptoms as *effects* of an underlying bodily (physiological-psychological) condition, which is the *cause* of the eating disorder. By deliberately bypassing all proximate mechanisms (causes) of eating disorders, explanations drawing on the symbolization model treat disordered eating as a *sign* that is not an *effect*. Because in constructing a language any term could be used to refer to what we are talking about, it is tacitly assumed that anything can be a sign, representation, or symbol of anything else. This assumption takes a truism about language and applies it to pathology, a problematic move. Consequently, an eating disorder is treated as a symbol of an underlying problem in coping with life that the individual could in principle express in any way, i.e., either in eating behavior or in behavior that has nothing to do with eating. Thus, eating disorders are about self-esteem.

In contrast, according to the reproductive suppression model the relation between self-esteem and eating behavior is literal and nonarbitrary, specifically, on the ground that everyday problems are crucial to how high a value as a mate one can attain. "Evolved psychological mechanisms" are defined, roughly, as *causal chains* from mental states to behaviors that persist because they serve adaptive *functions*. In human females, eating behavior is the outcome of evolved psychological mechanisms related to specific, well-known factors of biological fitness. Ecological food resources are ingested/metabolized in patterns that affect health and body image, which affect social status. Physical attractiveness is "sexy," making a girl's or a woman's self-confidence about her appearance also "sexy." Rank in a dominance hierarchy governs access to high quality mates. All these are factors of self-esteem. Hence, high self-esteem in general positively correlates with high mate value. Thus, eating disorders are about mating strategies.

My answer, that eating disorders are both about both self-esteem and mating strategies, has both substantive and methodological dimensions of relevance. Substantively, I agree that low self-esteem, internal distress, and social emphasis on appearance and thinness in adolescent girls are quite relevant factors indicating what eating disorders are all about. However, people do not realize that the nonevolutionary explanation in terms of these factors is based on the symbolization model. *Symbolization does not count as a proximate mechanism*; as the initial quote in this chapter about why individuals develop eating disorders states, it is neither a *cause* nor a reason. Therefore, symbolization cannot rule out any causal claim, such as, disordered eating behavior causes distress and low self esteem rather than vice versa, or that both occur in a feedback loop. Reconceiving self-esteem in evolutionary terms explains why self-esteem is normally related to whatever the "primary problem" is: Self-confidence is highly sexually attractive, so the low self-esteem and depression found in eating disordered people *functions* to delay in partnering and hence reproduction. This sort of hypothesis yields testable predictions, e.g., the risk factors and primary problems associated with eating disorders that generate low self-esteem are those related to mate value and biological fitness. Methodologically, if we take loss of ovulation seriously and want an explanation in terms of *causes* and *functions* (mechanisms), we see that our preconceived bias in favor of the symbolization model gives us the wrong kind of explanation. But this criticism cuts both ways. The case for the relevance of the evolutionary psychological explanation remains unpersuasive until it is developed to connect the gist of the reproductive suppression to each of the causes of eating disorders identified in the quote that begins this chapter. My chapter is devoted to this novel task.

So, to construct a relevant evolutionary explanation, I propose three generalizable strategies. First, we eliminate the kind of explanation inherent in the nonevolutionary explanation that constitutes a key source of irrelevance by extracting the potential causal explanatory factors (distress, low self-esteem, body image) from the symbolization model. Second, we produce relevance by constructing a different kind of explanation that puts the preceding causal factors into a model that connects proximate mechanisms (lack of desire to develop sexual traits, dislike of one's body size and shape, fear of becoming fat, seeing control as a virtue, desire to conform) with ultimate evolutionary processes and evolutionary history (sexual selection, past adaptations triggered by current thought, emotion, and behavior). As a criterion of adequacy for preventing explanatory loss, we must do justice to cognitive aspects of eating disorders associated with the prior non-evolutionary model (symbolization). To do so, we connect faulty thought patterns in these proximate mechanisms to both domain-specific cognition (self-assessment of one's mate value) and domain-general cognition (costs of inhibiting behaviors, such as inhibiting eating, as well as the ironic control of mental processes that produce system information-processing overload and leads to breakdown). This chapter justifies and executes these strategies, proposing the ensuing novel hypothesis about what constitutes the systemic breakdown in a genuine "disorder". We need to find out why some cases of inhibiting eating by dieting lead to eating disorders and others do not, in order to do a better job of preventing eating disorders and designing therapies for helping individuals manage or cure their eating disorders.

The first step in hypothesis formation is to define key evolutionary concepts. My "decreasing mate value hypothesis" posits the mutual relevance of the factors cited in the (nonevolutionary) symbolization model and the factors cited in the (evolutionary) reproductive suppression model. Sexual selection favors those traits that make one the best available mate (see Gangestad, this volume) as well as decision-making that uses environmental information to find the best available mate one can get given one's own traits (see Ketelaar and Todd, this volume), where "best available" is counted in the currency of biological fitness. Mate value is defined as "the relative attractiveness of one potential mating partner as compared to another based on their differential reproductive value" (Mealey, 2000, p.81). Reproductive value is "the statistically expected number of offspring of an organism of a specific age, sex, health, and status." Mealey (2000, p.81) emphasizes: "Overall, mate value increases with reproductive value and with status. That is, *an individual's desirability as a mate is positively related to his or her reproductive value and status as compared with same-sex others*. Mate value is thus strongly related to age and to what we humans would call wealth (Buss, 1992; Low, 1993)." Because reproductive value is a statistical concept and because we cannot tell what our own fitness is, we cannot perceive our own "mate value." So we need to introduce "self-perceived mate value" as a psychological concept, which we define as, "one's assessment of one's own mate value (attractiveness) as compared to potential competitors."

A second step of hypothesis formation is to embed it into background knowledge about natural processes. Sexual selection is posited to (1) make attractive mates those with high mate value (Jones, 1996; Thornhill, 1998). In the case of an eating disorder, I hypothesize that it will (2) create a positive correlation between increases and decreases of one's self-perceived mate value with one's actual mate value, with exceptions where self-deception is advantageous, and (3) favor those individuals whose self-esteem and valuation of one's body image incorporate one's self-perceived mate value. The cumulative result of this process is to (4) cause individuals with high mate value (and normally high self-perceived mate value) to be choosier and to exhibit more discrimination than individuals with low mate value, and (5) make individuals with low mate value more likely to use strategies other than advertising reproductive value or status (Mealey, 2000, p.81). With respect to eating disorders, I hypothesize that the result is to (6) cause individuals who experience cues of low mate value involving their body image to change eating behaviors to eliminate that source of their fitness disadvantage whenever possible, and (7) make individuals of high social status, high mate value, and high self-perceived mate value more susceptible to eating disorders because they have more to lose if their attractiveness is lowered than do individuals of low status.

A third step in hypothesis formation is to clarify the type of mechanism involved. In a trenchant criticism of the field, Mealey (this volume) charges that evolutionary psychology has been "more often descriptive than explanatory" because it has not yet gotten "down to the nitty gritty level of mechanism". I propose a genuine mechanism. The decreasing mate value hypothesis is that eating disorders often arise from earlier normal eating behaviors when there is a positive feedback loop between stimulus and response in which the stimulus is a cue for perceiving a threat to one's own mate value and the cognitively chosen response induces a behavior that triggers adaptations whose ironic effect is to increase the intensity of that stimulus. The feedback loop explains why once some people start dieting, they cannot stop, but continue to lose weight either by starving themselves (anorexia) or by inhibiting eating, then binging, and purging (bulimia). The nonevolutionary and evolutionary factors are unified by the two-way causal loop in which the lower the self-esteem (induced by perceiving oneself at a mate value disadvantage, a fitness threat), the more disordered the eating habits become and the more disordered the eating habits, the lower the self-esteem (induced by perceiving oneself at a greater mate value disadvantage, an even greater threat, due to unintended effects of disordered eating).

This mechanism gains relevance as the type of explanation we are looking for because it forms a "deviation-amplification feedback loop," one in which the normal state of eating habits is followed by successfully more deviation (disordered eating habits). To understand this type of loop by analogy, consider the U.S. military arsenal. Our economy works by producing goods, which involves destroying nature by turning it into commodities, i.e., paper is produced from trees. We extract raw materials from other countries. In order to back up our demands for the raw materials we want, we have created an arsenal of conventional and nuclear weapons. But a nuclear war would destroy nature where the bombs fall. So, nuclear war threatens the very existence of raw materials. To sustain our technology's capacity to destroy, we require increasing amounts of raw materials from other countries. The vicious circle continues, exponentially, creating risk of worldwide destruction deriving from our twin desires to destroy and to possess (which arise from selection for the most control of the most resources in order to maximize reproductive success).

The proximate mechanisms of the feedback loop involve a complex cycle. Complexity arises from features of food in which independent evolutionary mechanisms merge. There are evolutionary scenarios in which food intake involves both (A) previously described processes of sexual selection that create a connection between self-esteem and eating behavior and (B) processes of natural selection that create a connection between ecological resources and reproductive suppression. Food scarcity lowers food intake, which lowers the female's body fat, which endangers her ability to make healthy babies, and so natural selection prevents ovulation. That sets up selection pressures against all reproductive effort devoted to mating, creating an upward cascade of effects through all levels of organization in the organism via mind-body correlation, e.g., depression decreases mating activity and attractiveness. That cascade in turn changes the situation in which sexual selection operates on self-esteem and eating behavior.

Mealey (this volume) rightly emphasizes that the next scientific frontier in evolutionary psychology is to understand the internal and external triggers that activate the genetic, physiological, and psychological mechanisms of behavior. Now we have the background to understand in terms of "cues and triggers" how and why each step in the complex cycle leads to the next step of the feedback loop. Self-perceived mate value is a cue in the young female's assessment of her probable success of reproduction in the current situation versus future reproductive success. Self-perceived mate value reflects a female's representation of her own mating attractiveness status that includes how well she satisfies not only male preferences for female mates (e.g., youth, health, physical beauty) but also female opinions about the relative mating attractiveness and social status of females in the group. In female-female competition for mates, cues (e.g., being told "You are too fat" or thinking "I'm too fat") are cognitively processed as mating competition disadvantages. The cues warn her of a threat to her self-perceived mate value (an important component of self-esteem) that creates stress because she experiences a fitness risk. To relieve stress, decision-making defends the female organism by motivating dieting behavior designed to eliminate the threat. Severe dieting behavior "fools the body" by mimicking effects on food intake in conditions of scarce or fluctuating food resources. Such dieting behavior activates reproductive suppression adaptations (e.g., suppressed ovulation, which tracks body fat because that is needed to sustain pregnancy, healthy birth, and lactation) useful in the relevant environments of evolutionary adaptedness (EEA) when food resources are scarce or fluctuating. These adaptations not only have benefits (delaying reproduction when at a disadvantage) but also have costs, including a cascade of physiological changes (health problems). These adaptations also induce psychological changes that may be adaptively beneficial (distress and depression reduce male attention and therefore mating), or adaptively costly (because self-confidence is a factor of attractiveness, the woman may lose social status in the female hierarchy). These changes trigger new cues of one's mate value (including self-esteem, which is more affected by the opinions of other females than those of other males). In sum, dieting changes the female's self-perceived mate value so that her decisions and behavior can solve the adaptive problem of reducing the threat (perceiving herself at a mating disadvantage) by eliminating the disadvantage, or it can fail to solve the problem by triggering changes in her body and psyche that put her now at a greater disadvantage than before.

Change in mate value reflects a ratio of costs to benefits of the effects of these adaptations that varies with the individual, their patterns of thought and emotion, and their circumstances. When the change in mate value from initial to subsequent cues of self-perceived mate value is for the better, eating disorders are functioning as adaptations should (are designed by selection to) function, and are adaptive. When the response to initial self-perceived mating disadvantage leads to an even lower self-perceived mating disadvantage, the "adaptations have gone awry" as a positive feedback loop leads to more distress that motivates disordered eating habits. When this cycle repeats many times, a "downward spiral occurs" (analogous to well known downward spirals involving pain and inactivity), so that normal eating behavior becomes disordered eating habits and, eventually, a pathological eating disorder whose severity reflects the number of days or times the cycle occurs.

How can we argue that the decreasing mate value hypothesis is relevant to what eating disorders are all about? A definitive feature of evolutionary psychology is the effort of psychologists to incorporate an evolutionary perspective into their work. Unfortunately, there is a tendency to bifurcate psychology into traditional psychology and evolutionary psychology. Being committed to an evolutionary perspective on human life (Holcomb, 1993), I think that all psychology should be evolutionary in principle. Even so, it is up to traditional researchers to decide when evolutionary ideas are relevant to their work in practice because one can justifiably acknowledge that human minds evolved without believing that evolutionary research methods are relevant to one's specialized topic and the particular type of questions asked about it.

The burden of proof lies with evolutionists to demonstrate relevance, for three reasons. First, biologists distinguish "evolutionary biology" from other biological fields, and they often ask questions about organismic structure and function that do not require them to bring evolutionary biology into their research; the same goes for psychology. Second, people understand what is unfamiliar by relating it to what *is* familiar, and so the new science of evolutionary psychology has to show how it connects to traditional psychology in principle and in specific cases before people can understand and accept it. Third, a property of good explanations is that they fit with established principles and facts, a property based on conservatism about belief changes, and so the new paradigm must prove its worth to adherents of the old paradigms before they can accept it.

Because relevance (conceptual relatedness) is not merely a matter of facts (accumulated truths), trotting out more empirical tests of evolutionary hypotheses that confirm or corroborate them is insufficient to establish relevance when traditional researchers begin with preconceptions that render evolution irrelevant to their specialties. Relevance is established by arguments that combine theory construction, empirical testing of theoretical hypotheses, and conceptual work. A subtle distinction often missed is that we should not expect evolutionary psychology to unify all psychology because different fields utilize different kinds of explanation, even though phenomena across psychological sub-disciplines have evolutionary explanations. The usual explanation of eating disorders as symbols of subjective distress is a kind of explanation different from one in terms of proximate mechanisms of reproductive suppression and their evolutionary origins. The symbolization model delivers an "intentional-intelligibility" explanation: the eating disorder expresses the mind-set of the individual, and so the individual's intentions make intelligible the presence of the eating disorder, whatever its causes or reasons may be. In contrast, the reproductive suppression model yields a "functional-historical" explanation: that eating disorders are present because they are currently adaptive or currently maladaptive effects of adaptations that function to suppress reproduction when prospects of future reproduction are better than prospects of present reproduction. These are two different kinds of explanations; they differ in what facet of eating disorders they are explaining and in the types of explanatory factors invoked to do the explaining. People's preconceptions about relevance are relative to the kind of explanation they have in mind. To the extent that intentional-intelligibility explanations that cannot be integrated with functional-historical explanations are found throughout psychology, evolutionary psychology cannot unify psychology as a whole.

An answer to "What are eating disorders all about?" should show that explanations of eating disorders meet standard criteria of relevance that limit which explanations are worthy of scientific testing. In terms of general scientific method, I shall develop the "decreasing mate value hypothesis" with the aim to satisfy such criteria: testability, plausibility, conservativeness, usefulness, simplicity, theoretical generality, partiality, and objectivity. Because these criteria involve connections of explanations to background beliefs, if the criteria are satisfied then some degree of relevance is achieved. People normally think of all tests as "empirical tests," but conceptual tests are also pertinent to establishing relevance. The task here is to develop the "decreasing mate value" hypothesis so that it meets a conceptual test specific to evolutionary psychology: it makes evolutionary sense of all six proximate mechanisms of eating disorders.

2. PROXIMATE MECHANISMS AND THE REPRODUCTIVE SUPPRESSION MODEL

Anorexia can involve reproductive suppression in a variety of possible ways. These vary according to the evolutionary process at work and whether it is considered currently adaptive or maladaptive. Crawford (1989) has described a method for constructing and testing several hypotheses in terms of evolutionary and nonevolutionary predictions about anorexia nervosa. He states two (nonevolutionary) hypotheses that make no assumptions about behavioral adaptations in ancestral populations: the "sociocultural" and "developmental psychobiology" hypotheses. His "sociocultural hypothesis" is that anorexia is an attempt to conform to the cultural ideal that thin women are beautiful; this runs together the nonevolutionary hypothesis that I have shown to be based on the symbolization model and the proximate mechanism of the desire to conform (do what others do) that I shall later place into evolutionary perspective. His "developmental psychobiology hypothesis" is that anorexia results from fear of attaining adult weight and sexual function; that runs together the proximate mechanisms of "not wanting to develop sexual characteristics" and "fear of becoming fat" that I also later place into evolutionary perspective.

Such conceptual differences arise because definitions of anorexia as a syndrome build into it more (as Crawford does) or less (as I do). Crawford (1989, p. 17) states, "Anorexia is a syndrome characterized by a relentless pursuit of thin body shape and an exaggerated dread of fat and weight gain." In my broader definitions, anorexia nervosa is characterized primarily by self-starvation and excessive weight loss and bulimia is characterized primarily by a secretive cycle of binge eating followed by purging. My analysis has greater conceptual clarity because proximate mechanisms of eating disorders should not be incorporated into the definitions of the eating disorders themselves: that would make saying "the proximate mechanism of disorder X is Y" circular because X is defined in terms of Y.

Crawford (1989) also states two (evolutionary) hypotheses that do make assumptions about behavioral adaptations in ancestral populations: the "female-female competition" and "inclusive fitness" hypotheses. His "inclusive fitness hypothesis" is that anorexia decreases resources given to the anorexic, thereby increasing resources given to close relatives, which improves the inclusive fitness of the anorexic; this adaptation could be currently adaptive or maladaptive. The inclusive fitness hypothesis predicts that the incidence of anorexia is negatively correlated with the socioeconomic status (SES) of the family because low-status families have restricted resources (Crawford, 1989). However, the evidence is that anorexia is positively correlated with high SES: Anorectics typically come from high SES families, and their fear of cultural failure reflects their extremely high social (cultural) and educational (neocultural) aspirations of themselves and their families (Bailey and Roswell, 1994). That evidence is a counter-example to the inclusive fitness hypothesis.

Crawford's "female-female competition hypothesis" is that anorexia serves to delay reproduction in situations involving severe female-female competition for mates, which is his way of stating the reproductive suppression model. I would distinguish two versions of his "female-female" competition hypothesis because when we say that something "is adaptive" we need to ask: adaptive for which organism involved? We need to ask whether reproductive suppression is adaptive for the female experiencing it or for other females competing for mates with the female experiencing it. The "emergency personal fitness hypothesis" is that anorexia is an outcome of an evolutionary mechanism for enhancing fitness in emergency conditions that drastically lower the probability of producing and raising healthy offspring; this adaptation could be currently adaptive (adaptive self-suppression of ovulation) or currently maladaptive (adaptation gone awry due to mismatch of ancestral and current environments). The "manipulative pathology hypothesis" is that anorexia is due to intense female-female competition in which dominant females socially manipulate other females who might compete for their mates to suppress their ovulation, which increases the fitness of the dominant females and reduces the fitness of the eating-disordered females. These are all variant ultimate mechanisms of how something evolves in the life history of a species, not proximate mechanisms of how something arises in the life history of an individual.

The SES evidence does confirm the reproductive suppression model's versions in the form of the "emergency personal fitness hypothesis," because high SES females aspire to higher mate values and may need to delay reproduction until they can gain the needed social and educational skills, and in the form of the "manipulative pathology hypothesis," because those very aspirations of high SES females can be manipulated through visual media by other females (who can function at very low weights as models) to try to attain that ideal female figure even though they incur serious health problems in doing so. (In reply to the objection that the mass media are controlled by men, see proximate cause 5 in section 3.5 "The Desire to Conform.") The emergency personal fitness hypothesis holds that anorexia evolved because it was adaptive for the anorectic and therefore maladaptive for other females and predicts that in similar conditions it is currently adaptive for the anorectic, whereas the manipulative pathology hypothesis holds that anorexia evolved because it was adaptive for other females and predicts that in similar conditions it is currently adaptive for other females and maladaptive for the anorectic. The point is not to make evolutionary psychology unfalsifiable because, whether anorexia is adaptive or maladaptive today, we can think of a hypothesis that explains it—that is the gist of the charge that the field merely gives us speculative "Just-so stories" that ignore evidence. The point is that the total evidence indicates that anorexia is sometimes adaptive (e.g., when it eliminates male attention and reduces the intensity of female-female competition so that the anorectic female has time to solve problems and develop social and physical skills) and other times maladaptive for the anorectic (e.g., when it leads to death, or the delay goes on so long that women are childless, or the delay goes on ten years or long enough that future reproduction does not make up for present loss of reproduction, as found in !Kung women). So, we seek to understand how it evolved by positing different evolutionary mechanisms and circumstances for its adaptive form and for its maladaptive form, and then test whether our theoretical expectations fit the facts. Although I am not conducting tests here, my point is that we use evidence to construct more refined hypotheses: Scientific method involves a cycle of hypothesis construction then hypothesis testing, then more construction using test results, followed by more testing, and so forth in a self-correcting manner.

The SES evidence provides a rationale for pursuing the "sociocultural hypothesis" as well as the "developmental psychobiology hypothesis", both of which shall be developed later in this chapter. That evidence does provide an unexpectedly novel confirmation of my "decreasing mate value hypothesis," which is broad enough to connect ultimate and proximate mechanisms for the "emergency personal fitness hypothesis" and for the "manipulative pathology hypothesis." If someone has high aspirations, they probably want a partner whose mate value is high and to do so will develop a very high self-perceived mate value, whether honestly or self-deceptively. If so, then cues signaling that they have a mate value disadvantage will be acted on, because such cues constitute a great threat to their self-perceived mate value. When these cues are acted on through extreme dieting, leading to a better body image and higher self-esteem and/or to health problems, their method for solving their adaptive problem may solve it (it raises their mate value, and is adaptive) or it may not solve it (it decreases their mate value, and is maladaptive). This is typical of nature, as we all know that adaptations for avoiding predators sometimes work (the prey lives) and other times fail (the prey gets eaten).

To use Darwin's term, a variety of "conditions of life," which include not only internal but also external conditions that affect reproductive success, can trigger anorexia or bulimia because the conditions in which a female perceives herself to be at a reproductive disadvantage can be more varied than people realize. They may be: chemical (e.g., low brain serotonin levels), physiological (e.g., immunosupression dysregulation), life-historical (e.g., late sexual maturation), behavioral (e.g., dieting behavior), psychological (e.g., low self-esteem), sexual (e.g., lack of available sexual partners or desirable sexual partners), familial (e.g., dysfunctional family relations), social (e.g., lack of social supports), ecological (e.g., scarce food resources), or cultural (e.g., inability to attain cultural norms of ideal femininity). This categorization extends the reasoning of Wasser (1990) that individuals under social and psychological stressors should trigger similar physiological reactions as physical stressors, because females under social or psychological stress are just as unsuited to conceiving, bearing, and raising offspring as those under physical stress. A testable prediction is that these conditions would be risk factors for acquiring an eating disorder when they impose fitness risks.

This model subsumes the previous model's emphasis on self-esteem, treating it as one of many antecedent conditions. It corrects the symbolization model and subsumes it under a more general, circumstantially and theoretically complex, empirically supported and explanatorily powerful model. Advocates of the symbolization model argue that reproductive suppression cannot "get going" in some types of anorexia without being triggered by low self-esteem. This objection shows only that the reproductive suppression model is partial rather than a complete explanation of anorexia, not that it is wrong. The objection suggests a principled basis for treating lowered self-esteem as a "cause" (causal standing condition) of the eating disorder and reproductive suppression as an "effect" of the eating disorder, thereby reconciling the two models in terms of a single nexus of proximate causes and effects of behavioral events.

There are three grades of depth in evolutionary explanation in developing the reproductive suppression model using the "decreasing mate value hypothesis." By recognizing that these varied conditions of life provide extreme fitness risks, a variety of evidence potentially combines to support the superiority of the reproductive suppression model over the symbol model in explaining: (A) why eating disorders have the reproductive effects they do have (e.g. suppressed current ovulation); (B) why the proximate causes of eating disorders occur in the antecedent internal and external conditions they do (e.g., the influence of family environments on the relation between proximate causes and eating disorder effects); and (C) why the features of the whole cause-effect nexus are what they are (e.g., eating disorders occur in conditions of extreme fitness risks, breakdown of existing proximate mechanisms).

Another step in hypothesis formation is to clarify what a disorder is. What counts as a breakdown of existing proximate mechanisms? Evolutionary psychology claims to put rigor into the definition of disorders, by defining disorder as follows: Dysfunction occurs when an evolved psychological mechanism (strategy), which has been described and its proper function to enhance biological fitness is identified, as not performing as it was designed to perform in the contexts in which it was designed by natural or sexual selection to perform (Buss, 1999, p.399). According to Buss, dysfunction occurs in three possible ways: activation failure (the mechanism fails to become activated when the adaptive problem it was designed to solve is confronted), context failure (the mechanism is activated in conditions in which it was not designed to become activated), and coordination failure (the mechanism fails to coordinate as it was designed to with other evolved mechanisms). Dysfunctions are due to genetic mutations (genes for various diseases), genetic variation (heritability of the mechanism), or developmental insults (accidents causing brain damage).

Hence, we must distinguish problems correctly thought to be disorders (these three possible dysfunctions and their sources) from three problems wrongly thought to be disorders. First, mismatch theory permeates evolutionary psychological thought, namely, the fact that differences between modern and ancestral environments create currently maladaptive behavior. Buss claims an important, counter-intuitive insight that "eating disorders" due to mismatch are not genuine disorders (dysfunctions), because the evolved mechanisms are still operating in the manner in which they were designed to operate. This occurs when the mechanism of self-perceived mate value poses an insurmountable fitness disadvantage because it takes as input the bombardment of media images of extremely attractive women, which are a tiny percentage of millions of women in today's population, rather than the few attractive women in the local pool of potential mating competitors in ancestral bands of 20 to 300 people. Second, he implies that the "subjective distress" in eating disorders does not qualify aberrant eating behavior as a clinical disorder if it is produced by the normal operation of evolved functional mechanisms, such as the distress induced by comparing one's physical attractiveness to those of actresses and fashion models, which induces low self-esteem. Third, he implies that mistakes in a female's assessment of her future versus present chances of reproductive success do not qualify as genuine disorders if they are normal mistakes accompanying the "on average" functioning of the mechanism mate value assessment or time-related expectation of reproductive success.

The decreasing mate value hypothesis incorporates the ideas of environmental mismatch, subjective distress, and on average successful functioning within the previous sources of genuine dysfunction cited by Buss, namely, coordination failure. The hypothesis, briefly, was that social pressures for elevating thinness to a huge determinant of a female's mate value leads her to diet, which mimics behavior in ancestral environmental conditions of food scarcity, reducing fat to a level that triggers reproductive suppression as evident in lack of ovulation of anorectics and inconsistent ovulation of bulimics. When this occurs, these aberrant eating behaviors and associated features count as genuine disorders due to coordination failure: The mechanism for assessing reproductive timing in terms of mate value assessments is not coordinated, as it was in ancestral conditions, with the mechanism for assessing reproductive timing in terms of ecological food scarcity assessments due to mismatch of modern and ancestral environments. That is, eating disorders fail to solve adaptive problems posed by mate competition and by variable food availability, even though food is plentiful now and the female does not misperceive the amount of food available, because the two mechanisms are improperly coordinated.

Neither the symbolization model of anorexia nor the reproductive suppression model can explain much about anorexia, I submit, without being combined with more information about the proximate causes of the condition. Let us consider the six commonly identified proximate causes bypassed by the symbolization model as revealed in the initial quotation from the National Live Stock and Meat Board in their guide for school professionals to use with adolescents. I shall devote the bulk of my essay to this list of causes in a manner that integrates our nonevolutionary convictions about the causes of eating disorders with the new evolutionary science.

3.1 Lack of Desire to Develop Sexual Traits

Proximate Cause (1) is that young females do not "want" to develop sexual traits.

The reproductive suppression model gives a host of conditions in which delaying normal sexual maturation can be fitness enhancing, depending on the exact ratio of costs to benefits in the conditions of life. Hence, quantitative modeling is crucial to hypothesis construction. There would be little point to do the hard work of quantitative modeling if one believes the objection that the suppression of reproduction is just another health effect of eating disorders, with nothing evolutionary about it. Other bad health effects besides amenorrhea (cessation of menstrual cycling in premenopausal women) include physical emaciation, partial diabetes insipidus, endocrine dysfunction, abnormal hypothalamic function, and variety of hormonal dysfunctions (Bailey & Roswell, 1994); also, fat and muscle loss, hair loss, slow pulse and low blood pressure, decreased ability to concentrate, depression, chronic fatigue, insomnia, stress fractures/osteoporosis, overuse injuries, and cardiac arrest (Beck, 2000). It is objected that there is no reason to single out amenorrhea among other bad health effects of anorexia and treat it as a reproductive effect that requires explanation by its own special strategy. If we just say that anorexia causes a systemic breakdown, we would require a proximate cause of each malfunction without the need to posit an ultimate, evolved strategy.

The general problem with objections that arise from a completely nonevolutionary perspective such as this is that they ignore a crucial sort of evidence: evidence of reproductive suppression in nonhuman animals. The governing principle of the comparative method is that when nonhuman animals appear to be doing things similar to what humans appear to be doing, there is reason to explain this coincidence by positing strategies similar in humans to those already established in nonhuman animals, perhaps as analogous or homologous. Why is anorexia so common in girls as compared to boys? Established theory states that mammalian females invest far more heavily in offspring (due to internal fertilization and lactation) than mammalian males, so that reproductive mistakes are more costly for them. It is documented that mammalian females appear to possess mechanisms either to delay or to accelerate puberty and reproduction when warranted by the conditions of life (Wasser & Barash, 1983; Surbey, 1987; Anderson & Crawford, 1992). Mammalian females slow down sexual maturation through such mechanisms as suppression of ovulation, embryo resorption and fetal miscarriage. There are causal mechanisms that connect maturational delay, infertility, and amenorrhea (which suppress present reproduction) with other health effects (which harm survival chances), e.g., if a potential mother has too little body fat and hence insufficient caloric resources to sustain reproduction of healthy offspring, the hypothalamus reduces its hormonal output, which precludes ovulation (creates amenorrhea).

If the physiology of humans and nonhumans overlaps so much in terms of proximate mechanisms for "deciding not to reproduce," and the proximate mechanisms at hand operate in environmental conditions in which the body needs to reserve its energy for non-reproductive purposes, then the proximate mechanisms should lead us to posit evolved strategies in humans just as they do for nonhumans. This is especially true when established evolutionary theory of life histories (i.e. individual development over one's lifetime) contains a key principle of tradeoff between allocating reproductive effort to reproduction or to somatic growth and development (survival). Strong evidence that humans are among the species that exhibit reproductive suppression is found in the report by Wasser and Barash (1983): "Lowered self-esteem and the lack of social support from family and friends are commonly associated with elevated rates of reproductive complications, including infertility, delay of ovulation, habitual spontaneous abortions, complications at and following parturition, abandonment, and even child abuse ..." and "... that the physiological effects by which these processes are mediated appear to be quite similar to those experienced in other female mammals who are subjected to aggression by consexuals" (p. 531). Moreover, laboratory technicians who work with rats and other animals are routinely given guidelines pertaining to anorexia in other species induced by their techniques. This fact provides independent evidence that the condition should not be defined in terms that restrict it to humans and that the truth in the view that "our culture creates anorexia through media images of women," far from implying that evolution is irrelevant to anorexia, captures one of many biologically understandable ways to induce anorexia, some new and some old. Anorexia has a long evolutionary past. Selection has probably operated on it long enough to create strategies related to it in all the organisms whose fitness is strongly affected by it: those who experience it, those who induce it in others, and potentially kin, reciprocators, and group members of those who experience it or induce it.

The adaptive reproductive self-suppression hypothesis (Surbey, 1987; Voland & Voland, 1989; Anderson & Crawford, 1992) illuminates features of anorexia that cannot be explained by a purely nonevolutionary viewpoint. Why does anorexia develop mostly in girls and younger women (approximately 90% of cases) rather than in boys or men or older women? The trade-off between effort spent on reproducing versus surviving is most acute during adolescence, when girls (unlike boys) metamorphose in terms of body shape and fat distribution when they change from a prereproductive state to a reproductively mature state. For human males, sperm production requires little energy and no fat reserves. For older women, further delay of reproduction (which usually leads to no further reproduction) is far less costly in terms of fitness than in adolescent girls (who tradeoff risky early reproduction for less risky later reproduction, enhancing overall lifetime reproduction). Why is anorexia often associated with a first sexual experience? Sexual experiences early compared to those at sexual maturity carry the cost of having offspring prior to developing successful strategies of mate choice. Often they are undesirable, e.g., an extreme example is that sexual abuse is quite costly in terms of fitness (of course, mate choice early in life need not include sexual abuse).

Further, any one type of fact to be explained can be given a detailed evolutionary answer. Why is anorexia increasing in prevalence so rapidly at this point in history? Age to menarche (onset of menstruation) has declined recently, from an average age of 16 to 17 based on data from hunter-gatherer populations thought to represent our evolutionary past (Howell, 1979), with considerable decline this century (Tanner, 1962), and now typically hovering around age 12 to 13 in North American girls. Because girls are able to become pregnant far earlier than when many of the physiological adaptations originated in the human species, there has been an uncoupling of reproductive development from cognitive and social development (Surbey, 1998). Adolescent girls are at high fitness risk when they engage in early reproduction; the news is full of social problems due to early pregnancy, and such girls have a harder time than other girls in doing what it takes to maximize lifetime reproductive success. Early maturing girls not only have less time to choose a high quality mate, but they have less time to develop physical, emotional, cognitive, and social skills needed for successful group living and hence successful reproduction. That creates sexual selection pressures to delay reproduction for as a long as it takes to develop physical, emotional, cognitive, and social skills. It should be obvious that the varied conditions of life that span levels of analysis act as standing conditions that constitute kinds of risk factors for eating disorders. This is an example in which a Darwinian perspective does not merely "support the status quo," but suggests that rather than treat eating disorders merely with drugs we should change public policy to lessen risk factors and help these girls develop needed skills.

It may be objected that extended development toward maturity is well known and has nothing to do with the decreasing mate value hypothesis. But think again. The historically short time to menarche puts all girls at higher risk than before for sensing cues of fitness disadvantage relative to older females, because older females tend to have more access to resources required for child rearing. It puts girls into more intense female-female competition for mates among their age group. Studies of adaptiveness must consider costs in addition to benefits. The strategy, "Do not delay reproducing," has the obvious benefit of avoiding an unnecessary delay in reproduction, but that is outweighed in general by the cost of troubles dealing with sexual attention from peripheral males, e.g., in school. So, cues signaling "I'm at a large reproductive disadvantage now" are common, which puts them into the evolutionary scenario inherent in the decreasing mate value hypothesis. Early physical maturity in U.S. females compared to traditional hunter-gatherer societies is due to improved nutritional regimes, growing childhood obesity, lack of physical activity, and genes (the CYP1B1 gene speeds up the female body's breakdown of the male sex hormone and hastens breast development). As a result, maturity in other areas (emotional, cognitive, social) pivotal for mating and parenting lags behind, so that early maturity is a source of high fitness risks. If females "sense" that they will develop more skills as they learn more, they calculate that future reproductive prospects are better than present ones. So, temporary and non-life threatening anorexia may be adaptive, with subjective distress and low self-esteem reflecting warranted assessments of the adolescent female's reproductive status. Surbey's (1987) empirical findings show that, "In general, developing anorexia nervosa appears to alter a girl's developmental trajectory from that of an earlier maturer to that of a late maturer without decided ill effects".

This is not to deny that there are conditions in which early reproduction will be adaptive; there are costs and benefits of reproducing too late just as there are costs and benefits of reproducing too early. When young mammals encounter conditions of environmental risk and uncertainty reflected in emotional stress during development, such conditions are not favorable for survival, and it will generally be adaptive for them to reproduce early (Chisholm, 1996). Given the learning of reproductive strategies during critical periods, such as the first five to seven years of human life, a number of psychosocial circumstances influence the timing of puberty (Belsky, Steinberg, & Draper (1991). This evolutionary model of pubertal timing makes the novel prediction that early timing of puberty in girls is associated with negative health and psychosocial outcomes, on the grounds that high risk to survival would result in early reproduction in order to enhance reproductive success. The variable "age to first reproduction" is subject to intense selection pressures. When age to first reproduction gains more of a fitness risk related to the individual's ability to survive, she should reproduce early; whereas when it gains more of a fitness risk related to the individual's ability to reproduce successfully, she should suppress current reproduction and reproduce later. Both signals, "speed up reproduction" and "delay reproduction" have costs as well as benefits, and both kinds of signals are error-prone, especially when evolving. So, to know whether a mechanism is plausible requires a detailed consideration of costs. Is it plausible that there *are* such signals? How reliable would the signals have to be? How low would the error threshold of the mechanisms have to be? These are avenues of significant further research.

3.2 Dislike of One's Body Size and Shape

Proximate Cause (2) is that young females dislike the size and shape of their bodies.

This proximate route to anorexia obviously is related to sexual selection, e.g., male preference for females with good looks (as dependent on body size and shape) overrides other mate preferences when males seek a short-term mate and physical appearance is a crucial indicator of mate value when males seek a long-term mate as well. Thus, dislike of one's body size and shape, when that violates male and female assessments of attractiveness, would count as a cue that "I'm at a severe fitness disadvantage," which would be expected for evolutionary reasons to induce depression. Depressed mood has two functions: it helps us to disengage from a hopeless course of action that causes losses (sadness occurs when we anticipates losses, e.g., of money, mates and good mating opportunities, reputation, etc.) and consider other courses of action, and it deflates "blind" human optimism to allow us to more objectively reassess our goals and our self-perceived mate value (Nesse & Williams, 1994; Stevens & Price, 1996).

The symbolization model begs the question: Why should dislike of body shape and size be so distressful that it causes an eating disorder? The reproductive suppression model answers this question, using sexual selection theory. Of course, those standards of beauty have huge economic and social benefits for anyone who meets them. But what we are asking is not why gaining benefits causes enjoyment and why losing benefit causes suffering. We are asking why the suffering inherent in disliking one's body shape and size is connected to the physiology of reproductive suppression, given the complex multi-level interactions involved. The ideal female body shape has a waist-to-hip ratio of about 0.7, and men prefer women of this shape as mates (Singh, 1994). Hips need to be larger than waist in order to provide a birth canal large enough for the baby to evacuate the womb without injury. Women deposit fat in their breasts, hips, and thighs in order to carry a supply of food sufficient for a healthy fetus and baby. Women with higher ratios than 0.7 have been shown to incur a variety of medical problems.

Now, the problem arises here that it could be adaptive to dislike your body shape only if you could change it, and at a reasonable cost. How plausible is this, now and in ancestral environments? It is difficult to attain and retain a waist-to-hip ratio that approximates 0.7, or more precisely, the difficulty of doing so is highly age-related. As girls reach sexual maturation during the teenage years, the 0.7 ratio is not difficult to attain. As they age through adolescence, through early adulthood, and onward, the difficulty increases. So, to the extent that reproductive suppression was adaptive in the EEA it would be found in teenagers rather than early or late adulthood (EEA stands for the "environment of evolutionary adaptedness," which is not a single environment, but a sequence of ancestral environments in which the trait at hand evolved and was shaped by a corresponding sequence of selection pressures). Moreover, in the EEA the problem of too few resources (because they are scarce or fluctuate) was far more prevalent and significant for fitness differences than the problem of too many resources—especially if one thinks of the EEA in terms of hunter-gatherer ways of life, because they were not able to store food as agricultural or pastoral societies could. Females subject to hunter-gatherer EEA selection pressures would not have had the choice of how many or how few food resources they should put into their bodies to a degree comparable to present day grocery-store choices. Hence, they would have had more adaptations for dealing with too little food and emaciation than with too much food and obesity. Disliking one's body image—including depression, a sense of lowering of one's self-perceived mate value, and anxiety concerning what to do to raise it—could motivate changing it adaptively, mainly at youthful ages. Afterwards, disliking one's body image would tend far more often to be maladaptive in the EEA because one usually is "stuck with it," but it would be adaptive if it triggered a decision to go for a lower quality mate than one had been trying to attract or else to advertise other mate preferences held by males, the precursors of the following preferences shared by men and women today.

Age to first reproduction in females is subject to intense selection pressures in ways relevant to eating disorders. This prediction has been confirmed in ways related to food, body fat, body image, and depression, all of which are associated with eating disorders. Ellis and Garber (in press) cite the relevant literature: "In particular, early–maturing girls are at greater risk for breast cancer (e.g., Vihko & Apter, 1986; Kampert, Whittemore, & Paffenbarger, 1988), obesity (Ness, 1991; Wellens et al, 1992), and teenage pregnancy (Udry, 1979; Udry and Cliquet, 1982), and tend to show more disturbances in body image, to report more emotional problems such as depression and anxiety, and to engage in more problem behaviors such as alcohol consumption and sexual promiscuity (e.g., Caspi & Moffitt, 1991; Flannery, Rowe, & Gulley, 1993; Graber, Lewinsohn, Seeley, & Brooks-Gunn, 1997; Susman et al., 1985)." Ellis and Garber go on to test more novel predictions using a sub-sample of larger longitudinal study of maternal depression and adolescent development. They derived hypotheses about proximate mechanisms relevant to three classes of stressors that function as environmental antecedents of pubertal timing variations: physical stress (e.g., malnutrition, disease, economic hardship), socioemotional stress (e.g., harsh and neglecting family relationships, parental psychopathology), and divorce (e.g., presence versus absence of the biological father). These are conditional reproductive strategies, i.e., environmentally triggered processes that direct an individual toward a given reproductive strategy. They tested the predictions that early (i.e., up to a year) pubertal timing is predicted by: (a) a history of psychopathology in mothers, as mediated by stressful family relationships and father absence/stepfather presence; (b) higher levels of interpersonal stress in the family; (c) father-absent homes, (d) greater exposure of girls to unrelated adult males in the home, i.e., via mother's romantic partner (e) separate paths of interpersonal stress and father absence/stepfather presence. All these predictions were borne out by statistical analysis of facts, although the discussion becomes involved, as is true in science generally. These environmental triggers connect to internal physiological mechanisms, e.g., exposure to pheromones (chemical substances) delivered by axillary sweat glands of unrelated adult males could influence pubertal timing, as they are known to reduce variability in women's menstrual cycles (Cutler et al., 1986).

In sum, the vast evidence for numerous physiological, psychological, and social antecedents of variation in pubertal timing vindicates the basic postulate of the model that human females adaptively control the timing of their reproduction. Thus, it is reasonable to think that human females can adaptively delay time to menarche, and hence to reproduction, in response to variation in those antecedent conditions.

This postulate becomes even more reasonable once one looks at the literature on the timing of reproduction in traditional societies. For instance, empirical tests of hypotheses derived from precise quantitative optimization models reveal that the four-year interbirth interval among !Kung women leaves more surviving offspring than does either a shorter or a longer interval, and this novel result of optimal birth spacing has been defended against criticisms (Blurton Jones, 1997a, 1997b). The adaptively variable relation between eating behavior, ecological conditions, and birth timing posited in the decreasing mate value hypothesis has been empirically documented using quantitative models in the Paraguan Ache (Hill & Hurtado, 1997), e.g., in good years of abundant food resources women had more pregnancies than in bad years of food scarcity.

Still, different people have different intuitions about whether early or late maturation should be expected to be adaptive or maladaptive and whether eating disorders should be expected to be adaptive or maladaptive. These intuitions motivate opposing analyses of the same data. Let us note a principal reason why it will be hard to resolve these issues by doing quantitative cost/benefit models, although they can only help. Adaptive self-suppression is quite costly, and so the standard model posits that it occurs as an "emergency strategy" (Voland & Voland, 1989). That means it is supposed to occur in, or only in, conditions in which there are already high fitness risks. These risky conditions put the fitness-enhancing systems of the organism into extreme stress, i.e., the very ability of these systems to function is threatened. If we find that an eating disorder turns out to be maladaptive in the internal and external conditions modeled, e.g., anorectics do not make up for present lack of reproduction by reproducing much later, two things could have gone wrong. We might have an inappropriate comparison, i.e., of anorectics to anorectics in terms of reproductive success when we should compare what natural selection acted on, namely, anorexia in emergency conditions versus normal eating in emergency conditions (the former should outreproduce the later). Or we might have forgotten that in an emergency condition, there is no optimal strategy; even the best strategy will work sometimes and fail other times in an unpredictable manner.

This type of debate over opposing intuitions is common in evolutionary analyses. For instance, the reduction in fertility found in eating disorders raises broader issues of fertility reductions. Over the last two decades, evolutionists have developed several approaches for explaining the reduction in fertility in modern industrialized populations. Just as some see anorexia as maladaptive, some see reduced fertility during the last few hundred years as a maladaptive reaction to novel circumstances. Just as some see anorexia as adaptive, others see reduced fertility during industrialism as adaptive for those in the lowest social stratum for whom favoring offspring quality over offspring quantity leads to higher long-term fitness. Boone and Kessler (1998) develop mathematical models in which reduced fertility is an evolved strategy to maximize long-term fitness in response to periodic calamities (such as climatically induced resource shortfalls) that result in demographic crashes. If such strategies exist, then it is quite plausible that there are related strategies of the sort we are considering: eating disorders reflect strategies for dealing with emergency conditions, particularly fluctuating or scarce food supplies.

3.3 Striving to Meet Demanding Bodily Standards

Proximate Cause (3) is that young females acquire eating disorders because they strive to meet demanding athletic or other standards with special bodily (physical) requirements.

These may be standards for traditional female roles in the U.S., such as dance, modeling, or gymnastics. Or, they may be newer female roles in the U.S., such as wrestling or soccer; contrary to popular expectations there are anorexic soccer players. Anorexia is seen in female sport climbers, young and adult, because the high ratio of strength to weight aids in climbing and climbers with that high ratio tend to have low body fat. Each sport has its own route to anorexia, because susceptibility to anorexia depends on levels of body fat, and, as I have stressed, ovarian function tracks body fat. It is estimated that the enactment of the Title IX law is largely responsible for the change from young females being one of every thirty high school or college athletes to being one of every three such athletes. The symbolization model begs the same sort of question as before: Why should the need to meet these standards be so distressful that it causes an eating disorder? Evolutionary reasoning would suggest that if a female engages in these activities in order to acquire resources for her own survival or the survival of her kin, or in order to attract a male sexual partner, then failure to meet these standards can result in others preventing her from continuing to engage in these activities. Her survival or that of her family is then in peril.

Notice that the activities of dance, modeling, and track running all involve high calorie loss, where a thin physique is preferred. Such activities would lead more often to anorexia than bulimia, because by definition anorexia involves a body weight of 15 percent or more below the recommended national average, whereas bulimics may be low weight, of normal weight, or higher than normal weight. Where the sports activities involved can be carried out with a thick physique, such as wrestling, then bulimia is a greater risk than anorexia. Many experts think that the percentage of men with anorexia is higher than the commonly cited figure of 10 percent. Both female and male runners, especially long distance runners, suffer from anorexia. Male anorexia and bulimia are underreported for a variety of reasons: there is the dogma that anorexia and bulimia are female problems; feminism has developed this dogma into an indictment of patriarchal culture; the male runner's closest allies having these background beliefs are less likely to be suspicious that a male as opposed to a female has an eating disorder and so less likely to detect it in the male runner; the definition of anorexia in terms of absent menstrual periods and of bulimia in terms of irregular menstrual periods precludes the possibility; it is socially acceptable for men to either eat huge amounts of food (who might be bulimic) or to be picky eaters (who might be anorectic); and both sexes live in a culture that may overvalue performance, low body fat and an unrealistic body shape, size, and weight (Beck, 2000).

The key tenet of the symbolization model—eating disorders are not the primary problem, but, rather, symbols of a hidden primary problem—is applied to male runners:

> It seems unlikely that running, even at the highest competitive level, can "cause" an eating disorder; rather, men whose backgrounds, psychological profiles and emotional status fit a certain mold are probably more likely to be attracted to running and other forms of intense physical exercise than the average person. Thus they are placed in a situation ideally suited to unleashing disorder-like symptoms ... Most of the underlying psychological factors that lead to an eating disorder—low self-esteem, depression, anxiety, and difficulty coping with day-to-day challenges—are the same for both men and women (Beck, 2000).

The attraction of such men to running is, in evolutionary terms, an example of "niche-picking." When competition is intense because organisms adopt the same strategy, organisms are selected to seek niches in which there is less competition or competitions they are suited to win, and hence a greater average pay-off. Sulloway (this volume) holds that birth order affects niche picking, with firstborns and laterborns developing different personality types according to their within-family niches of differing relationships to parents. Thus, men whose perceived mate value is low due to a variety of problems may try to raise their mate value by doing something in which they are likely to win male-male competitions, such as running. Risk factors for eating disorders in men partially overlap those in women, including: obesity as children, recent dieting, participation in a sport or profession demanding thinness, and growing up in dysfunctional families exhibiting physical or sexual abuse or alcoholism.

Should we posit an additional evolved strategy related to efforts aimed at satisfying demanding physical standards? As a way of making that gambit plausible, remember that there are usually additional evolutionary considerations left out of most hypotheses during their formation. So far we have ignored facts that make it plausible to posit an additional evolved strategy. These facts are general, about humans, and have implications for both men's and women's strategies. Humans have migrated all over the globe, and it would be adaptive to postpone reproduction until after moving from place to place rather than achieve it during that movement because reproduction during migration would be forced to rely upon calories left over after intense physical activity. The fact that ovarian function tracks fat would be even more important during times of migration than times of scarce resources, because using more calories during migrating and taking in less calories during famines have the same result: less fat, hence, less ability to get through pregnancy, birth, lactation, and carrying around offspring.

Here is one evolutionary scenario and associated evolved strategy for men's eating disorders. It is the male version of the deviance-amplification feedback cycle posited for females, minus the female physiology of fat-tracking ovarian function and reproductive suppression. Buss (1994, pp.38-39) posits an ancestral scenario in which larger, stronger men physically dominated women. This led to injury and sexual domination (even rape), which prevented women from exercising mate choice. Our primate heritage is documented to reveal recurrent male physical and sexual domination of females, e.g., female baboons exchange sex for protection. Given the problem of protection, an important benefit to women of permanent mating is the physical protection offered by her male partner. In our evolutionary past and today, a man's size, strength, and physical prowess are cues to solutions to the problem of protection. The sort of men who meet demanding athletic standards today send just those cues to women wanting permanent mates, and so their athleticism as evident in winning male-male contests of running and other physically demanding sports increases their mate value. In the decreasing mate value hypothesis for men, those who find themselves at a disadvantage as a long term mate, owing to their "primary problems" inherent in the previous list of risk factors, turn to such sports in order to make up for that self-perceived mating disadvantage. However, because in our culture female athletes are expected to know how to cook and prepare healthy meals, whereas males are not and claim a lack of time or knowledge, or involvement in balancing a career with training, these male athletes often do not eat well enough to maintain health. The fact that some men know how to cook, especially athletes, distinguishes athletes at risk for eating disorders due to eating habits from those not at risk due to eating habits. Athletes who eat poorly may try to balance career with training in ways that keep them from solving the primary problems inherent in the risk factors, especially dysfunctional families in which fathers did not train them to have the marital, work, and parenting skills needed to maintain the family and raise culturally successful children. So, their efforts to meet demanding physical standards incur opportunity costs for doing what it takes to be a permanent mate, thereby decreasing their self-perceived mate value even more.

Further development of this hypothesis for men's eating disorders lies in the fact that women choose men as long term mates for many qualities besides physical prowess and athletic skill: ambition, industriousness, kindness, empathy, emotional stability, intelligence, social skills, sense of humor, kin network and position in the status hierarchy. One would relate these mate preferences to the risk factors in order to empirically demonstrate that men with eating disorders exhibit correlations of these risk factors, which involve their "primary problems," deficiencies in these qualities or in the tendency to display them to women. The distinction between long-term and short-term mating is especially important here, as evident in Gangestad's work (this volume). The evolutionary scenario posited physical protection in relation to permanent mates. Gangestad argues that models of good genes selection suggest that males who exhibit markers of good genes, which such athletes would do, may provide fewer material benefits than males who do not display those markers. He also argues that males who exhibit markers for good genes may be selectively advantaged as extra-pair mates chosen for their genetic benefits only and less advantaged or disadvantaged than long-term mates who invest resources in their female partners and their children. This theoretical outcome bolsters the male decreasing mate value hypothesis just constructed because most of the qualities preferred by females in their partners involve the ability, willingness, and signal that they benefit their female partners through resource investment rather than through good genes. So, males whose proximate mechanism of eating disorders is in striving to meet demanding physical standards are behaving in ways that lower their desirability as long-term mates, so that their athletic behavior decreases their self-perceived mate value in the vicious cycle just hypothesized. I am not focusing here on hypothesis testing (as in Holcomb, 1998), but on conceptual obstacles to integration of nonevolutionary and evolutionary psychology.

What is an "evolved psychological mechanism" relevant to an eating disorder and how can we describe an eating disorder as a malfunction without defining it in evolutionary terms? The idea of an "evolved psychological mechanism" relies on the criterion of differential reproductive success to specify its values (e.g., mate value). An evolved mechanism is a set of cause-effect relations such that the sequence of events makes a difference to a valued state presupposing that criterion. To speak of psychological mechanisms involves fitness as the success-determining value, cause-effect relations between cognitive events (in the wide sense), and the event of reproducing. For instance, anorexia reflects evolved psychological mechanisms of two types: one to decide whether to reproduce now or later, and another to survive in ecological conditions of scarce or fluctuating food resources. Stress is the body's response to a threat. The concept of threat presupposes a systemic state that has a value; without a value, it wouldn't make sense to say that a threat exists, because a "threat" is defined as a source of danger, trouble, problems, evil, harm, unpleasantness, discomfort, etc.—all of which are value-laden terms.

In proximate terms, suppose there is a threat to one's self-esteem. Too much dieting to enhance self-esteem is a malfunction, in a sense defined independently of evolutionary theory, because caloric loss interrupts such tasks as having enough energy to work effectively, participating in a social life that centers on meals, etc. The decreasing mate value hypothesis posits a cycle involving interaction of evolved mechanisms on both physiological and psychological levels, because once ovulation ceases, any mental or emotional or motivational states that would lead to reproduction would be a great waste, and hence strongly selected against. Stressors create cognitions of lowered self-esteem, that information is processed to signal perceived a handicap in one's own mate value that may motivate extreme dieting behavior, dieting creates internal triggers that mimic cues of famine, the body goes into starvation endurance mode, fat reduction mimics ecological conditions of scarce resources, ovarian function tracks fat quantity and distribution relevant to pregnancy, current reproduction is traded for survival, and the physiological changes feed back into cognitions of increased threat to one's mate value. If the stress reaction is internalized (it becomes part of one's "concept of self" in proximate terms and one's "self-perceived mate value" in ultimate terms), it leads to inadequate or malfunctional attempts to keep things under control, leading to more stressors, more stress reactions, and ultimately to an acute breakdown in health, sometimes even to death (Jon Kabat-Zinn, 1990, p.262). Presence or absence of internalization is a key factor for whether dieting turns into an eating disorder, just as it is a key factor in the triggering of chronic conditions that lead to systemic breakdowns in general.

3.4 Fear of Becoming Fat

Proximate Cause (4) states that young females develop anorexia because of their fear of becoming fat.

We needn't belabor the well-known fact that, for women in the current U.S. culture, thinness is associated with power, high status, beauty, and a perfect body, whereas fatness is associated with lack of power, low status, ugliness, and a deeply flawed body. A central, unexpected behavior genetic finding is that unshared family environment plays a large role in personality development (Sulloway, this volume). This point complements the finding cited by Ellis and Garber (in press) that behavior genetic studies using twin and sibling designs have shown that genetic differences account for significant variation in pubertal timing (Farber, 1981; Kaprio et al., 1995; Treloar & Martin, 1990). Note in passing that high levels of exercise have been found to delay pubertal maturation in female dancers (Calabrese et al., 1983), which further confirms the evolutionary view of the "fear of fat" proximate mechanism as one that not only involves self-representations (as the symbolization model states) but links self-representations involving weight as a factor of self-perceived mate value to the physiology of reproductive suppression (the symbolization model cannot possibly explain physiology).

Bailey and Roswell (1994) give an evolutionary restatement of the stress-diathesis view of anorexia in which they summarize the physiology of reproductive suppression that is not explainable merely through symbolization:

> Copeland (1985) suggests that anorexia is a disorder of weight regulation that appears related to hypothalamic noradrenergic activity, and Garfinkel and Garner (1982) say that anorexics typically show low plasma levels of gonadotrophin. Moreover, their 24-hour patterns of gonadotrophin secretion appear 'immature' in that they resemble those of preadolescent girls. Condit (1990) concludes that anorexia seems to interrupt the reproductive process by returning the individual to a prepubescent state of sterility. Anorexia may also be related to low levels of hypothalamic serotonin activity, which disrupts hunger satiety response in the 'never satiated' direction leading to compensatory hunger inhibition. Low serotonin levels may help explain the high levels of depression and high suicide rates in anorexia as well.

Activation failure occurs when evolved mechanisms are activated inappropriately in relation to the adaptive problem they were designed to solve. For instance, in eating disorders, the decreasing mate value hypothesis implies that self-assessment of mate value fails to be activated when potential mates are available. Because earlier puberty in girls increases their reproductive life span, selection drives the age of menarche downward, and self-assessment of mate value is selected accordingly to operate at earlier ages. Because boys can produce sperm long before they develop morphological features used to win male-male competition for mates, males can conceive offspring before they show outward signs of maturity. Because females can develop breasts before they can ovulate, females show outward signs of maturity before they can conceive offspring. This dimorphism in sexual maturation rate creates complex, interacting selection pressures. In females, breast development attracts male attention and allows for careful selection of mates before they can conceive, either through observation of male-male competition or prolonged assessment of potential mates given that pregnancy incurs such a high female investment. If variation in girl's rates of sexual maturation exists, girls who do not monitor both their own mate value and the mate value of boys who are potential mates are at a selective disadvantage compared girls who do so. That would drive activation of mechanisms for self-assessment of mate value, monitoring of the chances of present versus future reproductive success, and assessment of male mate value down to earlier ages. When females see magazines of models who are extremely high in mate value as determined by attractive appearance, noticing at ages as early as 8 and 9 years old that they have much more body fat, these conditions set of a "fear of fat" that induces dieting. Between normal eating behavior and an eating disorder is a "borderline eating disorder." "About seven million American girls and women suffer from eating disorders, and experts estimate that ten times as many have milder versions," diagnosed through a checklist of "disordered eating habits" that are based on a fear of fat; a recent poll showed that 21 percent of teen girl readers of "Young and Modern" magazine aren't sure whether they have an eating disorder (Colino, 2000).

Studies show that children at age 3 eat the amount they need, even if their plate holds more, but children at age 4 will override their natural hunger sensations and respond to social commands to "finish your plate" and eat whatever amount fits the social (including familial) norms. Social pressures create disordered eating habits, and to maintain these habits people change their perceptions of their own bodies. Striving for the highest possible mate value is extremely susceptible to socially induced deception of others and of oneself. "Because if you can pull off convincing other people that you are more valuable than you really are, then you get all the advantages that come from that status. In fact, if you look at social life, there are all kinds of social displays that are essentially deception, where people are trying to act like they're richer than they are, or sexier than they are, or smarter than they are, or something, in order to convince people. If you believe it yourself, that works pretty well" (Nesse, 1999).

Evidence for the implication that eating disorders sometimes involve cognitive or perceptual misrepresentation can be found in the new identification of a body image disorder. Body dimorphic disorder (BDD) is said to be a brain disorder (read: mental disorder) such that when people see themselves in a mirror they see a distorted body image. Symptomatic individuals, typically but not exclusively females, say that everything they thought was wrong with their body "stuck out" when they saw themselves in a mirror. This symptom reflects the focus on the negative (i.e., defects), which leads to control failure according to the theory of ironic processes of mental control. They are obsessed with their body image. At first sight, that symptom reflects a selfish self-absorption, but a high level of sensitivity to one's body size, shape, and weight could be helpful in recognizing cues to one's health during times of famine, cues important for self-assessment of mate value and regulation of reproductive timing. Some symptomatic individuals are obsessed with the idea of people staring at them. Females may think that others are thinking, "She's ugly," "She's fat," and "I don't want to look at her." Males may be preoccupied with pimples on their faces. Males and females say that they can't go to school because other people will see them like this. Such symptoms might be extreme expressions of mechanisms for assessing the dimension of one's mate value that involves physical attractiveness. When these people are given funhouse mirrors with convex shapes that distort normal body proportions and make some body parts larger than others are, they consistently report that these mirrors simulate what it is like for them to see themselves when their symptoms are expressed. This dimorphic perceptual disorder provides good evidence that eating disorders of some types do involve misperception of oneself or one's environment or both.

The view that this mental disorder is a brain disorder arises from the fact that such individuals can be helped through treatment by serotonin reuptake inhibitors. These medications are also given to people in chronic pain and to people with eating disorders. For all these dysfunctional conditions, some people are helped by serotonin reuptake inhibitors (e.g., Prozac, Paxil, or Zoloft), some are cured, and some gain little or no beneficial effects. This may involve variation in emotional components of eating disorders, as the emotions are "seated in" the limbic system, which is connected both to the nervous system (as the brain disorder view implies) and the hypothalamus (which regulates "automatic functions" such as sleep, as will be discussed later). The hypothalamus also regulates perceived intensity of stimuli, which may be relevant to the fact that it sometimes takes just one comment, "You are fat," to send someone into an extreme diet and hatred of fat that eventuates in disordered eating habits. However, the greatest therapeutic implication of the emergency personal fitness situation inherent in the decreasing mate value hypothesis is that eating disorders are triggered by some "primary problem" that makes a difference to self-perceived mate value. Generally, we should treat eating disorders in terms of these primary problems, as understood in relation to self-perceived mate value, e.g., we should help young girls (or boys) develop social, mating, parenting, and physical skills or whatever skills are needed to solve the primary problem and not just give them a pill.

3.5 Desire to Conform

Proximate Cause (5) is that young females want to do what others do, i.e., they have an inordinate desire to conform to social norms for the ideal female body.

A new research avenue pertinent here is to take account of female-female competition as it is intensified in female dominance hierarchies. One dimension is for a female to use information gathered from other females about which males and which females have high mate value, what the determinants of mate value are, and how to employ strategies and mechanisms for executing those strategies for acquiring the highest value mate possible. Another dimension is for females to use information gathered about female dominance hierarchies to better compete with other females. The symbolization model, which regards low self-esteem as the primary problem in eating disorders, is normally combined with a cultural hypothesis: Anorexia is an attempt to conform to the cultural ideal that thin women are beautiful (Crawford, 1989). The cultural hypothesis is supposed to explain why body image is the arena in which the low-self esteem is displayed: Women feel bad about themselves in virtue of feeling bad about their bodies, i.e., their appearance is felt to be unattractive by reference to a patriarchal cultural ideal of thinness that few women can healthily achieve and maintain.

Mealey (2000, p.247) makes the important distinction, especially in the light of feminism, between culture and male-dominated culture. Noting that anorectics are typically young, unpartnered girls or women who have not borne a child, she asks where their dissatisfaction with their bodies comes from: "Culture? Yes. Male-dominated culture? No." There is vast evidence that men prefer women with enough fat deposits to be curvaceous, whereas women themselves regard women who are as thin as fashion models as most attractive. Women, not men, regard the ideal female figure as far below average weight. Men prefer women who are plumper than women want to be (Buss, 2000). Women's magazines, not men's magazines, first started to portray thin models (Barber, 1998). The prototypical men's magazine, Playboy, has retained the curvaceous figure of its nude models over the years, as measured in terms of waste-to-hip ratio. Women's culture and the desire to conform to its ideal of thinness by young females is a common causal route to anorexia.

Given these facts, Mealey (2000, p.247) has advanced the hypothesis that eating disorders have their ultimate causes in female manipulation, which is the ability of dominant females to manipulate subordinate females to do things against the fitness interests of those subordinate females. The idea is that high status females, such as beauty models and those females that benefit from their existence, manipulate young females to do things which lead to their fully suppressed ovulation (e.g., anorexia) or partially suppressed ovulation (e.g., bulimia) to eliminate them as sexual competitors. This hypothesis is a novel development of the reproductive suppression model. It posits that eating disorders are not autogenetically produced adaptations for eliminating unwanted pregnancies or adaptations gone awry due to mismatch between modern and ancestral environments. Instead, they are results of control of some females by other females that are adaptive for the controlling females and maladaptive for the controlled ones. Instead of trying to explain the existence of culture by biology, this hypothesis takes as brute fact about the circumstances in which selection operates the cultural circumstance that the modern mass media idolizes thinness to a fault.

A reason to think that this scenario creates a strong selection pressure in female-female manipulation as an extreme form of female-female competition for gene perpetuation is that males choose females for their indicators of fertility, of which physical attractiveness is an overriding preference. Female manipulation eliminates young, nubile girls at ages centering on peak reproductive value (about 18 years old) and peak fertility (about 22 years old) from conceiving offspring some of the time, or in extreme cases of pathology, ever. Unlike men, women have evolved forms of aggression that do not risk personal survival, because their reproductive success in ancestral environments depended on variations in success in their own survival and that of their offspring. Female aggression is subtler than male aggression (e.g., female proportion of killing is small), and creates hierarchies designed to gain access to high quality mates on the basis of indirect aggression, such as making derogatory remarks about another female's looks (Campbell, 1999). Thus, girls who read and discuss fashion magazines are willing at younger and younger ages (e.g., at ages of andremarche, 8 and 10 years old) to look at women's magazines and say, "Look at her abdomen! She's too fat!" This intense, but subtle, female competition is made likely if not inevitable in modern culture by the fact that most women today have far less chance of losing offspring during pregnancy, infancy, or childhood than did our hunter-gatherer ancestors or ancestors in traditional cultures.

The main force of inherited predispositions useful in female-female competition is focused on a narrow range of genetic variances in the number of surviving offspring (which leads to less intense but more subtle competition), as well as competition for high value males who mate with many females and whose sexy sons will mate with many females (which leads to intense but aggressive competition). Mild forms of eating disorders may be adaptive for the symptomatic female, if it is the case that the cost of losing early menstrual cycles is more than made up for by later reproduction, a fact yet to be established by estimates of actual reproductive rates of anorectics and bulimics. Extreme forms are obviously maladaptive for them, e.g., if they die or fail to find high value mates because they violate male preferences. This hypothesis can account for two facts our previous discussion has not accounted for. So far, we have developed the reproductive suppression model to explain why young females are especially at risk for eating disorders and we can put together various mechanisms for the initial proximate causes and the mechanisms for the onset and maintenance of eating disorders. The female manipulation hypothesis, if true, explains why females would develop traits (e.g., late sexual maturation, loss of ovulation, skinny body shape) that violate male preferences. It also explains why the incidence of eating disorders has increased so dramatically in recent years as a correlation with media-driven body images equating beauty with thinness.

Following Cummins (this volume), low self-esteem is a functional mental trait of someone who is likely to lose dominance contests. Thus, enabling them to accept subordinate states when challenging dominant individuals would lead to costly consequences. Further, thinking in terms of the environments of evolutionary adaptedness would explain the pathologically low-self esteem often found in eating disorders as partly due to computing one's mate value by comparing one's body to the technologically enhanced pictures of fashion models in the media (the privileged few out of millions of people), rather than to other actual people of one's own social group. We lived in groups of 20 to 200 in ancestral environments, and our mate value would be much higher in comparison to our few sexual competitors than to the millions. This is a classic case of mismatch between ancestral and current environments. The mismatch idea has been developed conceptually, theoretically, and empirically into a disorder profile for early phase anorexia (Bailey and Roswell, 1995).

This hypothesis deserves to be tested as an alternative in order to increase the severity of testing; the more alternative hypotheses there are, the more severe and nuanced the tests between them can be, facilitating falsifiability. It could be tested for its ability to explain the specific characteristics of the recently increased incidence of eating behavior problems.

There are many societies or cultures in which virginity or non-promiscuity is necessary or helpful for high mate value. A female in such a culture reduces her mate value by accepting a single copulatory proposal from a possible mate. The more acceptances from males who do not have high mate value given her family background, social class, economic position, and so forth, the more likely she is to be derogated as a "slut," "a loose woman," or a "whore." In such situations, a better satisficing strategy would be not to accept any proposals to have sex, and to get sample information from females with the highest recognized mate value, e.g., beauty models. This sets up conditions in which female-female interactions are quite important.

A female should not simply try to develop traits that prospective males find desirable in females. Instead, she should attend to which females have highest status in the culture. Thus, the emphasis on body image so pervasive in eating disordered females needs to be expanded theoretically to take account, not just of male choice, but of all four theoretical sexual selection processes: two forms of choosing a mate (intersexual selection), including the ability of females and males to differentiate and select a mate, and two forms of competing for a mate (intrasexual selection), including the differential abilities of male individuals and female individuals to acquire mates in competition with other individuals of the same sex.

3.6 Seeing Control as a Virtue

Proximate Cause (6) is that young females become anorexic because they see control as a virtue.

It is important here to distinguish the symbolization model from psychological mechanisms of repression, transference, and so forth familiar from Freudian psychology. It is plausible to explain some eating disorders by diagnosing the individual's mind-set as one in which they feel out of control, and in their need to control at least something in their life, they fixate on food, taking control of food intake to extreme levels of starving themselves down from 130 pounds to 70 pounds (anorexia) or losing their sense of self and self doubt during an orgy (binge) of eating as much as 5,000 calories at a time followed by purging (bulimia). But that sort of intentional explanation relies on the reasons/causes of eating disorders that are not strictly part of the intelligibility explanation based on the three postulates of the symbolization model. Our previous quotes from the beef industry showed that it explained eating disorders via distress and low self-esteem, regardless of what the causes/reasons of eating disordered behavior turn out to be—Freudian or otherwise. Intentional-intelligibility explanations stay exclusively at the level of language: eating behavior is symbolic behavior in the sense that behavior is a vehicle for symbolizing a subjective psychological state.

Eating disorders occur as emergency strategies that are pursued whenever "the sovereign handling of one's own reproductive potential is not possible because of socioecological or ontogenetic constraints" (Voland & Voland, 1989). Given that many patients alternate between anorexia and bulimia, we confront the hard issue of whether anorexia and bulimia are two distinct but related phenomena or two expressions of one phenomenon. This issue is vital to the decision of whether we need separate theoretical explanations of the two conditions, and of their relations, or just one basic theoretical explanation of both.

A standard view is that bulimics are just "failed" anorectics. Whereas the anorectic denies their perception of their own hunger, the perception of hunger gets through to the conscious mind of the bulimic, who then acts on hunger. The import of this view for evolutionary psychology is that there would be just one evolved strategy for explaining both anorexia and bulimia, where bulimics execute the strategy more poorly than anorectics. If this view of bulimia is incorrect, there may be two distinct strategies for the two distinct eating disorders. Alternatively, the emphasis on hunger suggests that anorexia and bulimia may be more closely related. Whereas anorectics feel less hunger than normal people do (or deny, override, or do not act on their hunger perception), bulimics feel more hunger than normal people (or deny, override, or do not act on their perception of hunger satiation). If so, there may be one evolved strategy for dealing with hunger, with two distinct sorts of errors in the way the evolved strategy is carried out by psychological and physiological mechanisms.

Methodologically, because we know so little about our behavioral evolutionary history and those unknown facts can falsify even the hypotheses with the best evidence of many types that we have available today, we try to make minimal assumptions about our environments of evolutionary adaptedness (EEA) so that our hypotheses have the least chance of being incorrect. One pertinent example of making assumptions about the EEA that are as undeniable as possible is the following scenario. In the ancestral conditions in which our mental adaptations were forged, some young women were forced to rely on themselves for resources. This occurred in a variety of conditions: if they were abandoned by their family; if they were expected to move as young women out of their natal group and marry into another family some distance away from their natal group; if they were ostracized by the group or husband to which they went for marriage from their natal group, especially if they resisted rape or abuse by that group or husband; if they were kidnapped by another tribe; if they lacked adequate support from their family or tribal members; if they experienced environments of high male mortality or other things that made stable pair bonding unlikely, or in which actual or potential mates would not invest much in offspring (Sabura Allen, unpublished ms.). Sabura Allen uses the term "the abandoned mother" to designate this evolutionary scenario as opposed to "the betterment" scenario in which reproduction is delayed until conditions for successful reproduction get better, although both involve suppressed reproduction.

In such situations, one could raise one's probability of reproductive success by engaging in behaviors that acquired resources for personal use, e.g., gathering of food and goods, engaging in sexual behavior for non-reproductive purposes (i.e., exchanging sex for resources). Bulimia would be an extreme expression of a strategy for reproducing more successfully in such situations by securing resources for oneself. This would explain hoarding food in one's body through bingeing. Bingeing would have been adaptive in the EEA to ensure the survival of both mother and child, not only in extreme emergency conditions, but also in mild emergency conditions in which social and physical resources were unpredictable or unreliable. Now, given the core definition of bulimia as an eating disorder of frequent bingeing followed by purging, this EEA scenario explains bingeing in a natural way, but it prima facie does not explain purging. However, one could invoke the mismatch of present and ancestral environments here. Bulimics know in some sense that present day environments in the U.S. hold abundant resources, e.g., there are grocery stores nearby. If a modern resource assessment mechanism were jerry-rigged on top of an ancient one, a person would act like a hunter-gatherer in emergency conditions by bingeing. With food stores all around us, however, the environmental information that one really doesn't need these extra food resources all at once would also get through to the brain, and so one would purge. A simpler, alternative mechanism for purging would be that one would feel sick to one's stomach after taking in so much food all at once, but the mechanism for retaining food in conditions of scarce resources would not kick in, resulting in purging to eliminate the feeling of sickness. Falsifiability increases by testing multiple hypotheses against each other, and parsimony dictates that we have good reason to accept the simpler one before accepting the one that requires a more complex physiological-psychological mechanism.

At this point the "Just-so story" dimension of the vulgar adaptationism charge sets in. Even if positing such EEA scenarios for bulimia is reasonable to make about our ancestral conditions, the ad hoc ways of motivating purging in addition to bingeing in the explanation seem like they were manufactured out of thin air just to save the explanation. If evolutionists quit at this point the charge would hit the target, but because they go on to test the facets of the explanation that are testable, the charge misses the mark. Sabura Allen derived several predictions from the abandoned mother scenario and tested them. One prediction is that bulimics perceive a lack of social resources. This implication was tested by using levels of kinship affiliation to assess lack of social resources in important relationships. The evidence shows that bulimics expect lower levels of kinship affiliation from both their family members and dating partners than non-symptomatic people do.

A second prediction is that bulimics (a) perceive themselves as having low levels of commitment, help, and support from family members, (b) have high levels of openly expressed anger and conflict, and (c) are less likely to foster independence than non-symptomatic people. The data also supported this hypothesis and is consistent with the literature regarding bulimia and family environment (Bailey, 1991; Dolan et al., 1990; Kog & Vandereydken, 1989; Scalf-McIver & Thompson, 1989; Stern et al., 1989; Shisslack et al., 1990). Also data regarding family members are consistent with previous findings that bulimics perceive less social support and more negative interactions in their family relationships (Grissett & Norvell, 1992).

The confirmation of these two predictions paints a picture of family life in bulimics that fits the evolutionary theory of socialization introduced by Belsky et al. (1991) and discussed previously in relation to the empirical data on pubertal timing by Ellis and Garber (in press). The agreement of data for various aspects of the reproductive suppression model is strong evidence for it. The mutual relevance is based on the fact that, whereas anorectic females fully suppress reproduction (via loss of menstruation, and hence ovulation), bulimic females partially suppress reproduction (ovulation is an on-and-off matter). Children form expectations of the availability and predictability of physical resources based on early childhood experiences. Children from distressed families would develop a mistrust of others and skepticism of adequate available resources later in life. Hence, bulimics who grew up in distressed families expect a future of unpredictable and unreliable physical and social resources based on early childhood experiences. They lower their expectations, which are reflected in lower level of desired kinship affiliation in order to avoid high levels of disappointment when family members are unreliable. Bulimics still desire close relationships with their dating partners, but their expectations reflect an underlying pessimistic attitude that their desires cannot be attained. Indeed, previous research on how romantic relationships may be disturbed shows that bulimics are more likely to be in complicated, controlling or abusive relationships. They are less likely to be married than non-symptomatic persons are (Lacey, 1992; Kaneer et al., 1993). The problem of whether this is cause or effect arises here: a problem that runs through much of the data on eating disorders.

The general point is that we need a more exact description of the EEA relevant to the reproductive suppression model and its relation to today's environment, especially in the light of contraception, from which predictions are drawn and tested. One such research method that addresses the unsolved, basic issue of how evolved mechanisms relate to consciousness is:

> The environment in which a reproduction suppression (RS) mechanism evolved is assumed to be the Pleistocene era which included reproductively relevant stresses such as negative mental state (i.e., depression or unhappiness), coercive sex or undesired sexual attention/harassment from males, change in group membership through moving, marriage, or loss of status, withdrawal of support such as emotional support or resources from family members, etc. Reproduction suppression can be manifested in a variety of ways and at different points along the reproductive cycle. This study investigates the mentality of both conscious (i.e., contraceptive use) and unconscious (i.e. control of body fat) RS prior to conception. We compared contraceptive use, body image, dieting, WHR (waist-to-hip ratio), and BMI (body mass index) in female undergraduates grouped by level of stress, and attitude toward prospective pregnancy. Predictions were: 1) low stress women will show less conscious and uncon-

scious RS than high stress women, 2) women confident of contraception effectiveness will show less unconscious RS than less confident women, 3) women who rate unplanned pregnancy as negative will show greater RS than women who rate it positively, 4) women who are stressed and non-confident in their contraceptive use will show the greatest unconscious RS. Results are discussed in terms of their support for the Reproduction Suppression Model and their implications for investigating how consciousness interacts with the operation of psychological adaptations. (Walters & Lindberg, 1994).

4. DOMAIN-SPECIFIC MATING STRATEGIES AND DOMAIN-GENERAL COGNITIVE MECHANISMS

Proximate cause (6) was that young females acquire eating disorders because they have a strong desire to conform. The fact that such evolved proximate mechanisms are partly conscious and partly unconscious is by no means a suspicious oddity of evolutionary psychology, e.g., other information-processing mechanisms such as visual perception are partly conscious and partly unconscious. The strong desire to conform explains why, in a culture where "the thinner the better" is the rule, some girls (those with high desire to conform) are motivated to diet to the point at which they have an eating disorder while other girls (those with low desire to conform) do not. Culture alone cannot explain such within-culture variation; cultural explanations need to be combined with psychological explanations to fully explain phenomena of interest. However, the desire to conform by itself, and probably also the other preceding proximate causes previously discussed, will not produce a disorder unless they are combined with appropriate cognitive mechanisms. What sorts of cognitive mechanisms can do justice to aspects of eating disorders that the symbolization model, with its domain-general capacity, captures?

Once we follow the basic premise of evolutionary psychology that domain-specialized mechanisms of behavior evolved and retain the basic assumption of traditional psychology that domain-generalized mechanisms evolved, we confront a picture of human nature in which the two sorts of mechanisms interact today and have interacted for a long, long time. Thus, in order to advance both traditional psychology and evolutionary psychology it may help, in the context of explaining eating disorders, to try to find a pertinent domain-general mechanism of eating disorders. It would provide a genuine mechanism involving conscious thought activity, which is lacking in the symbolization model and in versions of the reproductive suppression model prior to the work of Walters and Lindberg. We need to link that mechanism to the reproductive suppression model as developed here in relation to domain-specialized mechanisms of mating. Evolutionary psychology has always confronted difficulties in relating evolved psychological mechanisms to consciousness, and so further hypothesis development should address that issue.

As an example of this gambit, I shall point out one relevant reply to a recent article on the effects and costs of behavioral inhibition (Polivy, 1998), given that most forms of eating disorders reflect mechanisms of inhibition of eating behavior. Polivy (1998, p.199) concludes that, "The attempt to inhibit emotions (and particularly their behavioral expression) and other motivated behaviors seems to have predictable outcomes for the individual—heightened emotionality and distress, cognitive disruption, behavioral excess once the prohibition is temporarily lifted, and possible illness or disease in the long term." In commenting on Polivy's article, Richard Wenzlaff and Daniel Wegner emphasize the relevance of their theory of ironic processes of mental control (Wenzlaff & Wegner, 1998). The claim that the reproductive suppression model needs to be tied to their theory of ironic processes is meant to take seriously a standard objection to the claim of evolutionists to be able to explain anorexia by appeal to the reproductive suppression model alone: Without the use of symbols inherent in the symbolization model, advocates of the reproductive suppression model cannot tell us how the eating disorders ever get going. If eating disorders symbolize a primary problem, and the primary problem involves fitness and the timing of reproduction, then evolutionists must get more deeply involved in traditional cognitive psychology and the study of symbol use to achieve their explanatory aims. Let us grant that reproductive suppression is an effect, not a cause of eating disorders. The six proximate causes reinterpreted in evolutionary terms in the preceding sections usually or always depend on cognitive symbol use and intentionality in order to cause eating disorders.

The theory of ironic processes of mental control fits well with the rationale for moving from "optimizing" to "satisficing" methods of information-processing (Ketelaar & Todd, this volume), namely, that organisms can process only so much information in making decisions about mating before finding feasible solutions to adaptive problems becomes intractable. To anticipate its implications for eating disorders, selection for information-processing economy leads to cognitive processes that break down (become dysfunctional) when the cognitive system is overloaded with information due to the way it "frames the adaptive problem." How would endeavors to inhibit unwanted eating behaviors be cognitively mediated? Information economy is met when the cognitive system focuses on producing behaviors that meet adaptive goals. That limits the search space of thoughts, out of all possible thoughts, to the few positive ones that might work to inhibit hunger for enough food to create culturally acceptable body weights, forms, and images. Information economy is violated when the cognitive system focuses on failures to meet adaptive goals, which demands that the system monitor all non-positive thoughts, which includes not only negative thoughts but also a vast number of neutral thoughts. The decreasing mate value hypothesis posits a system of mental control that focuses on failures to achieve the desired mate value in the circumstances, when cues signal "I'm at a mating disadvantage," which leads to decisions to become more attractive by changing one's body by inhibiting eating through dieting.

By constantly watching for lapses in dieting, the cognitive system has to monitor all thoughts involving eating, body weight and form, whether self-induced or induced by what other people might think of one's body. This leads to information overload, internalizing of the stress reaction to threats to one's feelings about one's own body ("self-esteem," "self-perceived mate-value"), then to inadequate attempts to keep one's weight under control, then to more stressors and more stress reactions, so that the whole psychological-physiological system breaks down (Kabat-Zinn, 1990, 248-264). Stress is the body's response to a perceived threat. Short term stress causes increased heart rate and increased neural activity, eliciting the typically adaptive response to stress, namely to eliminate the threat The stress reaction is a very functional response for emergencies (short-term, acute threats), but it is very dysfunctional when retained for long-term threats. By implication, the more years one perceives oneself to be at an unacceptable mate value disadvantage, the more likely that the reproduction suppression mechanism will lead to chronic dysfunction, such as eating disorders. As organisms, we cannot escape stress in the mating game. The key is learning to manage stress. We have to be able to "let the stress go." Of course, people who internalize stress can end up with a variety of medical, psychological and behavioral problems; internalization is a domain-general process to be added to evolved domain-specialized mechanisms in order to capture the etiology of eating disorders.

The theory of ironic control of mental processes specifies mechanisms involved in thought suppression, mental control by an individual generally, and circumstances that produce cognitive, emotional, and motivational problems. It postulates two domain-general cognitive mechanisms: an intentional operating process and an ironic monitoring system. First, the human mind contains an intentional operating process, which seeks thoughts that promote the preferred state (the goal state). Second, it contains an ironic monitoring system that searches for mental contents that signal failure to achieve the desired state. The common sense feature of these postulates relevant to anorexia is that they refine the saying, "Focus on the good, and cope with the bad." Functional behavior occurs when the person is "focusing on the positive," and relegating the negative to be something in the background of consciousness, which intrudes into consciousness only when the positive is not attained. As an example, "When trying to be happy, the operating process searches for thoughts pertinent to happiness, whereas the monitoring process searches for thoughts that indicate happiness hasn't been achieved. Although the operating process is effortful, consciously guided, and relatively proficient, the monitoring system is usually unconscious, less demanding of mental effort, and thus less proficient" (Wenzlaff and Wegner, 1998). Dysfunctional behavior, with pathology in the extreme case, occurs when a person is "focusing on the negative," which makes the negative occupy both the foreground and background of consciousness, overwhelming the goal-oriented system and causing a breakdown of its ability to function. Ironically, "under mental load, then, intentions to control the mind unleash a monitoring system that not only searches for the failure of mental control but then also tends itself to create that failure (for reviews of research evidence, see Wegner, 1994; Wegner & Smart, 1997; Wegner & Wenzlaff, 1996)."

This theory explains the failure of inhibition, the role of stress/distress, and the role of self-esteem in eating disorders in a natural way. Why does trying to inhibit eating actually cause bingeing (bulimia, eventual weight gain of dieters generally)? When mental capacity is taxed and the effortful process is limited, the subtle sensitivity of the monitor ironically can create the mental state that corresponds to control failure. What is the role of stress/distress? The oft-heard saying that "Distress causes eating disorders" is too vague, even as it is incorporated into the reproductive suppression model. Stress taxes mental capacity. You cannot learn much when you are stressed out. Evidence in childhood development and education shows that the dendrites of neurons are actively involved in the response to stress, preventing them from use in learning. Hence, children living in dysfunctional families have difficulty learning. Stress interferes with the general inhibition mechanism, so the ironic monitoring process takes over. This idea explains many antecedent correlates of eating disorders involving stressful conditions that we used in defining or providing empirical evidence for the reproductive suppression model: dysfunctional families, situations of extreme physical or psychological stress, lack of available sexual partners or supports, sexual rejection, etc.

Polivy summarizes Wegner's and Wenzlaff's theory this way:

> ...when mental control of behavior is taxed or depleted (as in having to do more than one self-regulatory task), the monitoring system that permits suppression actually might create expression of the behavior instead. Wenzlaff and Wegner go on to propose that this might reflect the tendency to set self-control goals as avoidance tasks. If, instead, one sets the goal as an approach action, the monitoring process searches only for positive instances of self-control, and cognitive loads or burdens do not produce disruptions of restrained behaviors. This sort of cognitive solution to the problem of restraining oneself is reiterated in similar forms in several of the commentaries (Polivy, 1998, p.238).

Further, the theory of ironic processes of mental control explains why framing goals in an approach-oriented rather than an avoidance-oriented manner can attenuate cognitive system breakdown, reducing the amount of distress and self-esteem. Recall that the reproductive suppression model emphasizes stress to the biological system and distress to the psychological system, and the symbolization model emphasizes low self-esteem and the distress it causes. The models are reconciled in considering such examples of contrasting thought modes as the following. "I aim to lose 10 pounds and become thinner" is a positively framed, approach goal. Given such thoughts, the operating process focuses on attaining the preferred state and the monitoring process searches for failures among a wide variety of neutral or weight-irrelevant mental contents. Disruption in the operating process only introduces a focus on neutral or irrelevant contents, which leads to a small anxiety increase. Dieting so framed is predicted not to lead to eating disorders. In contrast, "I aim not to be thought of as too fat," is a negatively framed, avoidance goal. Given such thoughts, the operating process seeks both weight-neutral and irrelevant contents. It must check every perception by others to see if they regard oneself as too fat, which is prone to self-obsessive asceticism (anorexia) or obsessive self-indulgence (bulimia), whose cognitive counterpart is information overload. The result is large anxiety increase. Obsession with weight can lead either to overdieting (triggering anorexia) or dieting then bingeing (triggering bulimia). "Discouragement and self-reproach that follows a binge episode (for bulimics) would exacerbate matters and could lead the person to forsake the desired goal state, thereby allowing the negative monitoring process to proceed unabated and unchecked" (Wenzlaff and Wegner, 1998; for more on paradoxical effects, see Bealers, 1986).

The typical cognitive thought patterns that maintain failure of mental control in eating-disordered females include catastrophizing, inappropriate comparing, and perfectionism. Each of these thought patterns focuses on the negative, not the positive, resulting in cognitively induced errors in reasoning that give a young female a sense that her mate value is intolerably disadvantaged. The relevance of the ironic control processes just described to eating disorders is apparent in each of these three signature types of faulty thought that trigger them, because each overloads the cognitive system by focusing on negative monitoring, causing the permanent stress reaction and eventual breakdown that qualifies a set of disordered eating habits as a disorder.

Catastrophizing is a cognitive symptom of assessing one's state as being in an emergency condition, as emphasized by Voland and Voland (1989). This would be expected from the decreasing mate value hypothesis because it posits an emergency state in which one's self-perceived mate value is handicapped so far as to be unacceptable to oneself. Both anorectics and bulimics may employ the inference, "If I make my body suffer, then I will purify my soul (my true self)." This mind-body split is characteristic of eating disorders (Bordo, 1993; Gosselin 1998). It occurs because being in a catastrophic state is a permanent condition of stress that causes a cognitive breakdown. When the cultural ideal is difficult or virtually impossible to attain, the threat of being deemed by others to be unattractive puts the female at extreme fitness risk, namely, of never finding a satisfactory mate, and so of becoming an evolutionary failure. As the self tries to maintain esteem despite information to the contrary, the self splits, constructing a mind-body dualism of treating the body as not-self and the soul as self. Catastrophizing becomes both an effect and an ensuing cause of control failure in the form of daily thoughts and private decisions. An anorectic might think, "Because I ate too much yesterday and am even worse off than before, I'll just eat a cracker for breakfast today." A bulimic might think, "Because I pigged out at breakfast today, the whole day is ruined, and so I might as well stuff myself today and start over tomorrow" (Dionne, 1998). The solution to such distorted thinking is to view setbacks as opportunities to make oneself a more attractive mate, a positive approach goal-orientation suggested by the theory of ironic mental control processes.

Inappropriate comparing is a cognitive symptom of error in assessing one's mate value by comparison to some rivals, i.e., other females who might compete for one's own chosen mate, rather than to others. Both anorectics and bulimics may employ the inference, "If others are capable of experiencing pain, then I must also experience pain" (Gosselin, 1998). This may take the form of thinking: "If I will hurt others, then I must be punished; I will hurt others; so, I must be punished." Females from dysfunctional families live in world in which family members hurt each other, and so conclude that they will hurt others also. Females that hurt others violate the cultural norm that women, more than men, are caretakers who feed and raise children and enable their man to work everyday and thereby provide her and her children necessary resources. Fear of lack of control is evident in the reasoning: "Eating means having food; food means having life; I should not have life because I do not deserve it; I do not deserve life because I have failed humanity; so, I do not deserve food."

This sort of thinking fits with the proximate cause of eating disorders of seeing control as virtue. It produces a self-assessment of an intolerably low mate value, which could lead to the decision to choose an extremely low mate value sexual partner or the decision to delay mating until conditions improve. In fact, bulimics continue sexual activity, but often accept undesirable partners, and anorectics reduce sexual activity far below non-symptomatic individuals, presumably expecting to delay sexual activity until later, as the reproductive suppression model posits. An eating-disordered female may think, "If I look like Kate Moss, then my boyfriend will pay more attention to me" (Dionne, 1998). She may decide to eat less, and the less she eats now the less she eats later. She lowers the rate of food intake in a vicious positive feedback cycle that produces dysregulation of the biological system, stress to the psychological system, then more dysregulation of the biological system due to low food intake, then more stress to the psychological system, maintaining anorexia. Or, she may decide to eat less, but compensate by eating more later, but purging to continue emulating an inappropriate role model.

The biological-psychological interaction is well represented in the sleep patterns of eating-disordered individuals. "People who have anorexia get more stage 1 sleep and less slow-wave sleep than those who maintain a healthy weight. Some people with bulimia awaken frequently during the night to eat. Extreme weight loss is associated with interruptions of sleep and early morning awakening" (Ince, 1999). Fast brain wave sleep occurs in stage one and stage two of sleep. People in stage one are easily jarred into wakefulness. People in stage two show slow brain waves mingled with fast brain waves, called "sleep spindles." A K-complex of bursts in brain activity occurs, a built-in vigilance system that keeps a person poised to be awakened if necessary. This is adaptive in an emergency situation, in which stress leads to an aroused autonomic nervous system. When it occurs as a chronic condition, sleep deprivation occurs, i.e., deprivation of the stage three and stage four slow wave sleep (deep sleep) necessary to leave the person refreshed. The result of deprivation of sleep quantity and quality is depression and psychosis. This explains the association of eating disorders with depression; indeed some think of eating disorders as a form of depression because of the effects of low self-esteem. To stop this vicious cycle, a cognitive therapist would advise the person to explicitly write up notes about themselves, especially on defining one's body image. By showing the comparison of oneself to the cultural ideal to be inappropriate and helping the eating-disordered person realize that she is being unfair to herself by making that comparison, therapists can open up the opportunity to help patients recalibrate their assessment of their own mate value. That idea is expressed in nonevolutionary terms as reassessment of their potential and worth as a person. The evolutionary approach explains why such traditional cognitive therapy works, and can be tested by testing the prediction that reassessment of a symptomatic individual's potential and worth as a person will work better when the valued qualities correspond to those that increase mate value.

Perfectionism is a cognitive symptom of error in assessing one's mate value by framing one's attractiveness and desirability as a deficiency according to a perfectionist standard, which would lead to an unacceptably low self-perceived mate value. It could lead to selection of a deficient sexual partner, so that partners of like mate values become partners. Assortative mating leads adaptively to couples with the highest combination of mate values all the way down from the highest mate value couples to the lowest mate value couples. When females frame their own attractiveness by comparison to a cultural ideal or perfect mate, their perfectionism changes what might be a high, medium, or low mate value assessment into the lowest possible mate value assessment. An eating-disordered female might infer, "If I am not perfect, then I am a failure" (Gosselin, 1998). The youthful expectation of being either great or a failure with no middle ground can result from inappropriate parenting or the youth thinking of their same-sex parent as great and trying to be equally great in order to have self-worth.

In sum, we have linked the reproductive suppression model with commonly known proximate causes of eating disorders—about individuals not wanting to develop sexual traits, fearing becoming fat, disliking the shape and size of their bodies, striving to meet demanding physical standards, wanting to conform to what others do, and seeing control as a virtue. From the picture we have painted with broad brush strokes and some necessary details, initial suspicion of a sample evolutionary hypothesis as irrelevant gives way to plenty of good reasons to regard it as relevant; there must be something to it, after all. Having built the case for the decreasing mate value hypothesis logically bit by bit, we see that the reproductive suppression model potentially explains far more than the symbolization model could ever hope to do, e.g., about why there are multiple pathways toward eating disorders and how they might be interconnected. We have illustrated research methods for studying the proximate mechanisms of anorexia and connecting them to evolutionary strategies. These methodological decisions in hypothesis formation are sufficiently general to apply not only to clinical psychology, but also to evolutionary psychology generally. When evolutionary psychology codifies into a systematic methodology not only decisions made in hypothesis testing but also made during hypothesis formation, it will come of age as a science.

REFERENCES

Anderson J.L. & Crawford, C.B. 1992. "Modeling Costs and Benefits of Adolescent Weight Control as a Mechanism for Reproductive Suppression", *Human Nature*, 3, 299-334.

Bailey, C.A. 1991. "Family Structure and Eating Disorders: The Family Environment Scale and Bulimic-like Symptoms", *Youth & Society*, 23, 251-272.

Bailey, K.G. & Roswell, L.B. 1994. "Anorexia: A Disorder Profile", *The ASCAP Newsletter*, 7, 6-12.

Barber, N. 1998. "Secular Changes in Standards of Bodily Attractiveness in

American Women Different Masculine and Feminine Ideals", *Journal of Psychology*, 132, 87-94.
Batten, M. 1992. *Sexual Strategies: How Females Choose Their Mates*. New York: Tarcher/Putnam.
Bealers, J.O. 1986. *Limits of Scientific Psychiatry: The Role of Uncertainty in Mental Health*. Bruner/Mazel, 1986.
Beck, K. 2000. "The Thin Men", *Running Times*, April, 61-65.
Belsky, J., Steinberg, L., & Draper, P. 1991. "Childhood Experience, Interpersonal Development, and Reproductive Strategy: An Evolutionary Theory of Socialization", *Child Development*, 62, 647-670.
Blurton Jones, N. 1997a. "Bushman Birth Spacing: A Test for Optimal Interbirth Intervals", in Betzig, L., ed., *Human Nature: A Critical Reader*. New York: Oxford, pp. 73-82.
Blurton Jones, N. 1997b. "Too Good to be True? Is There Really a Trade-off Between Number and Care of Offspring in Human Reproduction?", in Betzig, L., ed., *Human Nature: A Critical Reader*. New York: Oxford, pp. 83-86.
Boone, J.L. & Kesler, K.L. 1998. "More Status or More children? Social Status, Fertility Reduction, and Long-Term Fitness", *Evolution and Human Behavior*, 30, 257-277.
Buss, D.M. 1999. *Evolutionary Psychology: The New Science of the Mind*. Needham Heights, MA: Allyn & Bacon.
Buss, D.M. 1992. "Mate Preference Mechanisms: Consequences for Partner Choice and Intrasexual Competition", in Barkow, J., Cosmides, L., & Tooby, J., eds., *The Adapted Mind*. New York: Oxford University Press.
Buss, D.M. 1994. *The Evolution of Desire: Strategies of Human Mating*. New York: Basic Books.
Buss, D.M. 1985. "Human Mate Selection, *American Scientist*, 73, 47-51.
Calabrese, L.H., Kirkendall, D.T., Floyd, M., Rapoport, S., Williams, G.W., Weiker, G.F., & Bergfeld, J.A. 1993. "Menstrual Abnormalities, Nutritional Patterns, and Body Composition in Female Classical Ballet Dancers", *Physician and Sports Medicine*, 11, 86-98.
Campbell, A. "Staying Alive: Evolution, Culture, and Women's Intrasexual Aggression", *Behavioral and Brain Sciences*, 22, 203-252.
Caspi A. & Moffitt, T.E. 1991. "Individual Differences Are Accentuated During Periods of Social Change: The Sample Case of Girls at Puberty", *Journal of Personality and Social Psychology*, 61, 157-168.
Chisholm, J.S. 1996. "The Evolutionary Ecology of Attachment Organization", *Human Nature*, 7, 1-38.
Colino, S. 2000. "Fear of Fat", *Young and Modern* (*YM*), October, 48-49.
Cosmides, L. & Tooby, J. 1992. "Cognitive Adaptations to Social Exchange", in Barkow, J. et. al., eds., *The Adapted Mind*. New York: Oxford University Press, 163-228.
Condit, V.K. 1990. "Anorexia Nervosa: Levels of Causation", *Human Nature*, 1, 391-413.

Copeland, P.M. 1985. "Neuroendocrine Aspects of Eating Disorders", in Emmett, S., ed., *Theory* and Treatment of Anorexia Nervosa and Bulimia: Biomedical, Sociocultural, and *Psychological Perspectives.* New York: Bruner/Mazel, 51-72.

Crawford, C.B. 1989. "The Theory of Evolution: Of What Value to Psychology?", *Journal of Comparative Psychology*, 103, 1, 4-22.

Cutler, W.B., Preti, G., Krieger, A., Huggins, G.R., Garcia, C.R., & Lawley, H.J. 1986. "Human Axillary Secretions Influence Women's Menstrual Cycles: The Role of Donor Extracts From Men", *Hormones and Behavior*, 20, 463-473.

Dionne, L. 1998. "Body Image Breakthrough", *Shape*, 73-77, 131.

Dolan, B.M., Lieberman, S., Evans, C., & Lacey, J.H. 1990. "Family Features Associated with Normal Weight Bulimia", *International Journal of Eating Disorders*, 9, 283-288.

Duhem, P. 1998. "Physical Theory and Experiment", in Kourany, J.A., ed., *Scientific Knowledge: Basic Issues in the Philosophy of Science*, Albany, NY: Wadsworth, pp. 187-194.

Ellis, B.J. & Garber, J. (in press). "Psychosocial Antecedents of Variation in Girls' Pubertal Timing: Maternal Depression, Stepfather Presence, and Marital and Family Stress", *Childhood Development.*

Farber, S.L. 1981. *Identical Twins Reared Apart: A Reanalysis.* New York: Basic Books.

Flannery, D.J., Rowe, D.C., & Gulley, B.J. "Impact of Pubertal Status, Timing, and Age on Adolescent Sexual Experience and Delinquency", *Journal of Adolescent Research*, 8, 21-40.

Garfinkel, P.E. & Garner, D.M. 1982. *Anorexia Nervosa: A Multidimensional Perspective.* New York: Bruner/Mazel.

Gosselin, A. 1998. (unpublished manuscript). "On the Logic and Metaphysics of Anorexia".

Graber, J.A., Lewinshon, P.M, Seeley, J.R, & Brooks-Gunn, J. 1997. "Is Psychopathology Associated With the Timing of Pubertal Development?", *Journal of the American Academy of Child and Adolescent Psychiatry,* 36, 1768-1776.

Grissett, N.J. & Norvell, N.K. 1992. "Perceived Social Support, Social Skills, and Quality of Relationships in Bulimic Women", *Journal of Consulting and Clinical Psychology*, 60, 293-299.

Hill, K. & Hurtado, A.M. 1996. *Ache Life History.* Hawthorne, NY: Aldine-de Gruyter.

Holcomb, H. 1993. *Sociobiology, Sex, and Science.* Albany, NY: SUNY Press.

Howell, N. 1979. Demography of the Dobe !Kung. New York: Academic Press.

Ince, S. 1999. "Sleep Disturbances: A Special Report", *Harvard Health Letter*, Boston, MA: Harvard Medical School Health Publications Group.

Jones, D. 1996. "An Evolutionary Perspective on Physical Attractiveness", *Evolutionary Anthropology*, 5, 97-109.

Kabat-Zinn, J. *Full Catastrophe Living: Using the Wisdom of Your Body and*

Mind to Face Stress, Pain, and Illness. New York: Delta/Bantam.
Kampert, J.B., Whittemore, A.S., & Paffenbarger, R.S. 1988. "Combined Effects of Childbearing, Menstrual Events, and Body Size on Age-specific Breast Cancer Risk", *American Journal of Epidemiology*, 128, 962-979.
Kaneer, A., Bulik, C.M., & Sullivan, P.F. 1993. "Abuse in Adult Relationship of Bulimic Women", *Journal of Interpersonal Violence*, 8, 52-63.
Kaprio, J., Rimpela, A., Winter, T., Viken, R.J., Rimpela, M., & Rose, R.J. 1995. "Common Genetic Influence on BMI and Age at Menarche", *Human Biology*, 67, 739-753.
Ketelaar, T. & Ellis, B.J. 2000. "Are Evolutionary Explanations Unfalsifiable? Evolutionary Psychology and the Lakatosian Philosophy of Science", *Psychological Inquiry*, 11, 1-21.
Kog, E. & Vandereycken, W. 1989. "Family Interaction in Eating Disorder Patients and Normal Controls", *International Journal of Eating Disorders*, 8, 11-23.
Lacey, J.H. 1992. "Homogamy: The Relationships and Sexual Partners of Normal-weight Bulimic Women", *British Journal of Psychiatry*, 161, 638-642.
Low, B.S. 1992. "Sex, Coalitions, and Politics in Preindustrial Societies", *Politics and the Life Sciences*, 11, 63-80.
Mealey, L. 2000. *Sex Differences: Developmental and Evolutionary Strategies*. San Diego, California: Academic Press.
National Livestock and Meat Board. 1992. *Mirror-Mirror: A Resource Guide for Helping Adolescents Develop a Positive Body Image and Maintain and Healthy Weight*. Chicago, Illinois: Education Department of National Live Stock and Meat Board.
Ness, R. 1991. "Adiposity and Age of Menarche in Hispanic Women", *American Journal of Human Biology*, 3, 41-48.
Nesse, R.M. 1999. "Our Bodies are Imperfect, For Good Reasons", *Human Ethology Bulletin*, 14 (4), 1-6.
Nesse, R.M. & Williams, G.C. 1994. *Why We Get Sick*. New York: Times Books Random House.
Polivy, J. 1998. "The Effects of Behavioral Inhibition: Integrating Internal Cues, Cognition, Behavior, and Affect", *Psychological Inquiry*, 181-204.
Sabura Allen, J. 1997. (unpublished Ph.D. dissertation). Eugene, Oregon.
Scalf-McIver, L. & Thompson, J.K. 1989. "Family Correlates of Bulimic Characteristics in College Females", *Journal of Clinical Psychology*, 45, 467-472.
Singh, D. 1994. "Is Thin Really Beautiful and Good? Relationship Between Waist-to-Hip Ratio (WHR) and Female Attractiveness", *Personality and Individual Differences*, 16, 465-481.
Stern, S.L., Dixon, K.N., Jones, D., Frensche, P.L., & Rodin, J. 1989. "A Prospective Study of Disordered Eating Among College Students", *International Journal of Eating Disorders*, 5, 499-509.
Stevens, A. & Price, J. 1996. *Evolutionary Psychiatry*. London: Routledge.

Surbey, M.K. 1987. "Anorexia Nervosa, Amenorrhea, and Adaptation", *Ethology and Sociobiology*, 8, 47-61.
Surbey, M.K. 1998. "Developmental Psychology and Modern Darwinism", in Crawford, C. & Krebs, D.L., eds., *Handbook of Evolutionary Psychology: Issues, Ideas, and Applications*. Mahwah, NJ: Erlbaum, pp. 369-404.
Susman, E.J., Nottleman, E.D., Inoff-Germain, G.E, Loriaux, D.L., & Chrousos, G.P. 1985.
"The Relation of Relative Hormonal Levels and Physical Development and Social-emotional Behavior in Young Adolescents", *Journal of Youth and Adolescence,* 14, 245-264.
Tooby, J. & Cosmides, L. 1992. "The Psychological Foundations of Culture", in Barkow, J. et. al., eds., *The Adapted Mind.* New York: Oxford University Press, pp. 12-136.
Thornhill, 1998. "Darwinian Aesthetics", in Crawford, C. & Krebs, D.L., eds., *Handbook of Evolutionary Psychology: Issues, Ideas, and Applications*. Mahwah, NJ: Erlbaum, pp. 543-572.
Treloar, S.A. & Martin, N.G. 1990. "Age at Menarche as a Fitness Trait: Nonadditive Genetic Variance Detected in a Large Twin Sample", *American Journal of Human Genetics*, 47, 137-148.
Udry, J.R. 1979. "Age at Menarche, at First Intercourse and at First Pregnancy", *Journal of Biosocial Science*, 11, 433-441.
Udry, J.R. & Cliquet, R.L. 1982. "A Cross-cultural Examination of the Relationship Between Ages at Menarche, Marriage, and First Birth", *Demography,* 19, 53-63.
Vihko, R.K. & Apter, D.L. "The Epidemiology and Endocrinology of the Menarche in Relation to Breast Cancer", *Cancer Survey*, 5, 561-571.
Voland, E. & Voland, R. 1989. "Evolutionary Biology and Psychiatry: The Case of Anorexia Nervosa", *Ethology and Sociobiology*, 10, 223-249.
Walters, S. & Lindberg, T. 1994, June. "Conscious and Unconscious Reproductive Suppression", paper presented at the 6th Annual Meeting of the Human Behavior and Evolution Society, University of Michigan, Ann Arbor.
Wasser, S.K. 1990. "Infertility, Abortion, and Biotechnology: When It's Not Nice to Fool Mother Nature", *Human Nature*, 1, 3-24.
Wasser, S.K. & Barash, D.P. 1983 "Reproductive Suppression Among Female Mammals: Implications for Biomedicine and Sexual Selection Theory", *Quarterly Review of Biology*, 58, 513-538.
Watzlawick et. al. 1967. *Pragmatics of Human Communications*. New York: Norton.
Watzlawick et al. 1974. *Change: Principles of Problem Formation and Problem Resolution*. New York: Norton.
Wegner, D.M. & Smart, L. 1997. "Deep Cognitive Activation: A New Approach to the Unconscious", *Journal of Consulting and Clinical Psychology*, 65, 984-995.
Wegner, D.M. & Wenzlaff, R.M. 1996. "Mental Control", in Higgins, E.T. &

Kruglanski, A.W., eds., *Social Psychology: Handbook of Basic Principles*. New York: Guilford.

Wegner, D.M. 1994. "Ironic Processes of Mental Control", *Psychological Review*, 101, 34-52.

Wellens, R., Malina, R., Roche, A., Chumlea, W., Guo, S., & Siervogel, R. 1992. "Body Size and Fatness in Young Adults in Relation to Age at Menarche", *American Journal of Human Biology*, 4, 783-787.

Wenzlaff, R.M. & Wegner, D.M. 1998. "The Role of Mental Processes in the Failure of Inhibition", *Psychological Inquiry*, 9, 231-233.

CHAPTER 7

PATERNAL INVESTMENT OR MATERNAL INVESTMENT? A CRITIQUE OF THE PARENTAL INVESTMENT HYPOTHESIS IN AN AMERICAN POLYGAMOUS COMMUNITY

WILLIAM JANKOWIAK
University of Nevada

CLAUDIA WOODMAN
University of Las Vegas

Abstract. In this paper we will argue that there are a multiplicity of motives that affect the formation of women's parenting strategies in an American polygamous community. While biological factors may be compelling, unusual circumstances can modify or supersede parenting propensities. In this setting a woman may work to elicit her husband's parental contribution less through developing a love bond and more through the achievement of other statuses such as occupation success or dominance over other co-wives. Both of these avenues provide a woman with access to additional resources that she may then extend to her children or keep entirely for herself.

In presenting this argument we will concentrate on the politics of namesaking as a primary means for examining the interplay between parental investment concerns, cultural beliefs, and identity formation in two polygamous communities. The motives behind men's and women's namesaking decisions will be examined in order to provide a richer, more complete explanation for competing sex-linked parenting strategies. We will primarily focus on Angel Park, the oldest and largest Mormon Fundamentalist polygamous community in North America. Supplemental data from a Mormon polygamous community in northern Mexico will also be evaluated. The data presented in this chapter were collected between 1994 and 1998, during a four-year study of

H.R. Holcomb III (ed.), Conceptual Challenges in Evolutionary Psychology, 271–290.

Angel Park, and in a one-month study conducted in 1995 in the Juarez colony of northern Mexico.

INTRODUCTION

Despite the reluctance of cultural anthropologists to explore certain interfacings between biology and culture, several pioneering researchers, working primarily from a behavioral ecological perspective, have sought to document the shaping power of evolution on human behavior. The focus of this new scholarship contrasts with that of earlier studies that categorically ignored, or rejected, biological explanations. The new evolutionary studies document the existence of underlying mental templates or innate propensities as found in mate selection criteria (Buss, 1994; Symons, 1979; Jones, 1996), parenting behavior (Hewlett, 1992), kin selection (Chagnon & Irons, 1988; Betzig, Borgerhoff Mulder, & Turke, 1988), homicide patterns (Daly & Wilson 1984), and infanticide decisions (Hausfater & Hrdy, 1984; Trivers, 1972).

From an evolutionary ecological perspective, sexual conflict arises out of men's and women's competing and, at times, conflicting reproductive interests and parenting concerns (Chisholm, 1999; Hrdy, 1999; Low, 2000). For a man, conflicts may arise from any number of sources, from striving to invest in existing children, to cultivating resources to satisfy and thus hold onto a spouse, or to utilizing resources to acquire another spouse or mistress. For a woman, conflicts may arise from striving to obtain resources for her children, or cultivate a relationship with her husband, or work outside the home and, in the polygamous family, maintain her social position with other co-wives. Women's conflicts often revolve around the issues of paternal investment, the reassurance of paternal certainty, or something entirely different—the assertion of status within the family and position within the community. To date, most researchers have focused on the factors that shape men's perceptions of their offspring as sources of quality or as resources of quantity. In addition, this research has explored women's motives primarily as they pertain to obtaining resources from men or providing care for their children. There are no studies that examine how a harem (or Big House) polygamous family environment may shape women's motives, strategies and, thus, parenting behaviors.[1]

LIFE HISTORY THEORY AND AMERICAN NAMING PATTERNS

When the life history theory or parental investment model was first introduced, mate investment (e.g., attracting, keeping, and guarding a spouse) and parental investment (e.g., providing emotional and material resources for the offspring) were used to analyze male behavior in primate and human populations (see overview in Hewlett, 1992). The groundbreaking research of Smuts and Gubernick (1992, p.21) found that male parental care giving is often motivated more by mating concerns than by interest in the offspring per se. Hewlett (1992, p.5) aptly noted that fathers, who expend more parental effort than mating effort, practice a "dad" reproductive strategy (i.e., focus on developing a close relationship with their children), whereas fathers with the reverse pattern engage in a "cad" reproductive strategy (i.e., ignore their children in favor of obtaining additional mating opportunities). Since then there have been numerous studies of human and non-human populations seeking to identify possible factors for engendering one or the other strategy.

The life history, or parental investment hypothesis, has been explored primarily, but not exclusively (the exceptions being the Aka pygmies and Kung Bushmen) in societies where females are entirely dependent upon men for material support. These studies overwhelmingly focus on the factors that form men's decisions, but not women's, to engage in reproductive or parenting behavior, despite the fact that parental investment has two components: paternal and maternal. This bias in the research has made maternal interest coterminous with paternal interest.

Recently, a few scholars have sought to apply the inclusive fitness model to the presence of a sex-linked bias in American naming practices (MacAndrew, 1995; Johnson et al., 1991). They have looked at naming as a way to understand motives behind parents' naming decisions. Although most cultures do not use names to reinforce the father-child bond, in cultures where naming practices are not institutionalized men and women tend to use names to reinforce the father's perception of parental certainty.[2]

From an ecological evolutionary perspective, names are interpreted as representing a form of parental input (Evans, 1990) that "enhances the sense of relatedness experienced by the parents and other relatives" (MacAndrew, 1995, p.5) and, therefore, constitutes a kind of indirect parental investment. Accordingly, naming serves as a form of reassurance to induce "others to reinforce and facilitate the parent's own parenting behaviors" (Johnson et al., 1991).

A pioneering study of naming practices of white middle-class women living in the Chicago area (Rossi, 1965), for example, found that 61 percent of all children were named after kin, with 65 percent of sons and 22 percent of daughters named after the father. This is consistent with Smith's historical study, which found that 60 percent of first-born sons were named after the father. In a study of 323 unmarried teenage mothers, Furstanberg and Tralvitie (1980) found that, when the mother named the child after a relative, more than fifty percent of the children were given the father's surname, with boys being more likely than girls to be so named. They also surmised that women use naming to encourage the development of future bonding between father and newborn. This view is supported by the views of a Sacramento, California social worker who stated that at least 70 percent of the more than one hundred unmarried women who made up her caseload had named their children after the father to facilitate future contact between the two (interview conducted April 11, 1998).

Building upon these findings, Johnson, MacAndrew and Harris (1991) compared the sources of the names of natural and adopted children. They found that adoptive parents were much more anxious about others' perceptions about genetic relatedness. Hence, in order to increase public "perception of the child's genetic relatedness to the parents," adopted children were more likely than natural children to be named after the father (Johnson et al., 1991, p.369).

Francis MacAndrew, in a separate study (1995), suggested that namesaking is most likely to arise whenever the issue of paternal certainty is acute, or when there is a high level of anxiety about a father's willingness to become emotionally, or materially, involved in the raising of his children. He further noted that mothers are more active in choosing a paternally linked name for their child as a strategy to insure continual provision of resources from the father. Because MacAndrew does not explicitly explore the maternal corollary of this concern, how an environment of secure economic support might impact a woman's naming decisions remains unclear. It is this half of the equation that we wish to explore below.

Multiple Motives/Multiple Models: Predictions and Implications

We will gauge which of these explanations makes the most sense of the data on polygamous naming practices. First, there is a parental investment hypothesis which holds that men and women seek to attract, hold, and link spouses, relatives, and friends in the care of their children. In particular, we will explore the parental certainty hypothesis with its assumption that males are more anxious about paternal certainty, or cuckoldry, and, thus, feel a greater need of reassurance about their offspring's paternal identity. The second is a maternal investment hypothesis, which holds that women pursue a more broad-based strategy to elicit support from both maternal and paternal kin. Concomitant to this is the notion that they use the naming of their children to assert a sense of their identity. This is particularly true where a woman's autonomy is undermined and a variety of strategies as compensation come into play. A cultural explanation posits that the mind is a byproduct of the relationship between cosmology, gender, and naming practices. Given the Mormon patriarchal ethos, a strong patrilineal child-naming pattern is indicated, with an especially strong father-son naming bias.

If the parental investment model, at least as it has been applied in other naming studies conducted in the United States, is universally applicable, the women of Angel Park should be as concerned with establishing namesake linkage between the fathers and their offspring as women in other studies. Three phenomena should occur. First, if paternal certainty is a concern, Angel Park women should use namesakes that link children to their fathers. Second, if the fundamentalist cosmological model is the primary motivational force, there should be a pronounced tendency toward patriarchal naming, especially among male children. Third, a trend should be seen whereby the more secure a woman's position within the family, the greater the likelihood that she will use non-paternal names as a source of her children's names. We further posit that each type of wife, whether she is assertive, favorite or independent, will demonstrate different naming patterns that reveal the success or failure of their competitive struggles for power, status, autonomy, and love within the plural family.

THE POLYGAMOUS COMMUNITY

It is necessary to discuss at the outset several elements of the fundamentalist Mormon religious creed because they provide significant and salient components of the culture's cosmology. Fundamentalist Mormon theology is grounded in the teachings of three books: The Bible, The Book of Mormon, and The Doctrine and Covenants. The latter two books are prescribed to members of the Mormon Church as Holy Scripture, the words of God as they were revealed directly to Joseph Smith (Musser, 1944).

There are several non-negotiable tenets at the core of Mormon theology. One, God is a polygamous man who loves all his children but confers on men, in particular, an elevated spiritual essence which ensures that men who live "righteously" will obtain a higher spiritual standing. Two, men occupy leadership positions in their families and on the church council and have the potential, in the next life, to become god-heads with dominion over all their descendants. Within this cosmological creed, the father is charged with the duty to constantly expand his kingdom by entering into the institution of plural marriage (Musser, 1944).

Women's official standing, on the other hand, is determined by their performance in the highly valued complementary roles of mother and wife. Like Southern Baptists, Mormon fundamentalists interpret the scripture literally, "A woman should submit herself graciously" to her husband's leadership and a husband should "provide for, protect and lead his family" (New York Times June, 1998, p. A1).

In fundamentalism, women achieve salvation through obedience, first to their fathers and, then, to their husbands by becoming sister-wives (i.e., co-wives) in a "celestial," or plural, family. The marital contract "seals" a man and woman together "for time and eternity" in the Heavenly Kingdom (Musser, 1944). Because this bond extends beyond the grave into an eternal world, it is in a woman's "best interest to advance her husband's interests" (i.e., she should bear a large number of children) (Bohannan, 1985, p.81), while at the same time striving to uphold her husband's authority, especially in front of his children.

A second tenet holds that the father-son relationship is the core axis for the transmission of cultural and spiritual essence. First articulated by Joseph Smith in 1832, this tenet is a "theme that predominates throughout the Book of Mormon" (Clark and Clark 1991, p. 286). It is based on the belief of a Melchizedek priesthood whose lineage, extending back to Adam, recognized by fundamentalism as the only legitimate religious authority. This belief represents the church's insistence that the only acceptable basis or form of religious expression is based on a patriarchal social organization.

The polygamous family's behavioral expectations are derived from these theological axioms, which uphold men as the religious center and authority in the family. From an organizational perspective, serious and consistent familial attention should be on the father as both the ultimate adjudicator of family affairs and the representative of spiritual authority. His centrality is routinely reinforced as he leads the family in Sunday school service, conducts daily family prayers, arranges the marriages of his children, and disburses the family income.

In contrast, mothers are seen as a constancy of personal warmth, individual nurturance, and everyday intimacy. They embody strength and continuity but are not perceived as forces to contend with. The church's teaching that reproduction and childcare are a woman's primary duties imposes this perception. The qualified value of a mother, her relativity of place in the official culture, quickly becomes an unqualified one whenever a male or female interacts with his or her own birth mother. In this domain, mothers, unlike fathers, are seldom feared or rejected. They are the emotional, rather than the symbolic, glue that holds the polygamous family together. More importantly, mothers are spared the difficult, and often troubling, pressures of adoration or reverence that fathers are subject to.

A woman's unofficial, or earthly status, is much higher than her theological status. This stems from the realities of the polygamous marriage system where there is always a shortage of women. The scarcity of women lends a communal readiness to forgive and welcome back both wayward young and mature women who have strayed or left the community. The fundamentalist community is less likely to allow back wayward young men because they are often regarded with suspicion and because of the perennial difficulty of finding a wife or wives from within the community.

Church doctrine dictates that women must be taken care of. Any woman who requests to remarry will be immediately assigned and accepted into another man's family. Since a man's social standing is based on the number of wives and offspring he has, the additional wife, and any children, serve to enhance his social status. It is also understood that women who work or receive public assistance bring income into the family. Women are, therefore, seldom perceived as an economic drain or liability.

Fundamentalists hold the conviction that they are God's chosen people born to live "the fullness of the Gospel" and, thus, to create God's ideal. The polygamous family (Baur, 1988) lies at the heart of the fundamentalist communities' communitarian impulses to create a spiritually unified and socially harmonious system of order. The maintenance of this order depends on the strength and vitality not only of father-son relationships and mother-children relationships, but also of the relationships between co-wives. The plural family is held together as much by a collective will or a communal effort to maintain a strong image of a harmonious family as it is by any other action or behavior pattern.

Despite communitarian dictates against status discrepancies, the status system of the polygamous family is often organized hierarchically around criteria that the women develop among themselves. These criteria can include a husband's favorite wife, a wife's cooking ability, the number of her offspring, her monthly income, the level of her education, the quality of her choir singing and a host of other attributes. For some women, the relationship with sister-wives can be more important than their relationship with their husbands. For others, the primary preoccupation is with their husband, and not with other wives who they learn, at best, to tolerate. At no time, however, is intolerance of any degree between wives officially sanctioned.

Because co-wives are often in competition for their husband's attention, they contribute to the process of idealization of the father. In this struggle for predominance, children are often used as an indirect route to their husband's attention. Because the wives focus attention on their husband, the children, wanting to please their mother, follow suit. By cultivating father adoration, mothers hope to establish their uterine (i.e., based on women and their biological offspring) family's theological superiority in the eyes of the community.

On the other hand, women tacitly seek to balance their duties to uphold their husband's public image, while at the same time looking for ways to clarify and demarcate their individual status and their own uterine lines as distinct and valuable entities within the often-contentious polygamous family. One way women seek to accomplish this is by resisting the doctrinal mandate to surrender their children into the melting pot of their husband's progeny. This struggle over balance is sharply manifested in the naming decisions made over the course of their marriages.

As one woman noted: "We compete over everything, why not names too?" For example, if wife A knows that wife B is planning to name her next child Jessie James and wife A gave birth first, she might give her child that name, not because she liked it, but rather to assert her dominance over the co-wife. Another possible scenario is that a first wife, upon discovering that her husband intends to take another wife, names her next child after him. One second wife reported that a first wife did this very thing "in order to prevent me from using it to name my son." In another case, a third wife's plan to name her daughter after her best friend unraveled when the fourth wife insisted that she had always wanted that name for her child. In this way and others, later wives are as interested in establishing linkages through naming as earlier wives, but are more restricted in their efforts.

Depending on personality type, different strategies are used in order to advance a woman's interests. In both Angel Park and the Juarez Colony, there is a broad continuum of personality types of wives that range from assertive to favorite to low profile. These ideal types are not mutually exclusive. They often overlap and may fluctuate over a woman's life course in that a wife may seek to or be required to play a different role.

The assertive wife (or, in the vocabulary of the Angel Park women, the Queen Bee) actively seeks to organize and manage the daily family affairs for everyone. She seeks, through force of will and domestic and emotional skill, to control not only her children and her husband, but also her sister wives and their offspring. A favorite wife monopolizes most of her husband's time and resources and, therefore, makes a show of her most prominent position within the family. This succeeds because, for the moment at least, she captures much of her husband's attention. Accordingly she adopts the persona of the completely obedient and devoted wife, a posture she can easily assume, since she will be granted almost everything she wants because of her husband's emotional bond with her. The favorite wife is ambitious, and while not as openly vocal and outspoken as the assertive wife, is highly motivated to guide and direct the family. The difference between the two types is in strategy of approach and the quality of relationship that is established and maintained with their husbands. The low-profile wife, on the other hand, is at the bottom rung of the family, the frequent recipient of criticism from co-wives. She prefers to be a trooper who follows conventions while at the same time yearning for greater contact and emotional intimacy with her husband. These personality types give rise to different motives, including motives for naming practices.

ETHNOGRAPHIC METHODS

Naming is a salient activity in the Mormon polygamous community. People of all ages enjoy talking about it and are aware of who they are named after and why. There is no normative expectation regarding naming in Angel Park or Juarez. The folk notion is that naming is completely arbitrary and individualistic. Everyone insists that parents can and do give a child name they want. Once we informed people about a particular namesaking pattern, however, there was a near consensus as to its reasons. This consensus suggests that, at the tacit level, there exits a cultural uniformity of understanding regarding the motivation behind naming practices.

Previous naming studies examined first names but not middle ones; we wanted to do both.[3] We chose not to incorporate girls' maiden names into our sample, as we wanted to understand the politics of personalization and kin affiliation. About a third of the girls in our sample did not have a middle name because it is expected that, after marriage, a woman will use her father's surname as a middle name. The principal investigator collected the information with the assistance of several members of Angel Park and the Juarez colony. Women were asked the sources of all their children's names as well as who chose the name.

The strategy for our sample consisted of the "snowball" technique of following friends to other friends and relatives. It allowed us to contact 40 of the 320 families or 12.5 percent of the families living in Angel Park and 7 of 68 families, or 10.3 percent, in the Juarez colony.

From these families we obtained a total of 580 names for 320 children. There are forty husbands and 100 wives in our sample. Kin names form 49 percent of the total names in our sample, with 58 percent of them associated with the paternal side and 42 percent with the maternal. Kinship is highly salient in this culture.

"Nice names" (i.e., personalized names with no specific affiliation), comprising 27 percent of all names, tend toward the idiosyncratic and thus constitute a highly personal form of identity construction. There is no gender bias in the use of nice names, with 38 percent of all females and 34 percent of all males being given one. The other name groupings in order of frequency are: namesaking a friend (n=58 or 9.5 percent), a celebrity (n=16 or 3 percent), a Mormon leader (n=12 or 2 percent), and a community leader (n=8 or 1 percent). Still, it is significant that almost 50 percent of all names are those of kin.

Results

Results follow in Tables 1 through 4.

Table 1: The First Wives Reproductive History (Total: 37)

Top birth cohort (children 1-3):		
	Sons	30 of 51 (or 59%) - father side
		11 of'51 (or 22%) - mother side
		9 of 51 (or 18%) - other
	Daughters	18 of 55 (or 33%) - father side
		21 of 55 (or 38%) - mother side
		16 of 55 (or 29%) - other
Middle cohort (children 4-6):		
	Sons	17 of 42 (or 40%) - father side
		19 of 42 (or 45%) - mother side
		6 of 42 (or 14%) - other
	Daughters	7 of 28 (or 25%) - father side
		13 of 28 (or 46%) - mother side
		8 of 28 (or 29%) - other
Bottom cohort (children 7-11):		
	Sons	11 of 25 (or 44%) - father side
		3 of 25 (or 12%) - mother side
		7 of 25 (or 48%) - other
	Daughters	1 of 17 (or .06%) - father side
		7 of 17 (or 41%) - mother side
		9 of 17 (or 53%) - other
Total:*(200 children)		

Table 2: Name choice: (T= 681)

Mother chose 31% (n= 214)
Father chose 38% (n= 261)
Both chose 29% (n= 193)
other chose 2% (n= 11)

Table 3: Wife Type, Child's Sex, and Descent Affiliation

	Sons	Daughters
Favorite:	P: 33	P: 8
	M: 15	M: 21
Assertive:	P: 36	P: 6
	M: 22	M: 23
Low Profile	P: 56	P: 26
	M: 19	M: 32
Total:	P: 125 (69%)	P: 40 (42%)
	M: 56 (31%)	M: 66 (58%)

Table 4: Descent and Sex of Child

Sons:	Paternal Kin: 167 (70%) Maternal Kin: 71 (30%) Total: 238
Daughters:	Paternal Kin: 36 (23%) Maternal Kin: 74 (67%) Total: 110

Discussion

Our study of naming patterns in the Mormon polygamous communities shows a strong association between gender, birth order, and the frequency in which the child is named after the father or a paternal relative (see table 1). More first-born sons (60%) compared to first-born daughters (36%) are namesaked after paternal kin, with 38 percent (n=66 out of 172) of sons named after their father and only 14 percent (n=19 out of 137) of daughters named after their mother. There is a close association of sons with paternal relatives (70%) and daughters with maternal relatives (68%) (see table 4). This suggests that the father-son namesaking bond is significant as representative of the patriarchal principle, a route to role and ego enhancement and as signifiers of possible parental investment concerns.

Although our findings are consistent with those of other namesaking studies, closer analysis demonstrates that the underlying motivation for many namesaking decisions cannot be attributed to the cultural tradition or derived from a woman's strategy to address a man's concern with paternal certainty. There are other motives present.

The best explanation for the naming patterns and their variations appears to be a maternal investment hypothesis with its emphasis on women's decisions to focus on procurement of resources through engaging in strategies that range from becoming a favorite wife to establishing economic self-sufficiency to dominating the tacit female-centered domestic hierarchy.

It is in the micropolitics of naming and kinship association that a wife's personality and status in the family hierarchy are revealed. In these struggles, women use children as a means to achieve several ends: To maintain their husband's attention to themselves and their children, to strengthen affective bonds with natal kin, and to demarcate themselves as distinct autonomous units within the polygamous family.

Support for the maternal investment hypothesis can be found in the shift in naming practices between the founder generation and the subsequent generations. Members of the founder generation (i.e., born prior to World War II) were raised primarily in monogamous Mormon households and named their firstborn sons after a paternal kin only 50 percent of the time. This frequency is less than the 1940s American national average of 60 percent. In contrast, 80 percent of all first-born sons born after 1950 in Angel Park or the Juarez Colony were namesaked after paternal kin.

The cultural explanation with its normative emphasis on social continuity cannot account for the post World War II shift in the namesaking pattern of first-born sons. We believe that the increased emphasis on paternal-son namesaking arises out of living privately and publicly within the polygamous family system, which produces an institutionalized anxiety causing women to become overtly competitive with one another for public status, independence, and autonomy. This family reality was absent in the founder generation, which grew up primarily within a monogamous family environment, suggesting that the father-son naming association is a response to the new social condition typical of a polygamous family environment within the newly formed theocratic communities of Angel Park and Juarez Colony.

Unlike their monogamous counterparts (see Rossi 1965), polygamous women do not appear to become more comfortable, confidant, or content with their status in the marriage over time.[4] We suspect the primary motivation for the reversal back to a paternal-son emphasis in namesaking does not arise from new concerns with enhancing "public perception of genetic relatedness" (MacAndrew, 1995, p.11), but rather from the desire to celebrate something polygamous women have dedicated much of their earthly and spiritual resources to: the fundamentalist religion. Also significant is another motivation, more individualistic or personal: defense of their family status against the potential threat, actual or imagined, of a new wife in the household. For example, we found that first wives from the founder generation (married prior to 1950) who did not have a son named after their husband tended to change their attitude as they approached the end of their reproductive careers, particularly when an additional wife had entered the family. In this context, women's motivation for namesaking is directed more toward the entering wife than toward her husband. The earlier wife wishes to assert power through reproduction and demonstrating her authority and superiority in the polygamous family. In this way paternal namesaking may serve as a kind of closure to a woman's reproductive career, which also began by highlighting the father-son relationship.

A wife's personality and position within the larger family also accounts for the variation found in namesaking practices (see table 3). Both the favorite and the assertive wife are much more dominant players in the family than the low-profile wives. The favorite wife, whose status depends upon the continuation of her emotional bond with her husband, defends, advances, and upholds her husband's self-image as the patriarchal head of the family. To this end, she actively lobbies for her sons to be named after her husband or his paternal kin. Her concern has less to do with parental investment and more to do with mate investment. She offers her sons as an extension of honor to her husband; in turn, she expects that he will honor and adore her. A secondary, albeit important, motive is to retain the support of her husband for her children. As she becomes more secure in her marriage, the favorite wife begins either to lobby for, or to choose her children's names so that they link the child to her natal family. (see table 2).

In contrast, the assertive wife, at least at the start of her reproductive career, breaks with the more typical namesaking emphasis on the father-son association. Over 42 percent of the assertive wives' sons are namesaked after their maternal kin (11 of 26). Clearly, the assertive wife is not interested in paternal reassurance or paternal investment. However, at the end of her reproductive career, the assertive wife, perhaps due to the marginality of her marriage, overwhelmingly prefers that her sons be named after a paternal relative.

On the other hand, the low-profile wife, already out of the running for a valued place within the family hierarchy, continues to demonstrate her commitment to her religion by namesaking her children after her husband's paternal kin. In this way, the low-profile wife, like parents of adopted children (Johnson et al., 1991), utilizes kin names to reaffirm paternal bonds, and continues to assert her dedication to the cosmology by having her husband name her sons (48 of 54) while she names her daughters (66 of 85).

Names, then, can be the currency by which autonomy is asserted and religious commitment is reaffirmed. The fact that 45 percent of all children in our sample are named after maternal kin indicates that women constitute key symbolic figures in these communities. Excluding Native American cultures, which are organized around principles of matrilineal descent, the maternal namesake association present in this polygamous community may be the highest percentage found in any American subculture. The degree to which women are valued, despite efforts to render them culturally insignificant, was further revealed in a content analysis of the community's calendar, which found that 45 percent (n=64 of 160) of all birth dates noted are of local matriarchs or famous historical women.

Contrary to the fundamentalist folk assumptions that bestow value on patriarchal authority, women are actively participating in the naming process. Table 2 shows that 35 percent of naming decisions are cooperative efforts. It can be assumed that cooperative efforts imply that their husbands agree to women's choices. In this way, wives are involved in the naming decision more than 60 percent of the time. As more wives come into the family, men become more flexible in adhering to the patriarchal ideal and, thus, more willing to delegate the naming decision to later wives. Our finding stands in opposition to the patriarchal ideal of absolute male leadership and dominance in all situations.

What is the man's motivation for allowing his wife to participate in the naming of his children or to choose the name outright? In the polygamous community, it is incumbent upon men to orchestrate cooperative behavior. Granting women some control over the naming process is a relatively easy way to appease, reassure, and otherwise honor them.[5] Still, it is not clear whether the motivation for a husband to name a child after a wife's maternal kin is part of his overall mating or parenting strategy, a means to maintain familial cohesion, or merely indifference. It is clear, however, that, from the women's perspective, the association of daughters with their maternal kin has a deeper symbolic value.

We suggest that the emphasis on daughters arises from the fact that daughters are theologically not as highly valued as sons (see table 4). The strength of this natal link suggests there may be a second degree of maternal investment strategy present within the community. Women invoke a broader strategy that includes using sons and daughters to establish wide-ranging linkages to all kin. This may arise out of concern for future support for children, or it may arise simply out of the love and continued affection that a mother feels for a particular relative. In the latter, the bond being established is not between the offspring and his or her maternal relative but, rather, between the mother and her natal kin.

Another factor behind women's namesaking of daughters is the opportunity to create a sense of personalization that is critical for the establishment of individual borders and thus psychological health. For a woman, the opportunity to select a non-kin or "nice" name constitutes another way in which she strives to demarcate her uterine family. "Nice" names, like maternal kin names, serve to establish a border necessary to maintain a woman's sense of autonomous identity within the cauldron of polygamous family life.

Because there is no evidence that naming increases sentimental bonds between a child and his or her namesaked relative, we suspect that the use of names is more an index of a matricentric focus than it is of a desire to establish a natal (or maternal) alliance. For most fundamentalist women, namesaking children is seldom, if ever, about alliance building; instead, it is symptomatic of natal affection, ego enhancement, and identity assertion. For women the dichotomy of competing solidarities is mediated through giving, symbolically, sons to the father and thus, in a way, to the religion, while retaining daughters for themselves.

There are at least two overlapping themes in the polygamous communities, with the potential for women's manipulation of the symbolic system of naming to illustrate conflicting motivations and varied sources and expressions of power. Given the communities' unique cultural and economic contexts, we found that women were less focused on men as a source of mere survival. Many wives use their husbands as a reference point from which to declare their commitment to the fundamentalist theology, over and above their attachment to the man himself. At the same time, they seek to maintain affective bonds within the natal family. Here cultural contexts encourage the reduction of women's dependence upon men, and both the religious ideology and competitive polygamous environment contribute to wives becoming autonomous agents. This is evidenced from naming practices that link the child to the mother rather than to the father.

CONCLUSION

Our study sought to understand how an evolutionary model might make sense of namesaking, clearly a symbolic behavior. We wanted to replicate earlier studies that used namesaking as an index for measuring male concern with parental certainty. We found that naming variations reflect a woman's social position within the larger polygamous family much more than a normative, albeit theologically driven, ideal. We also focused on the conflict between co-wives, especially as manifested in the naming of children, as an additional explanatory factor for the formation of differential niches within the larger family. We found that women, unlike men, pursued a broader parental investment strategy that included paternal and maternal kin, with daughters serving as a route to honor filial obligations to one's natal family while simultaneously contributing to the creation of psychological boundaries between a mother's uterine family and that of her co-wives.

Although previous studies focusing on naming sought to establish an underlying biological basis for paternal certainty concerns, we found that a more careful analysis revealed additional, albeit often, competing motives that arise out of female competition for dignity and gain **amongst themselves** in the polygamous family. In this way, fundamentalist namesaking patterns do not simply reflect official cultural prescriptions or address men's paternal certainty concerns. There are variations that are shaped, in large part, by a woman's personality, relative social and familial status, and her concern for, and effort toward, autonomy in the family.

ACKNOWLEDGMENTS

We would like to thank for rendering advice, encouragement, and good will: Paul Davies, Monique Dietrich, Sarah Hrdy, Barry Hewlett, Harmon Holcomb, Bill Irons, Jane Lancaster, Thomas Paladino, Katherine Peek, and Elizabeth Witt. We especially would like to thank Francis MacAndrew who freely shared his insights into some fruitful lines of analysis as well as providing his unpublished paper on naming. We owe inspiration for this paper to his valuable studies on naming which served as the basis for our work.

ENDNOTES

[1] Although anthropologists have explored women's efforts to maintain affective ties with their kin (Watson, 1984; Fortes, 1969; Wolf, 1972), there are few parental investment studies that have probed for nonreproductive motives behind women's behaviors.

[2] The sex differences found in parental investment studies may be human universals, but as cross-cultural studies of naming demonstrate, that concern is not always institutionalized through the naming process (Alford, 1988). It does appear, however, to be a factor in mainstream American culture and therefore serve as a source of evolutionary critique of parenting strategies.

[3] Mormon fundamentalists prefer not to give one child the name of another within the same polygamous family (though a name can be spelled differently or inverted so that a first name is the middle or vice versa).

[4] In a study of Chicago middle class naming practices, Rossi's (1965) findings differ from Mormon fundamentalist naming patterns. She identified a strong

maternal bias in the middle and latter birth cohorts. She believes that this pattern arose out of men's growing indifference to participation in the naming process.

We also found a middle cohort effect, whereby sons are namesaked after both paternal and maternal kin, while daughters continue to be overwhelmingly namesaked after maternal relatives. Unlike Rossi's American monogamous data, however, we found that in a woman's later reproductive years, or the 8-13 birth cohort, namesaking patterns revert back to a paternal-son linkage, albeit to a degree less pronounced than the top birth cohort (see Table 1).

[5] Women who are active participants in family affairs are more content than the women whose husbands rigidly adhere to the patriarchal ideal of absolute male authority. This interpretation is supported by our analysis of name choice, kin affiliation, and the quality of marriage (e.g., content or discontent). We found the most important factor that determined a wife's relative contentment was not becoming a favorite or an assertive wife per se, but rather whether more than 50 percent of her children's names were associated with her kin. In this way contentment, at least for women, appears to be based in issues of partnership. If both spouses are actively involved and not rigidly following the patriarchal ideal, the marriage, at least for most Angel Park women, is more satisfying.

REFERENCES

Alford, R. 1988. *Naming and Identity: A Cross-Cultural Study of Personal Naming Practices*. New Haven: HRAF Press.

Baur, H. 1988. *Utopia in the Desert*. New York: SUNY.

Bohannan, P. 1985. *All The Happy Families*. New York: McGraw Hill.

Bradley, M. 1993. *Kidnapped From That Land*. Provo: University of Utah Press.

Cannon, J. 1992. "My Sister, My Wife: An Examination of Sororal Polygyny in a Contemporary Mormon Fundamentalist Sect," *Syzygy*, 1, 4, 315-320.

Chisholm, J.S. 1999. *Death, Hope, and Sex*. Cambridge: Cambridge University Press.

Clark, A. & Clark, I. 1991. *Fathers and Sons in the Book of Mormon*. Salt Lake City: Desert Books.

Daly, M. & Wilson, M.I. 1984. "A Sociobiological Analysis of Human Infanticide", in Hausfater, G. & Hrdy, S., eds., *Infanticide*. New York: Aldine.

Evans, R.M. 1990. "The Relationship Between Parental Input and Investment", *Animal Behavior*, 39, 797-798.

Foster, L. 1992. *Women, Family and Utopia*. Syracuse: Syracuse University Press.

Fortes, M. 1969. *Kinship and the Social Order: The Legacy of Lewis Henry Morgan*. Chicago: Aldine.

Furstenberg, F.F. & Tralvitie, K.G. 1980. "Children's Names and Paternal Claims: Bonds Between Unmarried Fathers and Their Children", *Journal of Family Issues*, 1, 31-57.

Gilmore, David. 1982. "Anthropology of the Mediterranean Area", *Annual Review of Anthropology*, 11:175-205.

Hrdy, S.B. 1999. *Mother Nature*. New York: Pantheon Press.

Hrdy, S.B. 1987. "Sex-biased Parental Investment Among Primates and Other Mammals: A Critical Evaluation of the Trivers-Willard Hypothesis", in Gelles, R. & Lancaster, J., eds., *Child Abuse and Neglect: Biosocial Dimensions*. New York: Aldine.

Hewlett, B., ed., 1992. *Father-child relations*. New York: Aldine De Gruyter.

Irons, W. 1988. "Parental Behavior in Humans", in Betzig, L., Muller, M.B., & Turke, P., eds., *Human Reproductive Behavior*. Cambridge: Cambridge University Press.

Johnson, J.F., MacAndrew, F., & Harris, P. 1991. "Sociobiology and the Naming of Adopted and Natural Children", *Ethnology and Sociobiology*, 12: 365-375.

Low, B. 2000. *Why Sex Matters*. Princeton: Princeton University Press.

MacAndrew, F. & Jennifer King. 1995, June. "The Naming of Children as a Form of Parental Investment", paper delivered at the meeting of the Human Behavior and Evolution Society, Santa Barbara, CA.

Musser, J. 1944. *Celestial or Plural Marriage*. Salt Lake City: Truth Publishing Co.

Parker, S., Smith, J., & Ginat, J. 1973. "Father Absence and Cross-Sex Identity: The Puberty Rites Controversy Revisited", *American Ethnology*: 687-706.

Robertson, A.F. 1991. *Beyond The Family*. Berkeley: University of California Press.

Ross, M. 1994. *The Culture of Conflict*. New Haven: Yale University Press.

Rossi, A. 1965. "Naming Children in Middle Class Families", *American Sociological Review*, 30, 499-513.

Sered, S. 1994. *Priestess, Mother, Sacred Sister: Religion Dominated by Women*. New York: Oxford University Press.

Shipps, J. 1985. *Mormonism: The Story of a New Religious Tradition*. Urbana: University of Illinois Press.

Smith, D.S. 1977. "Child Naming Patterns and Family Structure Change: Hingham, Massachusetts, 1640-1880", *The Newberry Papers in Family and Community History*, Paper No. 76-75.

Smuts, B. & Gubernick, D. 1992. "Male-infant Relationships in Nonhuman Primates: Paternal Investment or Mating Effort?", in Hewlett, B., ed., *Father-child Relations*. New York: Aldine De Gruyter, pp. 1-30.

Trivers, R.L. 1972. "Parental Investment and Sexual Selection", in Campbell, B., ed., *Sexual Selection and the Descent of Man*, New York: Springer-Verlag.

Van Wagoner, R. 1986. *Mormon Polygamy: A History*. Salt Lake City: Signature Books.
Watson, R. 1985. Inequality Among Brothers: Class and Kinship in South China. Cambridge: Cambridge University Press.
Wolf, M. 1972. "Uterine Families and the Women's Community", in Wolf, M., ed., *Women and the Family in Rural Taiwan*. Stanford: Stanford University Press.
Zoellner, T. 1998. "Polygamy on the Dole", The Salt Lake Tribune, June 28: A1.

PART THREE

ADAPTATIONIST RESEARCH SRATEGIES AND EVOLUTIONARY HISTORY

INTRODUCTION TO PART III

HARMON R. HOLCOMB III

In the main introduction to this anthology, we emphasized the difference between narrow and broad visions of evolutionary psychology. Critics charge that the field has been guided by an overly narrow "innatist-adaptationist interpretation" of how evolution operates: The mind is a set of evolved adaptations, encoded in discrete modules under genetic control, inherent in the human species, and identified by evidence of adaptive design for reproductive success. This statement of adaptationism is ambiguous, having features of both a substantive thesis (a thesis about what is real) and a methodological thesis (a thesis about how to study what is real).

The substantive thesis is that all or most organismic traits (including human traits) are the products of their own selective efficacy, whether through natural selection or sexual selection or both. That is, descendent features of organisms exist today because of the selective efficacy of their ancestral forms. The methodological thesis is that we should, in the course of discovering the function of any trait, assume that it is a product of its own selective efficacy. That is, our best hope for understanding the proper role of any trait is by treating it as if it has been selected for quite specific tasks, using arguments from special design (Williams, 1966); evidence that a trait exhibits specificity, efficiency, and economy for producing a particular reproductively beneficial effect is good evidence that the beneficial effect is the function of that trait.

Part Three critically reflects on adaptationism as a research strategy that permeates practice in evolutionary psychology, including the previous chapters in Part One and Part Two. Is the adaptationist research strategy a viable scientific method for achieving the aims of evolutionary psychology to explain the nature of the evolved architec-

H.R. Holcomb III (ed.), Conceptual Challenges in Evolutionary Psychology, 293–308.

ture of the mind (the subject of Part One) and the way that architecture contours decision making, for instance, about mating and parenting (the subject of Part Two)? My main point is that Parts One, Two, and Three collectively demonstrate that we need a broader vision of evolutionary psychology than either "innatism" or "adaptationism."

THE INNATIST-ADAPTATIONIST INTERPRETATION OF EVOLUTIONARY PSYCHOLOGY

Due to the problem of multiple possible interpretations of evolutionary studies, the contributions in this volume could be interpreted as knee-jerk reactions in the narrow way of "innatism-adaptationism." But "can" does not imply "ought." We ought not to fall back on the innatist interpretation. I shall argue that these studies call for and embody a much broader interpretation, summarizing exactly why evolutionary psychology as found in this volume is broader than "innatism." Then, I shall highlight the significance of the chapters by Griffiths and Richardson; their critique of adaptationist research strategies is the best to date, appearing immune to all previous defenses of adaptationism. They point out the need for a methodology of testing and validation broader than adaptationism that combines adaptationist methods with those of phylogenetics. Again, "can" does not imply "ought." Although we can study the evolutionary significance of a behavior without tracing its phylogenetic origin, we ought not to, because phylogenetic information can falsify conclusions of special design arguments that a trait has been selected for one specific task rather than another specific task or a related, more general task.

One principled argument against the innatist interpretation calls attention to what evolutionary psychology studies: the items to be explained. Our original interpretation linking innatism with adaptationism caricatures evolutionary psychology because it ignores the way evolutionary psychology's domain is specified. To interpret evolutionary psychology (or sociobiology) as saying that "the mind is a set of evolved, genetic adaptations" attributes to it a substantive claim that fails to recognize that sciences have procedures of domain specification (Holcomb, 1993): Evolutionary psychology does not study the mind as a whole but, by definition, the evolutionarily significant features of mind and behavior. That is, it restricts its domain to features of mind and behavior that have evolutionary significance. Crawford gives as an example cultural changes in which some traits have an evolutionary history whereas others do not:

> During the last 30,000 years, human inventions ranging from agriculture to television have irreversibly changed human culture. Each of the changes may lead to stress and pathology. It will be possible to trace some of these pathologies to the overloading and distortion of ancestral adaptation. However, there will be others that do not have an evolutionary history. For example, the change from a hunter-gatherer to an agricultural way of life may have resulted in changes in behavior that reduced the level of exercise. If the level of physical activity in ancestral humans was involved in some aspects of metabolism, its reduction in modern humans may result in a slight biochemical dysfunction that may lead to a physical or behavioral pathology. Such a pathology does not have an evolutionary history in the strict sense because it cannot be traced to an adaptation for dealing with ancestral environmental conditions (Crawford, 1989, p.19).

Because it is not an easy task deciding which human traits have an evolutionary history and which do not, evolutionary psychology needs criteria for making that distinction. Adaptationism comes into play here. Crawford's criterion of evolutionary significance is (1) the trait has an evolutionary past, and (2) it can be traced to an ancestral adaptation. So, our initial statement that projected an innatist-adaptationist interpretation onto evolutionary psychology was oversimplified in its talk of "any trait," i.e., that to discover the function of any trait we should assume that it has been naturally (or sexually) selected for quite specific tasks. That would imply "innatism" and "adaptationism," as unlimited views that cover all human traits. Instead, evolutionary psychologists make the innocuous claim found throughout evolutionary biology that to discover the function of any trait having evolutionary significance, i.e., that has an evolutionary past and can be traced to an ancestral adaptation, we should assume that it has been naturally (or sexually) selected for quite specific tasks. And they qualify that claim further to note that not all human traits having evolutionary significance are functional, but may be byproducts of functional traits, e.g. "although writing is a beneficial effect of the anatomy of the human hand, it is not a function of the hand because differential reproduction with respect to writing did not occur when the human hand was evolving its present form" (Crawford, 1989, p.17).

Now we can see why in principle the innatist-adaptationist interpretation of evolutionary psychology as saying that "the mind is a set of evolved, genetic adaptations" is a caricature. If we amend that claim to recognize Crawford's vision of how the field specifies its domain with increasing precision, it reads: "the mind—insofar as it consists in traits whose study reveals that they have an evolutionary past, are traced to ancestral adaptations, and are traced not as currently beneficial effects of anatomy but as functions in the environments in which they evolved—is a set of evolved, genetic adaptations." That claim is trivially true. Hence, the field does not fit the innatist-adaptationist interpretation, but only an innocuous limited interpretation: If we study the mind, we find some traits that have an evolutionary past and can be traced to ancestral adaptations in a way that distinguishes functions from beneficial effects, and so the mind is constituted in part (not in whole) by evolved, genetic adaptations. An upshot is that evolutionary psychology and traditional psychology should be envisioned neither as competing ways of studying the same thing nor as separate studies of different things. We need to construct research methods to study relations between traits having no evolutionary past (the subject of inquiry *without* using an evolutionary paradigm) and traits having an evolutionary past (the subject of inquiry *using* an evolutionary paradigm).

We can also see why the innatist interpretation is a caricature by examining evolutionary psychology as it is practiced. Part One on the evolved architecture of the mind began with a kinship-oriented methodology proposed by Mealey that goes beyond both innatist and environmentalist stances to ask and answer questions about the extent to which human variation is genetic or environmental, about the extent to which it is adaptive or not, and combined studies of species universals with individual differences. Instead of elevating genetic control over environmental factors, Sulloway clarified the role of different within-family environmental niches on personality development, and then Cummins clarified the role of the social environment on mental development by positing a type of adaptive design that lacks predetermined mental modules. Part Two, focusing on the way this evolved environmentally sensitive architecture works, treated mental adaptations as decision-making tools. Gangestad hypothesized that evolution by good genes selection created a plurality of conditional strategies in both sexes—not a single innate human nature or one nature of all women and another of all men—to account not only for differences between men and women, but also for variation among men and among women, in their adaptive mating tactics. Gangestad also defended a sophisticated form of adaptationism that uses optimization models and arguments from special design very cautiously, one that successfully rebuts the usual objections to adaptationism. Ensuing studies in Part Two focus on the use of environmental information as inputs to evolved decision-making devices for reproductively successful mating and parenting. These studies are a far cry from "innatism;" practice in evolutionary psychology is far broader than that. These studies are premised on the persuasive idea that our generation would not be here unless our ancestors responded to cues in their environments to make decisions that enabled them to reproduce successfully. We have inherited the potential to develop these mental tools (neither the tools nor the behaviors is innate), and we use different tools conditionally as various circumstances change this potential into actuality: These studies collectively show how we differ in the ecological, individual, familial, social, and cultural conditions in which we engage in different mating and parenting tactics.

It is undeniable, however, that an adaptation, even if it results in behaviors that vary with environmental cues, is at some deep level under genetic control. Why? Adaptations, by rough definition, are traits shaped by a history of a natural selection (including sexual selection), and selection operates directly on those traits (phenotypes) by selecting indirectly on genes (genotypes) that underlie them in shaping individual development. The "innatist" interpretation is bothersome because it says that evolutionary psychologists see humans as "predetermined," and people have an immediate emotional response that we are not predetermined. This interpretation is rampant only because it is vague about what, exactly, is "predetermined." Any scientific study of humans that explains human behavior according to some theoretical factors could be interpreted as saying that we are "predetermined" by those factors. That interpretation is no basis for a negative evaluation because it is so vacuous substantively: If humans are "predetermined" by naturally selected genes, by gene-environment interactions, or by evolved adaptations, then, by parity of reasoning, we are "predetermined" by our "family," "social environment," "social class," or "culture." The "innatist" interpretation dissolves upon analysis as confused, because it applies an idea having such a wide scope that it encompasses even the critics' favored nonevolutionary paradigm (e.g., "environment or culture, not evolution, explains us") but is levied as a criticism against only the evolutionary paradigm. Although science is often said to involve causal determination, surely we can agree that scientific explanation itself is not "predetermination," in the sense that controversy over "innatism" is about genes and not about science in general. Clarity advises us to limit "predetermination" to a specific innatist substantive thesis, say, that internal hereditary factors in the body trump external factors in the environment in producing behavior. The evolutionary psychological studies in this volume are obviously not "innatist" in this sense. But are they "adaptationist"? And if so, is that a good thing or a bad thing? And what is adaptationism, anyway?

SUBSTANTIVE ADAPTATIONISM AND METHODOLOGICAL ADAPTATIONISM

The substantive and methodological theses of adaptationism are distinct. We can endorse one while rejecting the other. We can hold the substantive thesis that most traits having an evolutionary history are selected for, but deny the methodological thesis, insisting we would do better by adopting some other approach. Or we can hold that the best way to discover functional properties is by assuming that organisms and their traits are the products of selection, even while thinking that they are the products of some other process. Despite this independence, advocates of the adaptationist program tend to endorse both theses. This certainly is the case among evolutionary psychologists.

Contributors to this volume disagree about the importance of adaptationism in studying the mind. All agree that the mind is a product of evolution and that our cognitive mechanisms evolved, at least in part, as a consequence of selection. There may be differences about the scope of natural selection in shaping our cognitive capacities, and thus there may be differences about the exact formulation of substantive adaptationism, but the substantive thesis is not the central point of contention. The contentious issue is methodological adaptationism. Some theorists embrace these methods while others doubt their scientific credentials. In order to clarify the target of the critique, let me explain the relationship between these two types of adaptationism.

Griffiths' chapter emphasizes that in describing the tasks the mind was designed by selection to perform, assumptions must be made about the adaptive problem and adaptive solution relevant to the trait under study. When evolutionary psychologists say that various human behavioral traits have been designed by selection for certain specific tasks, they typically assume that the behavior of ancestral organisms evolved to increase their fitness. How are we to substantiate these assumptions? Because phylogenetic information about the course of hominid evolution and cross-species comparisons is hard to get, evolutionary psychologists devised numerous methods to identify evidence of special design that bypass phylogenetic methods. That was good, because adaptationism (in that sense) got the field off the ground. But now that we have these adaptationist methods that identify special design using fitness arguments about adaptive problems and adaptive solutions, critics of adaptationism like Griffiths say that we cannot substantiate these assumptions without a return to phylogenetic methods.

Some evolutionary psychologists derive substantive adaptationism from two simple premises: Complex traits such as the mammalian eye arose from one of three possible sources—divine design, extraterrestrial seeds, or evolution via natural selection; but neither creationism nor seeding theory are scientifically viable; therefore, complex traits are the products of evolution by natural selection (Tooby & Cosmides, 1992, p.50; Buss, 1995, p.2). Other evolutionary psychologists argue that complex traits must have arisen from selection, since non-selective processes such as drift are incapable of producing such complexity (Pinker, 1997). Methodological adaptationism is embraced on the grounds that it enables us to uncover the information-processing mechanisms that other methods are liable to miss.

Griffiths points out that methodological adaptationism involves two forms of inference. Informed conjectures about the adaptive problems imposed upon our ancestors enables us to speculate about the kinds of mechanisms that would have been selectively efficacious. We begin with claims about past selective regimes and conclude with predictions concerning the sorts of mechanisms that we may have inherited. We then design experiments in order to confirm the presence and efficacy of such mechanisms. Alternatively we begin with information about psychological mechanisms we in fact possess and develop selective speculations concerning their acquisition. An understanding of the current functions of such mechanisms forms the basis of hypotheses about the evolutionary history of those mechanisms. Both inferences—but especially the first—illustrate the centrality of methodological adaptationism in evolutionary psychology (Barkow, Cosmides, & Tooby, 1992, p.10).

To facilitate critical discussion, I shall argue that the second inference—from known mechanism to selective history—has nothing in particular to do with the aims of cognitive psychology. Learning the history of our psychological mechanisms is interesting in its own right, but it is neither necessary nor sufficient for discovering the architecture of our cognitive capacities. It is plausible, in fact, that we must first know the functional architecture of these mechanisms in order to correctly trace their selective history, in which case the second form of inference is clearly irrelevant to the aims of cognitive psychology. Specifically, must we assume that for each proximate mechanism of some psychological phenomenon, there is an evolved strategy special to that proximate mechanism? I argue against that assumption. Establishing relevance must be done in terms of the ideas already present in specialized research, and not merely through general arguments.

Evolutionary researchers in psychology typically think that evolution is always relevant in psychology. Typically, they invoke the general distinction between "proximate" explanations (in terms of immediate causes, individual development) and "ultimate" explanations (in terms of the ancestral origins and evolved functions). Take the usual example, "Why is sugar sweet?" On the proximate level, there is a causal chain from molecular properties of sugar, to taste receptors, to a mouth-sensation of sweetness. On the ultimate level, in ancestral conditions sugar-containing foods such as fruits were useful for survival and our ancestors, whose bodies contained this causal chain and were motivated to eat sweet foods, outreproduced those who lacked a sugar-sweetness causal chain and were not motivated to eat sweet foods. The two kinds of explanation are then said to be complementary, not rivals (Barash, 1977; Mealey, 2000).

But the general distinction alone does not warrant that conclusion, as a simple logical point reveals. It is true that for every ultimate cause there is a proximate cause; hence, evolutionary psychology is rightly deemed an improved successor to sociobiology through its focus not just on the ultimate adaptive significance of human behavior, but also on the proximate psychological mechanisms of human behavior. That truth does not imply that for every proximate cause there is an ultimate cause, which is an inference evolutionists unwittingly slide into. That slide violates the basic tenet of quantificational logic that, within a domain, statements of the form "For every X, there is a Y such that…." do not imply "For every Y, there is an X such that…" For example, "For every effect, there is a cause" does not imply that "There is a cause for every effect;" the scientific premise of universal causation does not logically imply that a single cause of the universe exists (e.g., God). Hence, the general lack of evolutionary training of psychologists together with "I don't see how evolution could be relevant for my specific research (say, on eating disorders)" leads to resistance to evolutionary explanations for the data on eating disorders. When evolutionists write about "resistance to the evolutionary paradigm," they give the impression that all resistance is due to ignorance, misunderstanding, prejudice, etc. That often is the case, but some resistance is justifiable when relevance arguments are fallacious.

One such fallacy lies in evolutionary psychology's main relevance argument in clinical psychology: that disorders are "malfunctions" and "malfunctions" are to be defined in "ultimate" rather than "proximate" terms. To achieve relevance, the field treats "disorders" as "malfunctions," and "malfunctions" are defined in relation to a logically prior identification of a function. As Buss (2000, p.399) claims, "Once an evolved psychological mechanism has been described, a clear criterion exists for determining dysfunction: Dysfunction occurs when the mechanism is not performing as it was designed to perform in the contexts in which it was designed to function." The key fallacious gambit here is to define disorder/malfunction purely in terms of past history of natural selection, as evident from asking: What concept of "function" is operative here? It is wrongly assumed that a proper cognitive function is "whatever natural selection favored in the past and created in it the appearance of design for reproductive success," which is a definition of an "adaptation." If proper functions were defined as adaptations, it is logically circular to explain functions in terms of adaptations: If X is defined in terms of Y then Y cannot explain X because that would be for something to explain itself, and self-explanation is unscientific. Remember Darwin's schema: Natural selection occurs whenever traits vary, those traits confer fitness advantages, and the traits are heritable. Natural selection explains adaptation. For that to make sense the fitness-conferring traits cannot be defined as adaptations; if they were, the explanation would be conceptually circular. So, the traits have to be defined independently of adaptations. That applies to all traits of organisms, and therefore to all traits regardless of whether they are functional or dysfunctional. So, it is logically fallacious to try to explain the persistence of dysfunctional conditions by natural selection when "dysfunctional trait" is defined in terms of natural selection.

In sum, if functions and malfunctions were defined exclusively in terms of past or present natural selection, that explanation would be empty because it would be circular. So, functions and malfunctions have to be defined independently of natural selection, i.e., in "proximate" rather than "ultimate" terms. The fallacy arises from confusing what functions/dysfunctions are (defining something) with which functions/dysfunctions we would expect to find (identifying the existence of something already defined). Being the result of a history of natural selection is a criterion for deciding which functions/dysfunctions would persist; the adaptationist stance implies that they would be the ones that contributed to reproductive success in the evolutionary past and are present today because they were adaptive in the past. Evolutionary clinical psychology's goal should not be to define "disorders" as "malfunctions" and "malfunctions" as "adaptations that no longer enhance fitness." Its goal should be to use natural selection as a criterion for deciding what kinds of functions constitute the architecture of the mind, to investigate whether disorders (defined independently of natural selection) are currently adaptive or maladaptive, and to explore their relation to adaptations and failure of evolved mechanisms.

A common reason for falling into this fallacy is that disorders are typically defined in terms of their effects, i.e., a psychological disorder is something that prevents someone from doing things successfully that people normally do successfully, i.e., whose effect is to disturb or derange some means to and end. Which tasks? Given that success is a value-laden concept, success in relation to what valuable state? At a proximate level, psychology describes the tasks that the psychological system accomplishes (Griffiths, this volume), and tells us why these tasks rather than other tasks are the ones that count in terms of success by reference to valuable states (e.g., being able to hold a job, being able to get along with other people). This proximate level of task description and criteria for success must be executed in conceptual independence of evolution, or else it falls into circular explanation that explains nothing (as the previous conceptual argument shows). Given that nonevolutionary account, evolutionary psychology provides an ultimate explanation of these tasks and their success, by showing that many tasks are those that enhance reproductive success (e.g., holding a job is a way of obtaining resources for survival, being able to get along is essential for group-living organisms). A type of organism, human or otherwise, to the extent that current traits do not have an evolutionary past, has traits expected on the basis of evolutionary theory not to count as adaptations; much of the human mind may involve traits whose task descriptions at the proximate level make them "functional/ordered" or "dysfunctional/disordered" in a sense in which past evolution is irrelevant to them. This point is not unlike the one psychologists make who say that evolutionary psychology is irrelevant to what they are studying, but it is a bit more subtle and better informed by evolutionary theory.

Psychological disorders are conditions that interfere with normal functioning, but it is a common mistake to try to reduce functions to effects. Instead, function talk is a way of talking about a nexus or chain of causes and effects with regard to attainment of a state presumed to have a value. For instance, if we presume that the ability to eat the amount of food that would ensure survival is a value, then we can say that eating the right amount of food is functional and eating way too much food (leading to obesity if metabolized or to bulimia if purged) or way too little food (leading to anorexia) is malfunctional. Whereas effects follow causes in time, "eating the right amount" and "surviving" are simultaneous and "choosing the best available mate" and "having high lifetime reproductive success" are simultaneous. It follows that these pairs—and, in general, pairs of traits and their evolutionary fitness-enhancing functions—cannot count as causes and effects, contrary to the widespread tendency to reduce functions to effects. The crucial conceptual point that most scientists and philosophers of evolution fail to see is that cause-effect and function/malfunction are categorically distinct. That is, cause and effect are relations between temporally successive events, whereas functions and malfunctions are qualities of how some state with a value is achieved. Function talk thus involves causal talk but cannot be reduced to it because cause-effect relations need not involve values, whereas function talk must involve values.

Richardson's chapter evaluates Pinker's influential book, *How The Mind Works* (1997). Pinker focuses on substantive adaptationism. He asserts that the only historical process capable of producing traits that are adaptive and complex is evolution by natural selection. Since the human mind is both adaptive and complex—or so it is claimed—we ought to accept substantive adaptationism with respect to the mind. That is the core of Pinker's view. This claim is defended against various forms of skepticism that have been raised about the power and scope of natural selection. It is claimed, for example, that natural selection is not nearly so potent as sociobiologists and evolutionary psychologists presume in altering the basic architecture of organisms, where the most central components are deeply entrenched in the development of the organism. The thought is that any variation among such components will probably result in forms that are not viable. This is one of the central themes of the well-known attack on adaptationism offered by Gould and Lewontin (1979). Pinker agrees that the sorts of developmental constraints discussed by Gould and Lewontin rule out sudden significant alterations, to be sure. They do nothing, however, to prevent selection from sorting among small differences that, over long stretches of evolutionary time, add up to significant differences.

Pinker (1997) also endorses methodological adaptationism. His defends this thesis by appeal to the authority of Ernst Mayr (1983). Mayr rightly points out that biologists often come to understand a given trait by inquiring into its functions. He asserts that "The adaptationist question, 'What is the function of a given structure or organ?' has been for centuries the basis of every advance in physiology." (Mayr, 1983, p. 328, quoted in Pinker, 1997. p.166) And Pinker, following Mayr, takes this as evidence that biologists are employing some form of methodological adaptationism.

I do not agree with Mayr or Pinker on this point. Inquiry into the function of a trait offers neither comfort nor aid to the adaptationist program. Selection is crucial only for which functions persist over evolutionary time, not what counts as a function in the first place. So, the nature of functions can and should be understood in non-selectionist terms, as argued previously in this introduction (for the full argument, see Davies, 2000).

Even if Mayr is correct, we still must consider the actual procedures involved in methodological adaptationism, as well as the sorts of evidential standards imposed. It is the issues of evidence, and hence the validation of evolutionary explanations in psychology, that generate the doubts raised by Griffiths and Richardson.

CRITIQUES OF ADAPTATIONIST METHODS BY GRIFFITHS AND RICHARDSON

Griffiths is wary of the methods employed by some of the more prominent evolutionary psychologists. Both forms of inference—from adaptive problem to evolved mechanism and from known mechanism to adaptive problem—are problematic. We may speculate about the adaptive problems imposed upon our ancestors and, on that basis, predict that descendent organisms are likely to have inherited specific kinds of cognitive mechanisms. But in the absence of significant evidential constraints, such speculations can be multiplied to a fault. Proper adjudication requires adequate empirical tests. The problem, however, is that all too often researchers are beguiled by their own selective speculations, to the point of failing to subject them to serious test. Research about parent/offspring conflicts and on theories of emotions illustrates this worry. The second methodological inference begins with known psychological mechanisms and speculates on the selective history of those mechanisms. The problem here is that the adaptive explanation of a given trait can appear highly plausible—better than any available alternative explanation—and nevertheless be false. Griffiths illustrates this with several examples, including attempts to explain low birth weight in bears and the evolution of pregnancy sickness.

What these and other adaptationist studies overlook, according to Griffiths, is the power of contingency in evolution and the importance of phylogeny in understanding evolution. The impact of contingencies is hardly marginal. Of course interactions between organisms and their selective environments affect the evolutionary outcome, but so do stochastic factors. Analyses of organism/environment interactions that leave out the effects of stochastic factors are overly simplistic and the only way to include such effects is to discover details of the actual selective history. For these reasons, both inferences are unreliable. Moreover, by focusing upon speculations concerning the fit between organism and a single environment, methodological adaptationism leaves out a much wider store of historical information. It leaves out information gleaned from phylogenetic studies. Such studies focus upon the multiple lines of descent through which an organism has passed, culling information from different ecologies through which the organism has passed, as well as information about taxonomically related organisms and their ecologies. Such information can correct and suggest hypotheses about the organism we wish to understand. We thus should forego the methods employed by evolutionary psychologists and instead develop the study of phylogeny.

Comparative phylogenetic studies, however, are problematic if our aim is to understand the evolution of capacities unique to a single lineage. Such is arguably the case for language and reasoning. There need not be a difference in kind; a difference in degree may suffice. Because language and reason are markedly different in humans, we cannot readily compare either capacity in *Homo sapiens* with homologous capacities in other hominid lines, since the latter do not exist. Griffiths addresses this worry, suggesting we can minimize the problems involved. Moreover, it is noteworthy that the focus of Griffiths (1997) is not the cognitive capacities involved in language or reason, but the affective capacities involved in the emotions. This may support Griffiths' claim that the relevant comparisons can indeed be made. We can draw some useful taxonomic comparisons in the study of mechanisms that underwrite at least some emotions.

Richardson's chapter is even less optimistic about the scientific credentials of evolutionary psychology. Indeed his thesis is that we have "no credible reason to embrace the explanations offered within evolutionary psychology." This is not to say that the explanations are mistaken, only that we have too little access to the relevant historical record. Evolutionary psychological explanations, according to Richardson, fail to meet reasonable standards of scientific evidence. It is assumed that explanations within evolutionary psychology are beholden to the same standards of evidence as explanations within evolutionary theory generally, and the charge is that the explanations offered by evolutionary psychologists fail to meet those standards. Richardson argues by elimination. If there is adequate evidence for evolutionary psychological explanations, it will come from (i) population genetics, (ii) reverse engineering studies, or (iii) phylogenetic reconstruction. Only the third option is a serious contender—and in this regard the views of Richardson and Griffiths converge. But even here, according to Richardson, we have only the thinnest evidence concerning the emergence of capacities such as language and reasoning. It is here, in assessing the extent of information provided by phylogenetic studies, that Richardson and Griffiths appear to differ.

Against the first option, Richardson insists that often it is difficult to determine whether or not small phenotypic differences resulting from genotypic variations are currently adaptive in the population. But when the task is discovering whether or not a trait was adaptive during its evolutionary history, and hence now is an adaptation, the difficulties increase. Richardson appeals to five types of evidence required for "ideally complete" adaptation explanations and he argues that we are unlikely to find such information for the evolution of traits such as language, reasoning, and so on. The second option is reverse engineering. Engineering analyses identify the function of a trait within the organism's current environment and then speculate that ancestral tokens of the trait were selected for the same (or a similar) effect. The problem with such studies, however, is that they leave out information about the organism's actual selective history. Such studies are essentially ahistorical. Typically there are two or more historical routes by which a population could have reached its current composition; tracing the current functions of current traits does not tell us which is the correct route. We can note the functions served by current tokens and speculate about the adaptive problems such functions may have solved. The question remains, however, which of the possible adaptive problems were, as a matter of fact, the real selective pressures. And as Richardson argues, this is a question that optimality studies studiously fail to answer. Reverse engineering studies do not address the central historical question.

The bulk of Richardson's discussion is devoted to the sorts of evidence required by phylogenetic studies. Such studies provide grounds for the claim that a given trait is an adaptation only if there is evidence that that trait enhanced performance or survival relative to related variants of that trait. A variety of information is required, including information about the character of the environment, the lineage of the trait under consideration, related taxa and homologous traits, the relative adaptedness (adaptiveness) of those homologues, and so on. The crux of Richardson's argument is that such comparative studies of the evolution of human language and reasoning are evidentially anemic. It is an interesting question whether or not the worries raised by Richardson apply with equal force to phylogenetic studies of our affective capacities. With respect to language and reasoning, however, we do not have, and we are not likely to find, the required kinds of information. Evolutionary psychological explanations of these capacities, therefore, cannot appeal to phylogeny in defense of their methods. This completes the argument by elimination.

This principled argument gains force through plenty of well-chosen examples of concrete evolutionary analyses adduced by Griffiths and Richardson to make their point. The temptation to waive one's hands, saying that the argument expects too much from evolutionary psychology, is difficult to maintain in the face of their examples. So, we must think carefully about the details of their examples before we are justified in conjuring up some general reason not to take the principled argument seriously. If the critics' argument is sound and their examples make their point, evolutionary psychological explanations typically fail by the standards of evolutionary theory in general.

ACKNOWLEDGMENT

For his contribution in introducing Griffiths' chapter and Richardson's chapter, I thank Paul S. Davies.

REFERENCES

Barash, D. 1977. *Sociobiology and Behavior.* New York: Elsevier.

Buss, D. 1995. "Evolutionary Psychology: A New Paradigm for Psychological Science", *Psychological Inquiry,* 6, 1-30.

Crawford, C. 1989. "The Theory of Evolution: Of What Value to Psychology?", *Journal of Comparative Psychology*, 103, 4-22.
Daly, M. & Wilson, M. 1988. *Homicide*. Hawthorne, NY: Aldine-de-Gruyter.
Davies, P.S. 2000. *Norms of Nature: Naturalism and the Nature of Functions*. Cambridge, MA: MIT Press.
Gould, S.J. & Lewontin, R.C. 1979. "The Spandrels of San Marco and the Panglossian Paradigm: A Critique of the Adaptationist Programme", *Proceedings of the Royal Academy of London* B, 205, 581-598.
Griffiths, P.E. 1997. *What Emotions Really Are: The Problem of Psychological Categories*. Chicago: University of Chicago Press.
Holcomb, H. 1993. *Sociobiology, Sex, and Science*. Albany, NY: SUNY Press.
Mayr, E. 1983. "How to Carry out the Adaptationist Program", *American Naturalist*, 121, 324-334.
Mealey, L. 2000. *Sex Differences: Developmental and Evolutionary Strategies*. San Diego, California: Academic Press.
Pinker, S. 1997. *The Language Instinct*. New York: Morrow.
Tooby, J. & Cosmides, L. 1992. "The Psychological Foundations of Culture", in Barkow, J., Cosmides, L., & Tooby, J., eds., *The Adapted Mind: Evolutionary Psychology and the Generation of Culture*. New York: Oxford University Press.
Williams, G. 1966. *Adaptation and Natural Selection*. Princeton, NJ: Princeton University Press.

CHAPTER 8

FROM ADAPTIVE HEURISTIC TO PHYLOGENETIC PERSPECTIVE: SOME LESSONS FROM THE EVOLUTIONARY PSYCHOLOGY OF EMOTION

PAUL E. GRIFFITHS
University of Pittsburgh

Abstract. This chapter contrasts an approach that stresses the reconstruction of phylogenetic trajectories and in which the role of adaptation is inseparable from that of history with the adaptationist program as outlined by contemporary evolutionary psychologists. The phylogenetic perspective I advocate will allow evolutionary psychology to conduct more rigorous, quantitative tests of adaptive hypotheses and hence establish properly corroborated theories concerning the evolution of mind. Furthermore, the adaptive heuristic, as found in the adaptationist program, has a tendency to send psychologists down blind alleys. I support both these claims with several case studies in actual evolutionary psychology, particularly of the emotions.

1. THE EVOLUTIONARY PSYCHOLOGY PROGRAM

The evolutionary psychology program is the successor to the sociobiology of the 1970s. Sociobiology tried to explain human behavior using tools from population genetics and game theory. In many cases it interpreted observed behaviors as evolutionarily stable strategies—traits that cannot be outcompeted by any mutant form. Evolutionary psychology adds to the tools of sociobiology an approach to the explanation of behavior derived from the computationalist school in cognitive science and the Chomskyian tradition in psycholinguistics. Evolutionary psychology also differs from at least some earlier socio-

H.R. Holcomb III (ed.), Conceptual Challenges in Evolutionary Psychology, 309–325.

biology in clearly distinguishing questions about the *adaptiveness* of a biological trait from the question of whether it is an *adaptation*.

Adaptiveness is a measure of a trait's current effect on an organism's reproductive fitness. An adaptation is a trait that can be explained as the result of natural selection. Adaptiveness is neither necessary nor sufficient for a trait to be an adaptation. Vestigial traits like the eyes of blind cave fish are adaptations, although they are no longer adaptive. The ability to read has been adaptive for much of human history but reading is generally thought to be a side effect of other, more ancient cognitive abilities. Evolutionary psychology claims that the human mind is a bundle of cognitive adaptations. It does not claim that these adaptations are currently adaptive. The environments in which most humans live are very different from that which they occupied during most of their evolutionary history. The behaviors we observe in modern humans may not be evolutionarily stable strategies in their modern settings.

Evolutionary psychology takes from cognitive science and psycholinguistics a particular sort of concern with mental mechanisms. It accepts the characterization of the levels of explanation in psychology proposed by David Marr in his influential book *Vision* (Marr, 1982). Marr proposed than an adequate approach to psychology must make use of three levels of description. The highest level describes the task that the psychological system accomplishes. In the case of vision, Marr claims that the system takes patterns of stimulation on the retina and produces an interpretation of the world in terms of moving, three-dimensional objects with color. The intermediate level describes how the system computes information in order to accomplish this task. In the case of vision, it details the algorithms by which retinal patterns are transformed into representations of the world. The lowest level describes how these computational processes are implemented in the brain. The computationalist school of cognitive science assigns psychology the job of describing the intermediate level: the information processing methods used by the mind to perform its tasks. Describing the physical machinery that implements those methods is the job of the neurosciences, with the result of psychology acting as an important and perhaps essential heuristic. What evolutionary psychology adds to this conventional picture is the idea that evolutionary theory provides a way of describing the mind at the highest level, the level of task description. Analyzing the adaptive problems facing hominids in the "environment of evolutionary adaptedness" reveals the tasks the mind was designed to perform.

Adding an evolutionary perspective to the "Marrtian" program for psychology is supposed to provide three payoffs (Tooby & Cosmides, 1992). First, since computationalists have always recognized that a task description is an important, and perhaps essential, heuristic for finding algorithmic level descriptions, evolutionary analysis becomes an important heuristic in the search for psychological theories. In some cases the task description derived from evolutionary theory coincides with that of common sense. Cognitive scientists already knew the task description of the visual system. In other cases it is novel. Cosmides and Tooby provide a novel task description for human reasoning based on game theoretic models of the evolution of cooperation (Cosmides & Tooby, 1992). The "adaptive heuristic" assists psychology in a second way, because knowing the problems that mental mechanisms must solve restricts the range of acceptable theories about how those mechanisms actually work. Cosmides and Tooby argue that any mechanism for negotiating social exchange that is vulnerable to people not honoring their bargains could not have evolved and so can be ruled out *a priori*. Finally, adaptive thinking has an explanatory and organizing role. It makes sense of all sorts of otherwise senseless detail uncovered by psychology. The idiosyncratic rules followed by the rat in rejecting food items after illness fall into place when understood as a poison-avoidance mechanism.

Evolutionary psychologists believe that these two innovations, clearly distinguishing adaptation from current adaptiveness and treating mental mechanisms rather than behaviors as the locus of adaptation, will allow them to escape the strictures leveled at first-wave sociobiology. These are indeed important advances, but there are still major problems with the treatment of adaptation in evolutionary psychology. In one important way the attempt to introduce a truly Darwinian perspective on the mind has gone backward since the days of classical ethologists. Konrad Lorenz identified the *phylogenetic perspective* on behavior as Darwin's primary legacy to ethology (Lorenz, 1965). Lorenz believed that ethology began with Oscar Heinroth's insight that behavior, like morphology, could be subjected to comparative analysis (Richards, 1987, p. 529). Behavior was revealed as a pattern of affinities across species, reflecting the evolutionary relationships of those taxa. The study of this pattern—comparative biology—remains the most powerful method we have for testing adaptive hypotheses (Griffiths, 1996). My concern in this chapter will be to contrast an approach that stresses the reconstruction of phylogenetic trajectories, and in which the role of adaptation is inseparable from that of history, with the "adaptationist program" as outlined by contemporary evolutionary psychologists. I argue that only the former model will allow evolutionary psychology to establish well-corroborated theories concerning the evolution of mind. I also argue that the 'adaptive heuristic' is as likely to hinder psychology as to help it. Although I will concentrate mainly on recent examples, Darwin's own work on emotion is a particularly good example of a properly phylogenetic, or adaptive-historical, perspective on the mind and I will return to it at the end of the chapter.

2. ADAPTIVE THINKING

Adaptationism is the view that selective problems are strongly associated with particular solutions to those problems. This belief licenses two sorts of inference, as illustrated in Figure 1. First, adaptationists think they can infer the solution from the problem. If they know what adaptive problems an organism has faced, they can infer what adaptations it will possess. Second, adaptationists think they can infer the problem from the solution. They think that by looking at the complex forms evolution has produced, they can infer the ecological interactions between ancestral organisms and their environment that produced those forms. I have argued elsewhere that both these inferences are unreliable (Griffiths, 1995; Griffiths, 1996; Griffiths, 1997). Problems and solutions do not correspond in this straightforward fashion. Evolutionary psychologists are aware of the most egregious error of adaptationist reasoning. They accept that the solution cannot simply

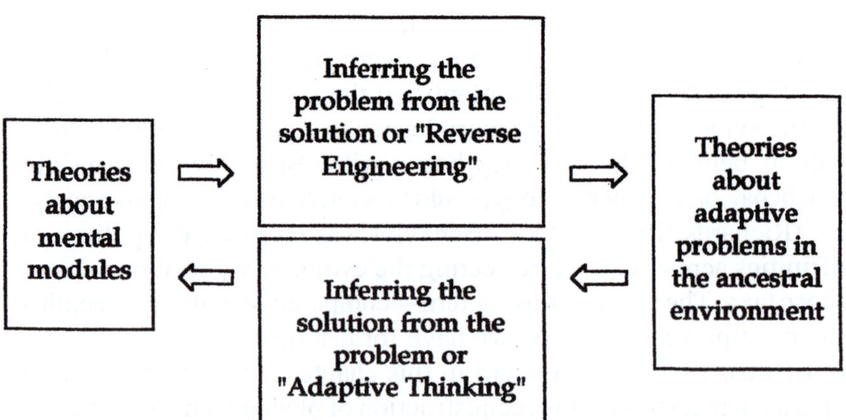

Figure 1. The 'adaptationist program' as described by evolutionary psychology.

be inferred from the problem. Merely producing an adaptive scenario in which that mental structure would be advantageous cannot prove a hypothesis about mental structure. An empirical demonstration that the mind is actually structured in that way is also required: "Although selectional thinking is an important source of inspiration for the evolutionary psychologist, nature always gets the last word" (Symons, 1992, p. 143-4). But although nature always gets the last word, many evolutionary psychologists think that the fact that a particular feature "makes adaptive sense" is a reason for taking data suggesting that it actually exists more seriously than the intrinsic strength of the data justifies. If nature disagrees with the adaptationist about what should have evolved, then she has to shout. If she agrees, then she only has to whisper. This is what is meant by saying that adaptive thinking has 'heuristic value': a hypothesis backed by a plausible adaptive scenario is more likely than the alternatives and hence more worthy of investigation.

In fact, adaptive thinking may have a substantial negative heuristic effect. It can induce complacency in the face of unreliable or ambiguous data, because that data or interpretation of the data "makes adaptive sense." At least two generations of evolutionary psychologists have been impressed by Robert L. Trivers' demonstration that the long-term interests of parents need not be identical with those of their offspring (Trivers, 1974). The parent wants to conserve some resources for the next generation, whereas the offspring wants as much as it can get. This model showed that observations of squabbling between parents and offspring around the time of weaning in primates, including humans, "make adaptive sense." It also created the expectation that offspring would deceive their parents about their needs in an attempt to get more resources. Evolutionary psychologists have seen this as an important insight. Parent/offspring conflict is "inherent to the human condition" (Pinker and Bloom, 1992, p. 483) and its "inevitability" is cited in support of the existence of psychodynamic mechanisms of deceit and self-deceit (Nesse and Lloyd, 1992). Yet the empirical evidence for parent/offspring squabbling over weaning is very weak. Patrick Bateson summarizes various studies which failed to find aggressive interactions at weaning in a wide range of species, studies which found offspring weaning themselves, and studies which found both parties engaging in reliable signaling in order to co-ordinate weaning (Bateson, 1994). This suggests that coordinating weaning is as important an evolutionary problem as obtaining an advantageous inter-generational distribution of resources. These empirical findings offer rich opportunities for adaptive explanation, and can probably be extrapolated to suggest new research on humans, but their discovery has been hindered by devotion to an *a priori* model of the evolutionary problem that behaviors surrounding weaning were supposed to solve.

An example of the ability of adaptive thinking to actually *drive out* existing empirical findings about the mind can be found in the evolutionary psychology of emotion. John Tooby and Leda Cosmides have predicted that emotions will turn out to be behavioral programs which are deployed in response to frequently recurring ecological situations: "Each emotion state—fear of predators, guilt, sexual jealousy, rage, grief, and so on—will correspond to an integrated mode of operation that functions as a solution designed to take advantage of the particular structure of the recurrent situation these emotions respond to" (Tooby & Cosmides, 1990, p. 410).

In an impressive program of research over many years, Paul Ekman and his collaborators have argued for the existence of a limited number of stereotyped, short-term emotional responses, corresponding very roughly to the short-term instances of the English emotion categories joy, anger, sadness, fear, surprise, disgust and, more tentatively, contempt (Ekman, 1972; Ekman & Friesen, 1986). These "affect programs" are complex responses integrating several physiological systems. They seem to be controlled by mechanisms that are significantly "informationally encapsulated" and hence conform to the "modular" conception of mind favored by evolutionary psychology. The affect programs display strong homologies with the emotional responses of other species, and in fact correspond remarkably well to the phylogenetically ancient expressions of emotion identified by Darwin in his efforts to show that emotions are not a uniquely human phenomenon (Darwin 1872/1965). Tooby and Cosmides treat the affect program as if it confirmed the predictions of their "adaptive thinking." But in fact, in many important respects what is known about the affect programs contradicts these predictions and points up a deep problem with "adaptive thinking" in psychology.

Tooby and Cosmides suggest that emotions are solutions to very specific ecological problems. They suggest that the emotion module will be programmed to respond to cues such as "looming approach of a large fanged animal" for fear and "seeing your mate have sex with another" for a postulated emotion of sexual jealousy (Tooby and Cosmides, 1990b, p.441). But although the *output* of an affect program emotion is stereotyped and pan-cultural, the *input* to the affect program—the emotional stimulus—is very flexible and varies between cultures and individuals. The "stimulus appraisal mechanism" controlling the affect programs simply did not evolve as a collection of adaptations to specific stimulus situations like those Tooby and Cosmides envisage. John B. Watson famously used the fact that babies do not show the fear response to classic phobic stimuli like snakes and spiders to support his view that nearly all human behavior results from individual conditioning (Watson, 1930). Newborn babies respond to loud sounds and loss of balance with fear, to prolonged restraint with rage, and to gentle forms of skin stimulation with pleasure. They respond consistently to very little else. Later research has added only one important thing to the list of neonate responses known in Watson's day. Newborn babies are extremely responsive to human facial expressions (Meltzoff & Moore, 1977; Izard, 1978; Trevarthen, 1984). In later life fear is produced by any stimuli an individual has come to associate with danger, sadness by stimuli associated with loss, and so forth. Contrary to the predictions of the evolutionary psychologists, affect programs are designed to cope with quite general evolutionary problems, and the affect program system is designed to redefine those problems as the environment changes.

J.B. Watson's extreme environmentalism was mistaken, not because there are more inborn or "maturing" emotional responses, but because learning is itself a complex ability with an evolutionary history. Associations between classic phobic stimuli and fear may be easier to acquire and harder to lose than associations between fear and other stimuli like flowers and colored shapes (Ohman, Fredrikson et al., 1976). The ways in which new emotion stimuli are potentiated are also significant. Associations are acquired through the child's own experience, but they are also acquired through observations of adult emotional responses to the situation (Klinnert, Campos et al., 1983).

The failure of adaptive thinking in this case makes two important theoretical points about the weakness of the adaptive heuristic. First, the affect program system seems to represent a subtle compromise between the need for flexibility in a changing world and the need to learn without many expensive trials. This particular compromise could not be predicted in advance by imaginatively reconstructing the evolutionary process. Second, the example of emotion stimuli exemplifies what Kim Sterelny and I have called the "grain problem" for evolutionary psychology (Sterelny and Griffiths, 1999), a problem also exemplified by the case of parent-offspring conflict over weaning described above. Tooby and Cosmides used an overly fine-grained analysis of the adaptive problem, treating "fear of predators" as a separate problem in itself. Evolution treated it as apart of a single larger problem—"fear in general." Knowing what constitutes one adaptive problem has a great deal to do with the number of degrees of freedom available to the organism to adjust its phenotype, since this determines whether various "problems" can be disentangled one from another. Such matters of the developmental structure of the organism are, of course, central to the literature on anti-adaptationism.

3. REVERSE ENGINEERING

So far I have concentrated on the pitfalls of adaptive thinking, or inferring the solution from the problem. The other form of adaptationist reasoning favored by evolutionary psychologists is inferring the problem from the solution, or "reverse engineering." Looking at the performance profile of a cognitive adaptation allows the evolutionary psychologist to deduce the evolutionary forces that have produced it, just as taking apart a piece of machinery allows a rival company to work out how it was produced. The problem with this strategy is simply that an adaptive explanation can be very plausible, but false. This is obvious when there are several well-developed adaptive explanations, as there are for the explosion in brain size in hominid evolution. The runaway selective process initiated by the invention of transmitted cultural skills, the "Machiavellian" needs of a social primate or the importance of thermoregulation on the savannah, or some particular combination of these and other causes may all have been the motor of evolutionary change in this case. The adaptationist normally recommends further development of each hypothesis so that its predictions become more detailed. The hope is that there will eventually be only one candidate that correctly predicts the observed traits. But an adaptive hypothesis can be falsified even when it adequately predicts the observed trait and has no rivals. If Mary McKitrick is correct, low birth weight in bears is not the cost they pay for the physiological ability to hibernate, as has been suggested (McKitrick, 1993). That hypothesis explains the trait and has no rivals, but is contradicted by the comparative data. Low birth weight occurs on the tree *before* hibernation and is preserved on branches of the tree where hibernation never originated, as in Giant Pandas. The adaptationist who does not standardly conduct comparative analyses of this kind presumably thinks that a really detailed adaptive scenario that predicts the observed traits is so likely to be true that it is not worth running the battery of available comparative tests. This is a quite extraordinary view. There are many other examples of comparative tests falsifying very plausible adaptive hypotheses (Brooks & McLennan, 1991; Harvey & Page, 1991). The adaptationist neglect of comparative testing in favor of testing the accuracy of predictions of the observed trait is profoundly unscientific. It systematically ignores chances to bring current presuppositions up against empirical data. That is a recipe for reinforcing existing assumptions and slowing down theoretical advance.

Although evolutionary psychologists are aware that their hypotheses make predictions that can be tested by the comparative method, the extensive range of quantitative methods that have become available in recent years and which are familiar to behavioral ecologists simply have not become part of the new discipline. For example, Margie Profet has made a major contribution to evolutionary psychology through her extensive work on the evolution of pregnancy sickness, which she hypothesizes may be an adaptation to protect the fetus from teratogenic substances (Profet, 1992). She makes a qualitative comparative prediction that pregnancy sickness or an equivalent should be most prevalent in exploratory omnivores like humans and rats, less prevalent in specialist herbivores, and least prevalent in dedicated carnivores. Although Profet does call for more work on this subject, she herself only discusses a few isolated cases that fit the prediction. Profet follows the same procedure in her work on the evolution of menstruation (Profet, 1993). Her "promiscuous primate" hypothesis suggests that menstruation flushes the uterus of sexually transmitted pathogens, and hence predicts that it will be associated with primate breeding systems in which females have a large number of different partners. Profet discusses three isolated comparisons between small groups of primate species and finds them generally in line with her hypothesis. Beverly Strassman has subjected the "promiscuous primate" theory to a proper comparative analysis (Strassman, 1996). Strassman treated menstruation and promiscuity as characters in a cladistic analysis, with increasing values of each as the different character states. She tested for an association between the evolution of promiscuity and menstruation in an independently estimated phylogeny for the primate species. She found no association between the two character states, a conclusion that was robust under several variations of the analysis, such as dividing the continuous character of menstruation into different discrete steps. Clearly, the same procedure could be applied to Profet's pregnancy sickness hypothesis, and the result would be a substantial, quantifiable, empirical test against independently collected data of a hypothesis that can otherwise only be debated on the grounds of plausibility.

One reason for the popularity of the adaptationist program is the mistaken impression that the alternative is to eschew evolutionary thought as a source of insight into the structure of organisms. Nothing could be further from the truth. The alternative to the adaptationist program is a richer approach to evolutionary explanation. A phylogenetic, or adaptive-historical approach recognizes the essential historicity of evolutionary explanation. This historicity has more than one aspect. First, there is the role of historical contingency in evolution. The outcomes of evolutionary processes do reflect the ecological relationship between organism and environment, but they also reflect stochastic factors in the selection process and the availability and order of occurrence of alternative forms. The role of these factors in determining outcomes makes it impossible to read back and forth from problem to solution and solution to problem in the manner of the adaptationist. Instead, empirical methods for the reconstruction of history, especially modern cladistic versions of the comparative method, are central to the study of adaptation. Adaptive thinking, even in its quantitative incarnations of optimization theory and game theory, does not yield explanations; it suggests adaptive hypotheses for comparative testing. A second aspect of historicity is the importance of phylogeny to thinking about evolution. Living organisms are at the end of lines of descent that pass through many different ecologies. The form that results from any one phase of adaptation reflects the previously evolved form of the organism as well as the ecology it encountered. Generalizations about the effects of particular ecological factors on the organism are likely to be richer and more reliable when confined within taxonomic groups that approach the "problem" with a common inheritance of resources with which to solve it. Hence adaptive analysis and phylogenetic analysis must go hand in hand. Both aspects of historicity support the view that taking an evolutionary perspective on the organism (or its mind) means seeking to reconstruct the phylogenetic trajectory by which the organism reached its current state, not analyzing its fit to some single environment. Natural selection operates repeatedly along this trajectory, with its agenda set not just by the environment but also by past history as embodied in the organism.

These phylogenetic-historical themes are borne out by the work on menstruation described above. Strassman found that menstruation has a strong phylogenetic signature. It is a character of the catarrhine group (old world monkeys, apes and humans). The endometrium in this group has a different vascular structure from that of other primates. By using more sophisticated comparative methods, Strassman was able to show that the distribution of menstruation in primates can be rather precisely accounted for by the hypothesis that the vascular structure of the endometrium in catarrhine primates does not allow full endometrial reabsorption at the end of the menstrual cycle in species which have a high litter mass to body mass ratio and therefore require a proportionally large endometrial structure (Strassman, 1996). One exciting thing about this alternative hypothesis is that it stitches adaptation and history together so seamlessly. Asking whether menstruation is being explained by adaptation or developmental constraint, or by how much of each, is obviously forced and quite irrelevant to the scientific project in which Strassman is engaged. Menstruation is being explained by evolution.

Kim Sterelny (pers. comm.) has suggested that claims about human cognitive adaptations will be more difficult to subject to rigorous comparative testing than many other aspects of evolution. Human cognitive adaptations are supposed to have evolved after the separation of hominids from other primates. *Homo sapiens* is the only living representative of the lineage involved. It is therefore not possible to test hypotheses about these adaptations by looking at the distribution of homologous traits in related taxa. This is a legitimate worry, but it should not be overstated. Natural history can creep up on the period in which our lineage became distinctively human from both sides (evolutionary psychologists tend to refer to this period as "the Pleistocene"). Many emotion theorists have crept up on the Pleistocene from behind. Darwin and his modern successors found extensive homologies between human emotions and those of other primates (Chevalier-Skolnikoff, 1973). New findings about the psychological and neurological bases of emotion can only enrich this collection of homologies. Similar results should be possible in many other areas of psychology. The discovery of primate homologies does not simply illuminate the history of the homologous structures themselves. Those structures and their history provide a framework with which evolutionary scenarios for uniquely hominid cognitive structures must be consistent. This consistency requirement creates extensive opportunities for testing those scenarios.

The possibility of creeping up on the Pleistocene from in front has been created by developments in molecular biology and historical linguistics. There is a discernible phylogenetic structure within the species *homo sapiens*, as there is in most widespread species. Modern molecular biology makes possible extensive investigations of the history and biogeography of human populations (Cavalli-Sforza, Menozzi et al., 1994). Further access to this structure comes from historical linguistics. Phylogenetic trees for human populations can be constructed on the basis of their languages. The congruence between the structures discovered by these two means is very considerable (Penny, Watson et al., 1993). These discoveries make it possible to discern homologies within *Homo sapiens* and to infer the "ancestral" condition of many traits (the condition which existed before the separation of current human populations). Obviously, only these traits can be explained as an inheritance from an ancient hunter-gatherer lifestyle. The access to history provided by these methods is limited by the fact that human populations only began to separate 150-200,000 years ago. Nevertheless, determining whether a trait was ancestral in humanity as a whole should be an essential precursor to any attempt to explain that trait by conditions in the Pleistocene. Hypotheses about the evolution of aesthetic preferences for landscape (Orians and Heerwagen, 1992) or the evolution of sexual attraction might well be eliminated by such a test. This test should also be applied to many of the putative emotional adaptations, such as the "sense of fairness" postulated by Robert Frank (1988).

4. THE PHYLOGENETIC PERSPECTIVE ON THE MIND

An adaptive-historical approach to evolution can play a heuristic role for psychology, although not quite in the way outlined by current evolutionary psychologists. The adaptive heuristic is supposed to take three forms, as outlined above. First, adaptive thinking is to provide the task descriptions for cognitive mechanisms. An adaptive-historical approach to psychology would be more skeptical of using *a priori* reflection on what seems likely to have evolved as a heuristic for finding cognitive adaptations. The new approach would suggest a phylogenetic heuristic. A phylogenetic heuristic would look for cognitive systems in humans which are elaborations of mechanisms seen in other species and which therefore may have existed in a common ancestor. In this way the adaptive-historical approach revives neglected elements of the evolutionary approach to behavior of classical ethologists like Lorenz and Tinbergen. The second role of the adaptive heuristic is to provide general "solvability constraints" that narrow down the class of possible behavioral mechanisms. Once again, the adaptive-historical approach would suggest a more phylogenetic perspective. Rather than ruling out certain classes of mechanism on general grounds we would take note of patterns in the comparative data. For example, the North American coyote's poison avoidance mechanism uses taste to potentiate food odors, while on the same prairie the red-tailed hawk uses taste to potentiate visual food cues. The two species solve the adaptive problem in a way that reveals their kinship to the laboratory rat and laboratory pigeon, respectively (Garcia & Rusiniak, 1980). It would have been very surprising if the mechanisms had been reversed. Finally, evolutionary psychologists look to evolution to make sense of the eclectic collection of mechanisms that make up the mind. Adaptive thinking is supposed to reveal the true purpose of these mechanisms. An adaptive historical approach would welcome the construction of adaptive scenarios for known behavioral regularities and/or cognitive mechanisms. But these scenarios are hypotheses waiting to be tested by the comparative method. Furthermore, well-constructed adaptive hypotheses will themselves have historical content. They will suggest that at a certain point in a lineage an ancestral form already equipped with such-and-such features acquired some new feature. The order imposed on the findings of psychology by this sort of adaptive reasoning will have a phylogenetic dimension. Tooby and Cosmides expect cognitive traits to make sense as suites of adaptations fitting humans to their Pleistocene "environment of evolutionary adaptedness." I expect them to make sense as a series of traits acquired, retained and modified as the lineage leading to *Homo sapiens sapiens* passed through many speciation nodes and many different selective environments. The origin of these traits will be reflected in their distribution in the nested hierarchy of primate taxa, hominid taxa, and human populations.

An adaptive-historical perspective is likely to be particularly valuable in the case of emotion. Darwin established a characteristic pattern of explanation for emotional expressions that is clearly adaptive-historical in form (Griffiths, 1997). Many of the communicative elements of emotional responses are arbitrary: Their form is not intrinsically suited to their function. There is no reason why a human being should use one facial expression to convey a particular signal, rather than another. Darwin's explanation of such traits can be disassociated from his Lamarckian theory of inherited habits to yield an explanation that holds up very well today. He argued that behaviors originally serving some more utilitarian function in the situation associated with the emotion, such as tooth-baring in primate anger, were recruited as signals of emotional state. They were retained, perhaps in a modified form, when their original functions ceased to be important. Classical ethologists referred to this as the "ritualization" of behavior (Tinbergen, 1952). If Darwin was right, then emotional expressions can only be understood as the result of successive modifications of a single structure through a number of different adaptive phases. The response of the lineage to each "adaptive problem" was not predictable unless the particular resources it brought to the problem were taken into consideration. Tooth-baring and pilo-erection would not have their current communicative function if it had not been for the earlier, agonistic functions of which they are vestiges:

> With mankind some expressions, such as the bristling of the hair under the influence of extreme terror, or the uncovering of the teeth under that of furious rage, can hardly be understood, except in the belief that man once existed in a much lower and animal-like condition. The community of certain expressions in distinct though allied species as in the movements of the same facial muscles during laughter by man and by various monkeys, is rendered somewhat more intelligible, if we believe in their descent from a common progenitor. (Darwin, 1872/1965, p. 12)

The role of history is also apparent in current hypotheses about the evolution of the general properties of emotions. Affect program theory explains the "passivity" of emotions in terms of modular information processing. We do not choose to produce an affect program response, and such a response may be at odds with our conscious evaluation of the situation confronting us. The existence of these information processing arrangements is hypothesized to be the result of incorporating into a highly flexible and intelligent cognitive mechanism an existing, less intelligent apparatus for responding to certain critical classes of event (Zajonc, 1980). A purely adaptive explanation of passivity would have to argue that there is a general selective pressure for responses to certain classes of ecological situations to become isolated from long-term, planned action. The "grain problem" discussed above suggests that such ahistorical explanations are unlikely to be forthcoming. The very notion of "a class of ecological situations" presumes an account of the cognitive resources available to the organisms to distinguish situations or to respond differentially to them. I have argued at more length elsewhere that Darwin's adaptive-historical perspective on emotion accounts for the great productivity of the research traditions in emotion theory derived directly from Darwin's work on emotional expression, as compared to those inspired by more adaptationist conceptions of evolution (Griffiths, 1998). Darwinian psychology should retain its evolutionary and phylogenetic perspective on the mind whilst shedding its adaptationism.

REFERENCES

Bateson, P. 1994. "The Dynamics of Parent-offspring Relationships in Mammals", *TREE* 9 (10), 399-402.

Brooks, D.R. & McLennan, D.A. 1991. *Phylogeny, Ecology and Behavior*. Chicago: Chicago University Press.

Cavalli-Sforza, L.L., Mennozzi, P., et al. 1994. *The History and Geography of Human Genes*. Princeton: Princeton University Press.

Chevallier-Skonikoff, S. 1973. "Facial Expression of Emotion in Non-human Primates", in Ekman, P., ed., *Darwin and Facial Expression: A Century of Research in Review*. New York and London: Academic Press, pp.11-89.

Cosmides, L. & Tooby, J. 1992. "Cognitive Adaptations for Social Exchange", in Barkow, J.H., Cosmides, L., & Tooby, J., eds., *The Adapted Mind: Evolutionary Psychology and the Generation of Culture*. NY & Oxford: Oxford Univ. Press, pp.163-228.

Cosmides, L., Tooby, J., et al. 1992. "Introduction: Evolutionary Psychology and Conceptual Integration", in Barkow, J.H., Cosmides, L., & Tooby, J., eds., *The Adapted Mind: Evolutionary Psychology and the Generation of Culture*. NY & Oxford: Oxford Univ. Press, pp.3-15.

Darwin, C. 1872/1965. *The Expression of the Emotions in Man and Animals*. Chicago: University of Chicago.

Ekman, P. 1972. *Emotions in the Human Face*. New York: Pergamon Press.

Ekman, P. & Friesen, W.V. 1986. "A New Pan-cultural Facial Expression of Emotion", *Motivation and Emotion*, 10 (2), 159-168.

Garcia, J. & Rusiniak, K.W. 1980. "What the Nose Learns from the Mouth", in Muller-Schwarze, D. & Silverstein, R.M., eds., *Symposium on Chemical Signals in Vertebrates & Aquatic Invertebrates*. NY: Plenum Press, pp.141-155.

Griffiths, P.E. 1995. "The Cronin Controversy", *British Journal for the Philosophy of Science*, 46, 122-138.

Griffiths, P.E. 1996. "The Historical Turn in the Study of Adaptation", *British Journal for the Philosophy of Science*, 47 (4), 511-532.

Griffiths, P.E. 1997. *What Emotions Really Are: The Problem of Psychological Categories*. Chicago: University of Chicago Press.

Harvey, P.H. & Pagel, M.D. 1991. *The Comparative Method in Evolutionary Biology*. Oxford & NY: Oxford University Press.

Izard, C. 1978. "On the Development of Emotions and Emotion-Cognition

Relationship in Infancy", in Lewis, M. & Rosenblum, L., eds., *The Development of Affect*. NY: Plenum Press, pp.389-413.

Klinnert, M.D., Campos, J.J., et al. 1983. "Emotions as Behavior Regulators: Social Referencing in Infants", in Plutchik, R. & Kellerman, H., eds., *Emotion: Theory, Research and Experience* Vol.2. New York: Academic Press, pp.57-86.

Lorenz, K. 1965. Preface to *The Expression of the Emotions in Man and Animals* by Charles Darwin. Chicago: University of Chicago.

Marr, D. 1982. *Vision*. New York: W.H. Freeman.

McKitrick, M. 1993. "Phylogenetic Constraint in Evolutionary Theory: Has it Any Explanatory Power?", *Annual Review of Ecology and Systematics,* 24, 307-330.

Meltzoff, A.N. & Moore, M.K. 1977. "Imitation of Facial and Manual Gestures by Neonates," *Science,* 198, 75-78.

Nesse, R.M. & Lloyd, A.T. 1992. "The Evolution of Psychodynamic Mechanisms", in Barkow, J.H., Cosmides, L., & Tooby, J., eds., *The Adapted Mind: Evolutionary Psychology and the Generation of Culture.* NY & Oxford: Oxford Univ. Press, pp.601-626.

Ohman, A., Fredrikson, M., et al. 1976. "The Premise of Equipotentiality in Human Classical Conditioning", *Journal of Experimental Psychology,* 105, 313-337.

Orians, G.H. & Heerwagen, J.H. 1992. "Evolved Responses to Landscape", in Barkow, J. H., Cosmides, L. and Tooby, J., eds., *The Adapted Mind: Evolutionary Psychology and the Generation of Culture*. NY & Oxford: Oxford Univ. Press, pp.555-579.

Penny, D., Watson, E.E., et al. 1993. "Trees From Languages and Genes are Very Similar", *Systematic Zoology,* 42, 382-5.

Pinker, S. & Bloom, P. 1992. "Natural Language and Natural Selection", in Barkow, J.H., Cosmides, L., & Tooby, J., eds., *The Adapted Mind: Evolutionary Psychology and the Generation of Culture*. NY & Oxford: Oxford Univ. Press, pp.451-494.

Profet, M. 1992. "Pregnancy Sickness as Adaptation: A Deterrent to Maternal Ingestion of Teratogens", in Barkow, J.H., Cosmides, L., & Tooby, J., eds., *The Adapted Mind: Evolutionary Psychology and the Generation of Culture*. NY & Oxford: Oxford Univ. Press, pp.451-494.

Profet, M. 1993. "Menstruation as a Defense Against Pathogens Transported by Sperm", *Quarterly Review of Biology*, 68, 335-386.

Richards, R.J. 1987. *Darwin and the Emergence of Evolutionary Theories of Mind and Behavior*. Chicago: Univ. of Chicago Press.

Sterelny, K. & Griffiths, P.E. 1999. *Sex and Death: An Introduction to the Philosophy of Biology*. Chicago: Univ. of Chicago Press.

Strassman, B.I. 1996. "The Evolution of Endometrial Cycles and Menstruation", *Quarterly Review of Biology,* 71 (2), 181-220.

Symons, D. 1992. "On the Use and Misuse of Darwinism in the Study of Human Behavior", in Barkow, J. H., Cosmides, L., & Tooby, J., eds., *The*

Adapted Mind: Evolutionary Psychology and the Generation of Culture. NY & Oxford: Oxford Univ. Press, pp.137-159.

Tinbergen, N. 1952. "Derived Activities: Their Causation, Biological Significance, Origin and Emancipation During Evolution", *Quarterly Review of Biology,* 27 (1), 1-32.

Tooby, J. & Cosmides, L. 1990. "The Psychological Foundations of Culture", in Barkow, J.H., Cosmides, L., & Tooby, J., eds., *The Adapted Mind: Evolutionary Psychology and the Generation of Culture.* NY & Oxford: Oxford Univ. Press, pp.19-136.

Trevarthen, C. 1984. "Emotions in Infancy: Regulators of Contact & Relationship with Persons", in Scherer, K., & Kerman, P., eds., *Approaches to Emotion.* New Jersey: Erlbaum, pp.129-162.

Trivers, R.L. 1974. "Parent-offspring Conflict", *American Zoologist,* 14, 249-264.

Watson, J.B. 1930. *Behaviorism (Revised Edition).* New York: W.W. Norton.

Zagonc, R.B. 1980. "Feeling & Thinking: Preferences Need No Inference", *American Psychologist,* 35, 151-175.

CHAPTER 9

EVOLUTION WITHOUT HISTORY: CRITICAL REFLECTIONS ON EVOLUTIONARY PSYCHOLOGY*

ROBERT C. RICHARDSON
University of Cincinnati

Abstract. Human language and reasoning are evolved capacities. Evolutionary psychology advances an explanation for them, following a typical pattern. This explanation assumes that relatively complex characters are adaptations to ancestral conditions, and evolutionary psychologists claim to explain their presence as a consequence of natural selection operating under Pleistocene conditions. As evolutionary explanations, they are subject to the same kind of expectations we would impose on any other evolutionary explanations. There are several broadly accepted ways to empirically assess evolutionary explanations—and particularly adaptive explanations. One of these is the comparative method. This requires that we compare related taxa for derived characters, evaluating whether analogous traits evolve within lineages under associated environmental conditions. In light of the sort of evidence that is actually available to us in the anthropological and paleontological record, we are not likely to be able to meet the standards we would otherwise require for adequate evolutionary explanations; in particular, the comparative method is not likely to apply in a way that would allow us reasonable confidence concerning the explanations offered within evolutionary psychology for human language and reasoning. Given the sorts of problems confronting evolutionary psychology, I suggest that this inadequacy is not likely to be overcome. A skeptical attitude is the reasonable attitude.

1. INTRODUCTION: EVOLUTIONARY PSYCHOLOGY

Human beings, like other organisms, are the products of evolution. Our psychological capacities are evolved traits as much as is our gait

H.R. Holcomb III (ed.), Conceptual Challenges in Evolutionary Psychology, 327–373.

or dentition. Human beings, like other organisms, embody traits that are the products of natural selection. Our psychological capacities are subject to natural selection as much, again, as is our gait or dentition. In this minimalist sense, there can be no real quarrel with evolutionary psychology; but evolutionary psychology does not settle for the minimalist claim. In practice, it embodies a far more ambitious, and more problematic, program. On the more ambitious program characteristic of evolutionary psychology, psychological processes are adaptations, not to present circumstance, but to our ancestral environment. That they are adaptations explains their presence, their design features and their failings. Advocates of evolutionary psychology often rely on very general arguments to show that these are adaptations. One begins with complexity of design. Features with complex functional designs must somehow have evolved, and must have provided a substantial advantage to our ancestors in virtue of that design. It is not enough that there be evolution or even natural selection. There must have been selection *for* that complex functional design. The presence of complex features must be the consequence of evolution by natural selection (Tooby & Cosmides, 1992, pp. 49 ff.; Pinker & Bloom, 1992). One goal of evolutionary psychology, correspondingly, is to provide evolutionary explanations of our natural psychological capacities in terms of natural selection. Leda Cosmides and John Tooby, to take the most prominent examples, claim that human reasoning consists of a set of mechanisms organized around social exchange. They describe the program this way:

> According to the evolutionary psychological approach to social cognition ... the mind should contain organized systems of inference that are specialized for solving various families of problems, such as social exchange, threat, coalitional relations and mate choice. ... Each cognitive specialization is expected to contain design features targeted to mesh with the recurrent structure of its characteristic problem type, as encountered under Pleistocene conditions. Consequently, one expects cognitive adaptations specialized for reasoning about social exchange to have some design features that are particular and appropriate for social exchange, but that are not activated by or applied to other content domains (1992, p. 166).

According to Cosmides and Tooby, the "evolutionary function" of human reasoning involves facilitating and monitoring social exchange and social relations. Human reasoning then would be a "cognitive adaptation" to social conditions encountered by our evolutionary ancestors, established and maintained by natural selection. The goal of evolutionary psychology, thus, is explaining psychological processes as biological adaptations to Pleistocene conditions, shaped and maintained by natural selection. According to evolutionary psychologists, this is true not only for social cognition but for a wide array of psychological mechanisms and social behaviors. Similar explanations have been offered by evolutionary psychologists for family structure, parental care, marital jealousy, sex roles, sexual preferences, familial affection, personality, and the moral sentiments, to mention just a few.

So what we are offered is an evolutionary claim. Such evolutionary hypotheses are subject to a variety of empirical tests, though rarely the kind of test that would warm the heart of more narrow experimentalists. The task involved in assessing an evolutionary explanation— whether of human psychology, or of clutch size in birds or of flower structure in orchids—is one of assessing the historical antecedents and context, given both the contemporary forms and the relevant historical record. These antecedents in turn provide evidence essential for a defensible evolutionary explanation.

Evolutionary psychology is a program within the general framework of evolutionary theory. The sorts of conjectures defended by evolutionary psychologists should at least be held to the same standard that properly is to be demanded of evolutionary explanations more generally. I will argue that if we apply those standards, evolutionary psychology does not fare well at all. What I will show is that the historical record, though wonderful in its detail and exciting in what it offers, is insufficient to support the explanations offered. If we were given an explanation for clutch size or flower structure based on similar evidence and reasoning, we would, and should, reject it as premature and inadequately supported. We should be no less uncompromising in our expectations for evolutionary psychology. We should correspondingly reject evolutionary psychology as unconstrained speculation. It is speculation portrayed as fact. It embodies evolutionary claims ungrounded in evolutionary history. It is important to understand that the moral I offer is a skeptical one. I will press that we have no credible reason to embrace the explanations offered within evolutionary psychology. I offer no alternative explanations. I do not claim even that these evolutionary hypotheses concerning human psychology are false. I claim instead that the historical record gives us no reason to accept them.

2. HUMAN REASON AS AN ADAPTATION

In his recent book, *How the Mind Works* (1997), Steven Pinker embraces the program of evolutionary psychology with characteristic and unrestrained enthusiasm, proclaiming:

> The mind is a system of organs of computation, designed by natural selection to solve the kinds of problems our ancestors faced in their foraging way of life, in particular, understanding and outmaneuvering objects, animals, plants and other people (p. 21).

In Pinker's view, like that of Tooby and Cosmides (see, for example, Cosmides & Tooby, 1987, pp. 294-302), the mind is modular in structure, embodying a variety of specialized "organs of computation," with each module specialized and adapted to our ancestral lives. The structure of each module, moreover, is "specified by our genetic program" (loc. cit.). As adaptations, the operations of these modules were molded to fit the "problems of hunting and gathering," which defined our ancestral environment. The modular structures, and modular functions, are selected for. Pinker suggests, furthermore, that the key to understanding human intelligence, unlike that of many other organisms, is its flexibility, that is, the human ability to modify behavior to fit variable and novel circumstances.[1] In a similar vein, Robert Nozick claims that natural selection should favor capacities whose exercise has increased fitness as a consequence of their reliability (1993, p. 113). In a locally stable and predictable environment, simple reflexes might suffice. Humans, however, respond to variations in circumstance, and this, Pinker and Nozick agree, is the key to understanding the evolutionary function of rationality (cf. Nozick 1993, p. 120). Rationality allows us to deal with "new and changing situations." On their view, that is the "proper function" of human reason.

Whereas Cosmides and Tooby see human reason as an adaptation for social exchange, somehow reflecting the social structure of primitive human social groups, Pinker offers a less specific portrait, derived from Tooby and Irven DeVore (1987). The "cognitive niche" occupied by humans, in Pinker's hands, portrays human intelligence as an adaptation shaped by a variety of features, including our gregariousness, the need for "information," tool use, hunting, and social coordination (cf. Pinker 1997, pp. 188 ff.). Human reason serves not only the purpose of social exchange, but many other purposes as well.

Robert Nozick is interested, similarly, in articulating the "evolutionary function" of rationality, which in turn depends on the processes that historically shaped and maintained it. Nozick is studiously vague about precisely what this function might be beyond the idea that rationality enables humans to cope with "new and changing situations."

In *Darwin's Dangerous Idea*, D.C. Dennett (1995) similarly defends the centrality of biological adaptation in understanding human psychology. In defending the view that "biology *is* engineering" (1995, ch. 8), Dennett proclaims that "Darwin's great insight was that all the designs in the biosphere could be the products of a process that was as patient as it was mindless, an "automatic and gradual lifter in Design Space" (1995, p. 188). Without denying that there are differences between the cognitive endowments of humans and other animals, and without denying the importance of culture, Dennett recognizes that "the brains we are born with have features lacking in other brains, features that have evolved under selection pressure over the last six million years or so" (1995, p. 371). The critical recognition, according to Dennett, is that anticipation has its advantages over trial and error from a Darwinian perspective, just as learned responses have advantages over fixed responses from a Darwinian perspective. It is at least true that in some contexts, anticipation and learning have advantages over their counterparts. The use of information—typically not information derived from our ancestors, but from our contemporaries —gives us a systematic advantage, and Dennett observes that that advantage has allowed humans to spread broadly across the Earth while ancestors apparently less adept did not, or if they did they were displaced. The similarities of Dennett's view to those of Pinker and Nozick are transparent. The conditions under which anticipation is better than trial and error, or learning is better than fixed responses, are the kinds of conditions under which flexibility would be favored. In an environment that is relatively uniform, in which a local sample is representative, the efficiency of fixed responses would favor them, and there would be no need for flexibility. Once evolved, this flexibility would facilitate our spread to diverse environments.

One key to the novel use of information characteristic of humans, Dennett thinks, is language. Neuropsychology has revealed structures in human brains that are unique, and implicated somehow in language use. He says, to my mind rather over-optimistically,[2]

> There is no question that the origin of language is theoretically a much easier problem than the origin of life; we have such a rich

> catalogue of not-so-raw materials with which to build an answer. We may never be able to confirm the details, but if that is so this will not be a mystery but only a bit of irreparable ignorance.

Dennett is following Pinker and Paul Bloom, who earlier defended the idea that human language is the product of Darwinian natural selection, designed to facilitate communication and coordination. Pinker and Bloom also assume that adaptive complexity requires an explanation in terms of natural selection (1990, section 3; 1992, pp. 454 ff.). Human language involves a complex capacity, governed by rules and representations that are not taught in any straightforward way: there are rules governing phrase structure that provide information about the underlying structure and meaning; grammars depend on lexical and phrasal categories that simultaneously limit what can be said and give language its remarkable expressive power. Pinker and Bloom exploit these general features in concluding that language shows "signs of design" (1992, p. 459). Language does seem to be a natural candidate for explanation in terms of evolution by natural selection. Nonetheless, if Dennett is right that the details may forever elude us, and that we may well be faced with an "irreparable ignorance," then whether the problem is the origin of life or the origin of human language, an answer advanced in the absence of a knowledge of the details is an hypothesis without a firm foundation. We would not accept an explanation of the origin of life without knowing the details. We should not accept an explanation of the origin of human language in ignorance. An answer that is offered with an "irreparable ignorance" of the details is no more than unbridled speculation. It does not warrant our assent.

We could, of course, remain optimistic about what we might come to know. Our ignorance might not be irreparable. The argument I will offer in what follows should not leave us optimistic about repairing our ignorance. If we ask whether the hypotheses characteristic of evolutionary psychology are even *likely* to be subject to the sort of treatment we should require of them if they are to be taken seriously as evolutionary explanations, then I think the most reasonable answer is not only that the sort of data we would want is not available, but also that it is unlikely to be forthcoming. This conclusion is certainly not demonstrated by anything I will say. It is theoretically possible that our ignorance is reparable. I do not think it is likely. The generality of the conclusion is also debatable. In focusing on human reasoning and language, I mean to fix on a case that in one way is a good case for evolutionary psychology. It is right that these are likely to be adaptations, in large part because they are complex capacities. The question is whether in these cases we have the kind of information we would need in order to know that some specific explanation of their presence is true. Again, I will urge that we do not. I conjecture that such knowledge is unlikely to be forthcoming. It certainly does not follow from this that evolutionary explanations lack credibility for, say, the emotions or passions. These have also been subjects of interest to evolutionary psychologists. In some ways they are better cases for evolutionary psychology, because there is a great deal of similarity between our tendencies and those of our primate kin. This was clear to Darwin. Perhaps these offer more plausible domains for evolutionary psychology. Again, I am skeptical but offer no argument to the contrary. On the other side, there are considerations that suggest that the negative conclusions I will argue for here are likely to be general. We already know a great deal about our human ancestors and our primate kin. We are not lacking information. One problem is that the kind of information we can glean from the record is not the sort that sheds substantive light on the origins of the human mind. Another problem is that the human case in many ways is not one hospitable to the kind of analysis evolutionary psychology requires. This will be my primary focus.

3. ADAPTATION AND HISTORICAL EXPLANATIONS

An *adaptation* is a trait that is present, or was maintained, because of the selective advantage it offered to ancestors; in this sense, to claim that something is an adaptation is to make an historical claim (cf. Brandon 1978, Burian 1983, Amundson 1996). Current use or advantage is not enough to show that a trait is an adaptation. To use an example from Darwin (1859), skull sutures are certainly adaptive in humans, since they facilitate passage through the birth canal. They have an important, and even essential, current use. They are adaptive. Skull sutures are nonetheless not an adaptation for parturition, since birds and reptiles also have skull sutures, though they hatch rather than being live born. As Darwin says, even though these skull sutures "no doubt facilitate, and may even be indispensable for" mammalian birth, they are not adaptations for that function. Adaptive features are not necessarily adaptations. What is required to show that a trait is an adaptation, and to provide an evolutionary explanation for a trait in terms of natural selection, is knowledge of historical antecedents and conditions. Without history, evolutionary explanation is empty.

Evolutionary psychologists not only recognize but also embrace the importance of history. It is no part of their argument that our psychological faculties are adaptations simply because they are adaptive in the current circumstance; indeed, when we turn to reasoning, our faculties may not be adaptive in our current setting, since we are systematically prone to errors in reasoning. Current adaptive value is not the point. What is crucial for the evolutionary claims at the heart of evolutionary psychology is that our psychological faculties were adaptive in our ancestral environment, or what is often, but rather misleadingly, described as "the Pleistocene environment." Their *current* presence must be a consequence of this *historical* adaptedness. The claim is explicitly historical. This raises the central challenge, namely, confirming the historical explanation. It is one that is certainly not insurmountable in principle. In practice, it may be difficult or impossible to meet. We cannot, except in unusual and limited circumstances, directly observe the action of selection. Instead, we must infer what processes were present on the basis of the products we observe. We must infer the historical process based on currently available variations and the fossil record. This is, as it turns out, not a trivial limitation, even though it is one that concerns practical limitations. How can the challenge be met?

The most direct approach to explaining adaptations draws on the extensive work done within population genetics. To show that something is currently adaptive, it is necessary to show that there is variation, and that there is some environmental feature that differentially favors one variant over another. A simple example illustrates the problem of showing that a feature is adaptive, that is, showing that its current function favors its persistence or presence. Hemoglobin is a protein that has been extensively studied from both a functional and an evolutionary perspective. Many of the features of the hemoglobin molecule do have some adaptive significance—such as its oxygen-binding properties—and are evolutionarily conserved. Genetically based variations such as hemoglobin S (involved in sickle cell syndrome) provide paradigm cases for genetic variations established and maintained by natural selection. There are, on the other hand, a number of hemoglobin variations within and between species resulting from amino acid substitutions whose adaptive significance and evolutionary importance are unclear. The overall structure, and thus the function, of hemoglobin molecules are largely conserved through a wide variety of amino acid substitutions. This suggests that many of these variations may be selectively neutral, with the differences between species established by drift rather than selection. We can in fact be reasonably certain that some blood type differences are the consequences of drift rather than selection, as the classic work of Cavelli-Sforza and his colleagues shows (cf. Bodmer & Cavelli-Sforza, 1976). However, unless population sizes are very limited, small selective differences can have dramatic effects over a longer time. That is, small physiological effects, resulting from genetically based variations, may have large evolutionary effects. With reasonable effective population sizes, we can expect selection coefficients as small as 10^{-3} or 10^{-4} to be evolutionarily significant. Differences of this magnitude are not impossible to detect in uncontrolled settings (Gillespie, 1991), but they are below what is likely to be detected in most natural settings. So, though we know that some variations do have adaptive significance, and though we know that some do not, there are a number of variations for which it is at the very least difficult and most likely impossible to assess their adaptive significance.

This moral concerns adaptive significance. Adding an historical dimension compounds the difficulties. It is here we can see the methodological problems in greatest relief. Showing that a trait is an adaptation involves showing not that it *is* adaptive, but that it *was* adaptive and that the adaptive features were critical in its evolutionary development. We need to fill in the details relevant to an adaptive explanation, inferring past states from current conditions. Robert Brandon (1990, ch. 5) claims that there are five conditions that we would like to see met by "adaptation explanations," or at any rate that an ideally complete adaptation explanation would meet. None are necessary, but all are relevant. How good a putative adaptation explanation actually is can, in his view, be seen as a matter of how nearly it meets these five conditions. The conditions Brandon defends are these:

(1) There must be evidence that selection has occurred: we need to be able to distinguish selection from other candidate evolutionary explanations, and to demonstrate something about the character and strength of selection; this will require that we have some information about the character and extent of variation in the ancestral forms.

(2) There must be an ecologically based explanation for selection: there must be some ecological factor explaining the adap-

tive differences, whether this is some factor in the physical environment or in the social environment.
(3) Differences between individuals must be heritable: there must be a correlation in the phenotypic traits of parents and offspring which is greater than would be expected by chance.
(4) There must be information concerning the environment, the population structure, and gene flow: effective population size, population structure, gene flow, interbreeding coefficients, and mutation rates all affect the rate of evolution; as a result, we need to be able to fix these parameters if we are to determine whether some observed effect is due to selection.
(5) We must know what traits are primitive and what traits are derived: Whether a trait is ancestral within a clade, whether it evolved within a single lineage, or whether its presence is the result of convergence among lineages is relevant to assessing whether something is an adaptation.

These five conditions reflect the constraints for showing that something is (currently) adaptive, but also recognize that there is an historical component to adaptation explanations. When the question is one of whether a trait is an adaptation, and not merely adaptive, the problem is one of inferring from observed patterns of genetic and phenotypic variation to the historical processes which yielded those patterns. Selection is but one of several possible explanations for existing patterns. If we are to distinguish natural selection as the explanation for some trait or pattern, and if we are to distinguish some particular adaptive scenario as the explanation for some trait or pattern, this is the sort of historical information required. Should selection be settled on as the explanation for some trait or pattern, alternative selective scenarios will also be possible. It is not impossible to show that selection is responsible for an observed evolutionary change. Likewise, it is not impossible to determine the selective agent. To take one important recent example, M. Kreitman has done elegant work on polymorphisms in the alcohol dehydrogenase locus of *Drosophila*, that shows unambiguously that these *are* maintained and established by selection (see Kreitman, 1983; Kreitman & Hudson, 1991). Nonetheless, it is notoriously difficult to distinguish the effects of selection from those of drift even in the simpler molecular cases (cf. Lewontin, 1974). As we will subsequently see, it is even more difficult when the traits of interest are phenotypically complex.

As I have argued elsewhere, the sort of information that is necessary for a complete explanation of either human intelligence or human language in terms of population genetics is not available, and it is not likely to be forthcoming (Richardson, 1996; cf. Lewontin, 1990; De Jong & van der Steen, 1998). We have no information concerning the strength of selection, or the extent of variation. We have at best a meager grasp on the ecological and social conditions that are relevant. Heritability values are unknown. The ancestral population structure is something we do not know, and the relevant ancestral traits are also unknown. This is exactly the sort of information that is necessary to show that a given evolutionary change is a consequence of natural selection rather than some other process, using the resources of population genetics. It is simply not available for evolutionary psychology. So if we are to defend evolutionary psychology, we must look elsewhere. There are alternative approaches to evaluating adaptation as an explanation.

4. REVERSE ENGINEERING AND THE STUDY OF CURRENT FUNCTION

The study of adaptation within an historical framework, as I have underscored, involves inferring historical process from contemporary products. Population genetics is a *transformational* or *dynamic* model especially suited to microevolution (Lewontin, 1969; Lauder, 1981). The focus is on historical sequence and causal antecedents, emphasizing prior conditions as determinants of contemporary patterns. Some biologists, especially evolutionary ecologists and many behavioral ecologists, focus instead on questions of current form and function irrespective of the historical paths that produced them. The question at the focus of investigation then is not historically based adaptation, but whether a trait is adaptive; that is, the question is the extent to which the trait currently improves or optimizes fitness among a specified set of variants and within a specific environment (cf. Reeve & Sherman, 1993). The focus is typically on equilibrium states, in which there is a balance between selective forces or in which a population is at a fitness maximum. The specific history is lost, and irrelevant, since any number of histories would lead to the same equilibrium state. Evolutionary explanations are constructed from an assessment of current utility.

The architects of evolutionary psychology generally resist an ahistorical approach to the study of adaptation, but the method has left its mark, and certainly has a salient place in biological research.

Dennett claims that in an important way biology is engineering and that "reverse engineering" is the critical step in the analysis of adaptation (1995, pp. 213 ff.). Pinker expresses a similar thought this way:

> A good adaptationist explanation needs the fulcrum of an engineering analysis that is independent of the part of the mind we are trying to explain. The analysis begins with a goal to be attained and a world of causes and effects in which to attain it, and goes on to specify what kinds of designs are better suited to attain it than others (1997, p. 38).

There is a long tradition within biology that recognizes the importance of something like reverse engineering.[3] To pick one prominent example, R.A. Fisher recognized in 1930 that life history evolution was likely to involve tradeoffs between growth and reproduction; and he provided the classic explanation for the prevalence of a 1:1 sex ratio. What Fisher argued was that under a wide variety of conditions, a 1:1 sex ratio is the stable equilibrium, and that a population which deviates from a 1:1 ratio would tend to evolve toward this equilibrium. The trick is to recognize that fitness effects can extend beyond one generation. If a population has a skewed ratio with more of one sex than another, then, other things being equal, an individual's fitness will be enhanced by investing preferentially in the sex that is less common. The result is a natural dynamic leading to a 1:1 sex ratio. It is possible to explain why some populations deviate from this expectation. Other things are not always equal. In the idealized case, Fisher assumes, for example, that the cost of bringing male and female offspring to maturity is equal. If they are not, then one result can be a deviation from the anticipated 1:1 equilibrium. What is important for our purposes is that the explanation of why populations typically exhibit a 1:1 ratio, and why some do not, is ahistorical: A 1:1 ratio is, under a wide variety of conditions, the equilibrium condition and an equilibrium toward which populations will tend to evolve. It is what is called an "attractor." The history of actual populations that exemplify a 1:1 ratio is obscured by the existing equilibrium. For all that Fisher argued, it could be that a given population at equilibrium had never deviated from that equilibrium, that there had never been variation from the equilibrium, and that it was not even subject to stabilizing selection. That would make the equilibrium no less of an equilibrium. There are a variety of purposes to which an ahistorical treatment might be put. The question is whether, or when, ahistorical analyses can shed light on evolutionary questions.[4]

Richard Lewontin and Stephen Jay Gould famously have attacked what they call the "adaptationist program," the view that natural selection is "so powerful and the constraints upon it so few that direct production of adaptation through its operation becomes the primary cause of nearly all organic form, function and behavior" (1978, p. 76). In place of a complacent commitment to adaptation as *the* explanation of organic form, Gould and Lewontin argued for a more pluralistic approach to evolutionary biology, in which a wider array of alternatives to natural selection needs to be systematically considered. Showing that some trait is the product of natural selection requires eliminating alternative explanations, or at least showing that they are less likely. Though each of the two has independently raised a variety of issues, with somewhat different emphases, the paper often taken to be representative of an anti-adaptationist perspective is "The Spandrels of San Marco and the Panglossian Paradigm" (1978). Gould and Lewontin there offer a variety of alternatives to natural selection, including genetic drift, but the image of the spandrels of San Marco suggests strongly the role that developmental or structural constraints might play in evolutionary processes (cf. Alberch, 1982: Gould, 1977: Maynard Smith, et al., 1985). The attraction of the idea is not difficult to see. The construction of a phenotype from a genotype is a complex affair, and the idea that the phenotype could be indefinitely molded to meet the demands of the environment is unrealistic. Pleiotropy and epistasis entail that there are nonlinear, and non-additive, interactions among genes. There are dependencies in ontogeny that entail that characters cannot be readily decoupled (cf. Schank & Wimsatt, 1988; Wimsatt & Schank, 1988). Suites of characters evolve in concert, in ways that are impossible to predict in the absence of specific developmental information. What has emerged in the twenty years since the "Spandrels" paper, though by no means solely as a result of that paper, is in fact a much more eclectic, more pluralistic, research program in evolutionary biology, with more sensitivity to problems involving developmental limitations, more awareness of the deep conservatism in underlying genetic mechanisms, the significance of phylogenetic analysis, stochastic effects on evolution, and at the same time a more nuanced understanding of the action of natural selection. Among other things, the more rigorous standards imposed on studies of natural selection have benefited the study of natural selection "in the wild" tremendously, illustrating its role and prevalence (cf. Endler, 1986).

Though Gould and Lewontin are often taken, with some cause, to be promoting a more ambitious agenda, the use of reverse engineering in the analysis of adaptation is perhaps the clearest example of the sort of research they criticize. They claim that the "adaptationist program" typically involves two steps. First, an organism is "atomized" into traits that are explained as structures optimally designed by natural selection. This often amounts to assuming that genetic variation is both extensive and additive, and that as a consequence phenotypes are almost indefinitely malleable. Evolution can mold organisms almost indefinitely. These are in fact exactly the conditions under which optimization would be expected. Second, since the attempt to so treat traits as independent is almost invariably unrealistic and since this lack of realism is almost immediately evident, interaction effects are acknowledged and incorporated as "trade-offs." Competing demands on organismal design are acknowledged, but without compromising the commitment to optimal design. Instead of simple optimization, we are offered optimization under some set of side constraints. Developmental constraints are acknowledged, and treated as limitations to adaptive design, though often as limitations that are extremely soft.

An example may help. John Maynard Smith et al. (1985) suggest that one consequence of developmental constraints may leave some "phenotypes accessible or inaccessible, given any developmental mechanism" (1985, p. 269). Developmental mechanisms may be such that some phenotypes simply cannot be grown. Reeve and Sherman transform the idea, and gloss it as the suggestion that phenotypes are developmentally constrained "when the alternative phenotypes have lower fitnesses because they depend upon developmental processes that would seriously disrupt the original developmental program, thereby reducing viability" (1993, pp. 21-2). There can be selection against some phenotypes because their ontogenetic predecessors are less viable, or inviable. Perhaps this explains some developmental conservatism, such as the presence of gill slits in early embryonic forms of terrestrial mammals. Whether this is so or not is a complicated issue. I want to insist on a different point. Understood in Reeve and Sherman's way, we do not have developmental constraints on evolution; rather, they would be developmental consequences of evolution by natural selection. It is one thing to emphasize limitations on growth; the corresponding developmental constraints would show up as limitations on the extent of genetic or phenotypic variation. These are developmental constraints. It is another thing altogether to emphasize limitations on the viability of intermediate forms as Reeve and Sherman do; in this case, the existence of variation is assumed and the "constraints" are treated as adaptive consequences of evolution orchestrated by natural selection. Reeve and Sherman assume (reasonably) that the presence of gill slits in early human embryogenesis is a consequence of the inviability of intermediate forms, rather than a result of some other explanation such as lack of variation. They assume we are explaining an adaptation. This is itself an example of the kind of assumption originally criticized by Gould and Lewontin. Reeve and Sherman think of developmental "constraints" as themselves the consequences of adaptation, constrained this time by developmental dependencies, but present as a result of selection. By contrast, Maynard Smith, et al., think of developmental constraints as limitations on the range of available organic forms. In either case, we are faced with the prospect that some phenotypes are simply unavailable to selection because of a lack of variation. If these are properly developmental constraints, then the observed form was not a consequence of selection since there was no antecedent variation. If they are now absent as a consequence of selection, then their maintenance is not a consequence of selection since the variation has been eliminated. In the absence of variation—from whatever cause—there can be no selection. As a consequence, the "optimal" design may simply not be accessible to evolutionary processes.

There are a number of forms that reverse engineering can take, with varying purposes (see Amundson, 1994; Amundson and Lauder, 1994). In one form, information concerning constraints is incorporated into evolutionary models as constraints on the available range of biological form, specified prior to the assessment of adaptation. Thus, marine water striders, *Halobates*, can move across the surface of seawater with amazing quickness, and also can jump and land on the surface of the water. There are two physical constraints that make this remarkable trick possible. First, like many insects, the water strider has a coating that makes it resistant to surface wetting. Second, there is a match between the total perimeter of its feet (which determines the contact line for the striders) and surface tension (which varies with salinity). For a water strider with a mass of roughly ten milligrams, the contact line would need to be roughly 1.3 millimeters total in order for it to move about and roughly 13 millimeters of contact line for it to sustain a jump (cf. Vogel, 1988). *Halobates* meets the expectation from the physical model. It has specialized hairs to increase its total contact, thus making both movement and jumping possible. In cases such as this, the constraints are easily defined beforehand, since they depend on physical parameters (body mass and surface tension). This use of constraints is, in itself, not objectionable. It amounts to using environmental, or developmental, information to structure the problem, constraining the alternatives systematically. It is, though, difficult to use this approach generally. George Lauder observes:

> The claim that relevant design criteria can be specified *a priori* to allow the analysis of biological design amounts to a claim that we can specify in advance the problem or problems that the design is supposed to solve. Although it is almost always possible to specify *some* design criterion, the more complex the design, the less likely it is that we will be able to determine what the relevant performance and mechanical functions are that any given structure needs to solve. And furthermore the less likely it is that we will be able to meaningfully weigh alternative performance goals (1996, p. 71).

The difficulty is simply that *a priori* design constraints are often not specified, and in many cases cannot be specified before the fact.

The alternative is to develop design constraints *a posteriori*, using the performance of the phenotype to determine the environmental factors that shaped that design (see Beatty, 1988). The point of MacArthur's (1957, 1960) early ecological work on the abundance of species in the "broken stick" model was that the existing data on relative abundance of species fit better with a model based on non-overlapping niches. These data, MacArthur concluded, fit better with an analysis assuming competitive exclusion as the cause of species distributions than with an analysis assuming species distribution patterns are caused by, say, abiotic factors. He argued on the basis of current distributions that competition was the evolutionary factor molding niche preferences. The analysis of pattern is, for MacArthur, a vehicle for understanding evolutionary history. MacArthur found a rough qualitative fit between the observed distribution of species and a competitive model. The fit is by no means perfect, though. Common species are more abundant than the models would predict, and rare species are less abundant than predicted. This led to a second stage in his analysis. Faced with a partial failure of fit between the data and the model, MacArthur shifted to an *a posteriori* use of design. MacArthur saw that if the environment were heterogeneous, then it would be possible to improve the fit between data and model. He went on to reverse the dependence, estimating environmental heterogeneity in terms of the failure to fit the model: "The divergence from the ideal curve may, in fact, be regarded on this hypothesis as a measure of heterogeneity. Experimentally, for bird communities, this appears to explain most of the 'steep' curves" (MacArthur 1957, p. 293). This is a standard strategy in handling a mismatch between model and prediction, in order to explain apparent lack of optimal design (see Kingsland, 1985). Given a preliminary analysis in terms of some putative causal factor that exhibits only a partial fit, the deviation from initial predictions is explained by superimposing a second application of the same optimization model, taking up the slack in the fit by *assuming* optimal design. In an *a posteriori* application such as MacArthur's, the degree of mismatch with the predicted optimum is used as a measure of the significance of the constraint.

In a similar way, optimal foraging models also get, at best, an imperfect fit between behavior and theoretical models. In experimental work with Great Tits (*Parus major*), Krebs and his collaborators provided meal worms of different sizes and manipulated the frequency of encounter. Though the models predict a discrete threshold, the actual preference function is a sigmoid one (Krebs, *et al.*, 1977). The general qualitative fit again is impressive, and there are a variety of ways to explain the deviation from the theoretical prediction; for example, there may be incomplete information concerning the value of prey or alternatives, or there may be undetected differences in prey quality. Faced with a failure of fit between data and model, we can restructure the adaptive problem and use this failure of fit *a posteriori* to estimate the values of other environmental parameters. This is exactly what is done. The slack between data and the optimal model is taken up by assuming foraging is optimal and retrofitting the model with constraints to compensate for the looseness of fit. As Gould and Lewontin observed, this kind of retrofitting does insure that there will be some optimal model for the behavior in question. Responses to the challenge of finding an analysis that makes a trait optimal depend primarily on the creativity of the theorist. Granting that, such an *a posteriori* analysis is consistent with a wide variety of hypotheses concerning the evolutionary origin of the traits in question. It is silent on the evolutionary question, as well as on any choice between alternative competing selective explanations. We need a method that incorporates history.

5. PHYLOGENY AND ADAPTATION: THE COMPARATIVE METHOD

Population genetics is not the only route to the historical study of adaptation, though it is one that is widely used in the study of microevolution. Another important approach incorporates what is often called the "comparative method," and has recently gained currency among evolutionary biologists in studying macroevolution. To propose that a trait, behavior, or character is *adaptive* entails that that trait confers an advantage relative to other organisms that lack that trait. To propose that a trait, behavior, or character is an *adaptation* entails that that trait is present because in the past it did confer an advantage relative to other organisms that lacked that trait. Adaptations are relative in several ways. A trait is adaptive relative to alternatives present. Similarly, an adaptation must have been adaptive relative to historically available alternatives. Moreover, a trait is adaptive relative to a given environment; what is adaptive in one setting may be maladaptive in another. An adaptation was adaptive in some past environment, and thus it was adaptive relative to its previous ecological setting. The comparative method accordingly requires that we compare a trait to phylogenetically related ancestors and environmental conditions. These constitute the evolutionarily relevant alternatives within related evolutionary lineages. If it turns out that a trait would not enhance performance or survival relative to the actual variants among its ancestors, then that would undercut the claim that that trait is in fact an adaptation. Similarly, if a trait originated in a lineage for which the current function would be irrelevant, then that would undercut the claim that that trait is in fact an adaptation, whether or not it is currently adaptive.[5] To draw again from Darwin's *Origin*, there are structures, hooks, on the branches of a climbing variant of bamboo. Darwin was not oblivious to the adaptive functions they serve for these climbing species, but nonetheless he rejects them as adaptations. The reason is simply that those same structures are present in other, nonclimbing, relatives. Since the structures precede their "function," that function cannot explain their presence.

The comparative method eludes the pitfalls noticed by Gould and Lewontin. Where engineering analyses are silent on history, the comparative method makes evolutionary history the centerpiece of the analysis of adaptation. Phylogenetic analysis is used to develop an historical perspective on the pattern of acquisition of traits within lineages, with an eye to inferring answers concerning the adaptive character of traits from the order of appearance of those traits. The pattern of acquisition or loss of a putative adaptation can then be mapped onto an independently established phylogeny.[6] Given adequate knowledge of the relevant selective regimes associated with the ancestral forms, it is then possible to compare alternative characters for their adaptive potential. The method imposes severe requirements: We need comparative data on related taxa, developmental information, information concerning the character of the environment, the trait family under consideration, and the relative adaptiveness of the traits characteristic of the several taxa (see Baum & Larson, 1991; Brooks & McLennan, 1991; Coddington, 1988; Emerson & Hastings, 1998; Endler & McLellan, 1988; Frumhoff & Reeve, 1994; Harvey & Pagel, 1991). In many cases, we may not be able to meet the requirements for a comparative study, and even if we do the method is fallible. The comparative method offers an approach to evaluating adaptation as an explanation. It is not the only method, and it is not guaranteed to yield a definitive answer. It is nonetheless a useful method.

Adaptationist hypotheses have implications concerning phylogenetic origin and the "utility" of characters relative to alternatives. Such hypotheses are inherently comparative, projecting an advantage for one character or trait relative to others; likewise, they depend on identifying one variant form as more fit than alternative competing forms. Phylogenetic analyses can be used not only to determine the ancestral conditions, but also to assess the range of ancestral variation and the adaptive significance of a trait. If a character did not provide an advantage relative to variants present at the time it evolved, or if it originated in a lineage for which its currently adaptive function would not be significant, then that trait could not be an adaptation. The relevant phylogenetic information can naturally be captured in a cladistic framework. A cladistic methodology is designed to expose an independent historical constraint on evolutionary explanations. Given a derived trait within a cladistic lineage, it is possible to identify both the primitive, or ancestral, traits and their functions. It should also be possible to obtain some idea of the range of relevant variation. When the data is adequate to do this, it is then possible to evaluate adaptationist hypotheses.

The basic idea of phylogenetic systematics is straightforward: Given a group of related organisms, it is possible to recover the phylogenetic relationships within that group on the basis of shared similarities—synapomorphies—in extant species. A phylogenetic tree is typically given as a branching diagram, depicting a series of speciation events; any given phylogenetic tree is essentially a hypothesis concerning the genealogical relationships among a set of taxa. If we begin with a monophyletic group, or clade, consisting of an ancestral species and all its descendants, the problem is to determine which of the possible phylogenetic trees is the actual one. With even a small number of species, the number of possible trees can be quite large. To use an artificial and simplistic example, if we have four species A, B, C, and D, each of the phylogenetic trees in Figure 1 represents a distinct genealogical hypothesis. In order to determine which tree is the correct tree, we consider the traits characteristic of the species within the clade. In theory, similar characters could be due to common descent (homologies), or to parallel or convergent evolution (homoplasies). Cladists, following Willi Hennig (1966), recognize that the patterns among shared derived characters can be used to construct a phylogeny, which in turn should reveal the pattern of common ancestry. They start with the conviction that genealogical influences should be sufficiently pervasive that homologous characters should predominate over homoplasous ones. Convergent and parallel evolution, that is, should be less common than similarity due to common descent. Whenever possible, then, we should assume that similar characters are due to common descent rather than convergent or parallel evolution. The correct tree can then be inferred by examining character similarities, with the best tree minimizing homoplasies. To follow our artificial example, suppose we have character distributions among our species for eight traits as in Table 1. For simplicity, we assume that the characters at issue are binary: A species either has or lacks the character, and possession is not a matter of degree.[7] If we assume that there are as few character transitions as possible—that is, we assume that homology is the rule and not the exception—then it is possible to determine which tree constitutes the best representation of the actual history of the clade. Based on the information concerning the five "species" in Table 1, it is possible to see that characters 1 and 2 define the clade and distinguish it from the outgroup O. Characters 3 and 4 identify a monophyletic group (consisting of B, C, and D). Characters 5 and 6 define a third monophyletic group (consisting of C and D); the loss of character 4 and acquisition of 7 distinguish the species within this group. The first tree is the correct one on cladistic principles.

Systematics certainly has had its share of controversies, and even among cladists, there are a number of issues within phylogenetic systematics that are currently unresolved. These often depend on how we establish character polarities (for good discussions, see Cracraft, 1981; also Maddison, Donaghue, & Maddison, 1984), and what counts as a single character (see Farris, 1983; and Kluge, 1989). A phylogenetic analysis depends critically on defining an appropriate outgroup, that is, a closely related species outside the clade used for comparative purposes.[8] In our simplified example, the outgroup O has only one of the characters in common with the clade. The only characteristics which provide useful information for an evolutionary phylogeny are those not shared (by descent) with the outgroup, since only these give information concerning the topology of the tree. Again in our example, character 8 provides no phylogenetic information, since it is ancestral and not derived. The issue of selecting an outgroup is critical precisely because the task is one of determining what the ancestral, or plesiomorphic, character state is; the choice of ancestral traits is critical for the construction of the phylogeny. The insistence on using monophyletic groups is, however, widely accepted even if the particular means for determining which groups are monophyletic is not.[9] The net result, if the data is sufficient to reach a definitive resolution, is that the resulting phylogeny can be used to reconstruct patterns of character diversification; and this in turn can be used as a check on adaptationist explanations.

	Characters							
Species	1	2	3	4	5	6	7	8
O	-	-	-	-	-	-	-	+
A	+	+	-	-	-	-	-	+
B	+	+	+	+	-	-	-	+
C	+	+	+	-	+	+	-	+
D	+	+	+	+	+	+	+	+

Table 1. An artificial and simplified character set for five species and eight characters. Assuming discrete characters, a + indicates the presence of the character within the species, and a - indicates its absence. O is the outgroup to the clade, exhibiting none of the features (1 through 7) used to determine the phylogeny of the group. Species A differs from the outgroup in only two traits (1 and 2), while B differs in four (1, 2, 3, and 4). Species C differs from the outgroup in five traits (1, 2, 3, 5, and 6), and D differs in all seven characteristic traits.

Adaptations are then those traits, or combinations of traits, correlated with environmental variables, which have arisen within a lineage as an adaptive response to common selective pressures. In evaluating the significance of a putative adaptation, comparative studies require systematic information concerning the ancestral environment, and the factors affecting its evolution. These factors may in turn be abiotic or biotic. If they are biotic, they may be interspecific, as would be, say, predation. In this case, the organism is involved in an "arms race" (Dawkins 1982). They may also be intraspecific, or social, as would

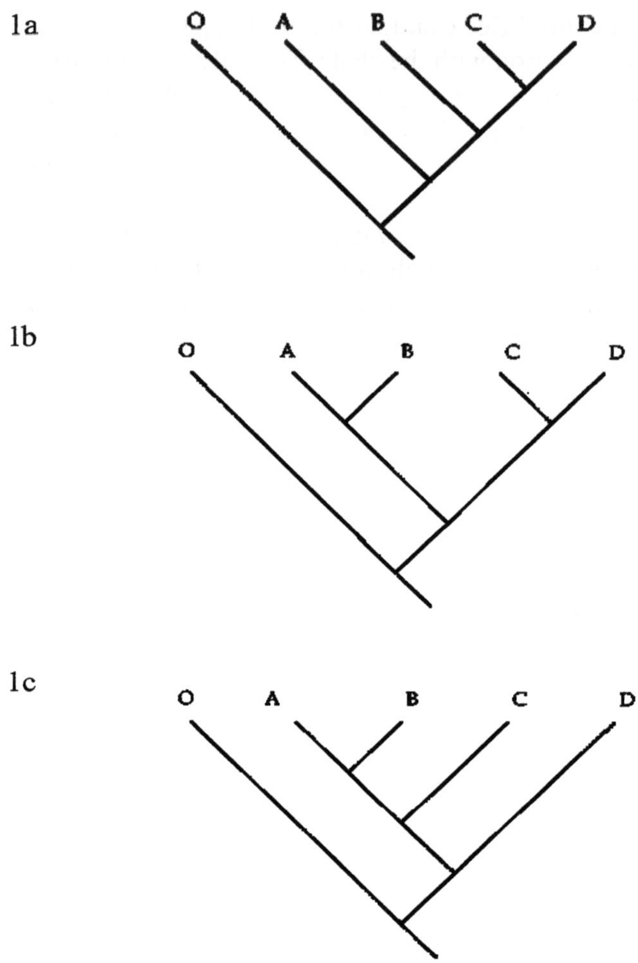

Figure 1. Alternative phylogenies. Cladograms describe the structure of monophyletic groups and not, strictly, ancestral relationships. A cladogram groups species in such a way as to reflect evolutionary descent, based on shared derived characters (synapomorphies). Trees with distinct topologies represent alternative phylogenies. With four species within a clade (A, B, C, D), each of these three trees represents a distinct evolutionary phylogeny. The correct tree is 1a, given the character distribution in Table 1 and given that the most parsimonious tree is the one that minimizes character transitions. The clade consisting of A, B, C, and D is defined by characters 1 and 2, because these are the shared derived characters within the clade. They distinguish the clade from the outgroup. Characters 3 and 4 in turn define a distinct clade consisting of B, C, and D. Characters 5 and 6 define a clade consisting of C and D. D is distinguished by character 7. C is distinguished by the loss of 4. (Within the clade including C, the ancestral state is positive for 4, so it must involve a secondary loss.) Character 8 provides no phylogenetic information, since it is ubiquitous within the clade and the outgroup O.

be sexual selection. In evaluating relative performance, one option is to turn to functional morphology, assessing performance directly. However this is done, though, a proper assessment requires comparing the character under study to alternatives, including in particular its alternatives present within the ancestral environment. If the selective regime offers a selective advantage to one variation within the lineage, that scenario is consistent with a hypothesis of adaptation.

Comparisons across phyla are useful in evaluating adaptation, and must take into account their respective environmental settings. With similar environments, similar traits across phyla will reflect common ancestry or stabilizing selection if the traits are homologous; similarity indicates convergent evolution if the traits are homoplasous. If the environments are not similar, then similarity of traits suggests phylogenetic constraints are critical or that there is convergence that is not environmentally driven. Let us return to our simple example. If we observe, say, maternal care in species D and biparental care in C, we may wonder how this difference arose. If maternal care is characteristic of the outgroup and of the other members of the clade, then an explanation of maternal care in D which treats it as an adaptation will be inappropriate. Biparental care is apomorphic, or the derived trait within the clade. Maternal care is plesiomorphic, or the ancestral trait. The proper explanatory focus would be on the evolutionary origin of biparental care in C. Of course, it may well be that maternal care is an adaptation for some ancestor to the clade. That would in turn require that maternal care be a derived trait within a wider clade; this does not compromise the point that a focus on maternal care as an adaptation within the clade, or an adaptation specific to D, would be misplaced. It is an ancestral rather than a derived trait within the clade. If, on the other hand, we were to observe maternal care in A and D, but biparental care in the other members of the clade and in the outgroup, then that is consistent with the view that maternal care is the adaptation. It is now a derived trait within two lineages—a homoplasy—and needs an evolutionary explanation. Coupled with an evaluation of the respective environments of the species, such comparisons can either support or disconfirm the hypothesis that the trait is an adaptation. An analysis in terms of reverse engineering, by contrast, may tell us a good deal about the adaptive character of maternal care, but it will tell us little about whether it is an adaptation or, if it is an adaptation, what it is an adaptation to or for. Indeed, reverse engineering by itself cannot even guarantee that we are dealing with a derived trait. It is in fact consistent with Fisher's analysis of sex ratios that a 1:1 ratio is ancestral and not derived; it might be, as many suggested early in the twentieth century, that a 1:1 ratio is the simple consequence of Mendelian segregation and a chromosomal determination of sex. If so, then in spite of Fisher's elegant equilibrium explanation, a 1:1 ratio may not be an adaptation at all, at least within lineages in which sex is determined chromosomally. Correspondingly, deviations from a 1:1 ratio are sometimes derived traits, and those may need an evolutionary explanation.[10]

Three quick examples may serve to illustrate the usefulness of the comparative method for evaluating adaptationist hypotheses, and to suggest some morals that might be derived from comparative studies. Some evolutionary scenarios can be definitively rejected using the comparative method. In criticizing the assumption of ubiquitous adaptation, Lewontin once (1978) suggested that the difference between the one- and two-horned rhinoceros is not adaptive. The two forms might be alternative, but equally good, solutions to an environmental problem arrived at independently; for example, both might be useful for predator defense, but the difference between them might be insignificant. J. A. Coddington (1988) subsequently pointed out that the simplest phylogeny supports the view that the ancestral form within the clade had two horns rather than none or one and that the Indian rhinoceros lost the frontal horn secondarily. The proper evolutionary question is why this species lost one, but not both, horns. The hypothesis that they arose independently as solutions to a common environmental problem, though, is not defensible given the phylogenetic analysis.[11] The result that these are not phylogenetically independent does not settle the question whether the loss of one horn was an adaptation, but it does effectively drive home the conclusion that the one-horned variation is an evolutionarily derived trait.

There are comparative studies that offer significant support for adaptive explanations. Anoline lizards have proliferated on Cuba, Hispaniola, Jamaica and Puerto Rico and occupy a diverse range of habitats. On each island, the range of habitats is similar and the lizards have developed similar morphologies. Some arboreal forms, in particular, have toepads that allow them to hold on to smooth surfaces. Others that live on the ground lack this particular feature. Functional studies have confirmed that these toepads are important for their ability to cling, and phylogenetic studies support the view that arboreality evolved before the toepads. Moreover, though there are morphologically similar types on the four islands, similar morphological types have evolved independently; cladistic relatedness is uncorrelated with morphological similarity. This supports the conclusion that toepads are an adaptation for the arboreal lifestyle of the lizards (cf. Larson & Losos, 1996), and not merely that they are maintained within the group because of their selective advantage. Generally, *the recurrent evolution of similar morphological types in these habitats suggests that adaptation rather than some other process*—for example, colonization or developmental constraints—*is the cause of the evolutionary changes* (see Losos, et al. 1988).

Finally, it is possible to see what kinds of considerations can make a case more ambiguous. The genus *Montanoa*, or so-called "daisy trees" because of the structure of their florets, includes thirty taxa. Twenty-one of these species are shrubs, five are trees, and four are vines. The shrub form is apparently ancestral, and the prevalence of this form suggests significant conservatism within the group. Roughly 15 percent of the species is associated with a change in habitat, and has given rise to tree and vine forms in roughly equal proportion. A study of convergent adaptation might be appropriate for these forms, following the lines of the previous example; and, of course, an adaptive explanation would be misplaced when applied to the shrub form, which is plesiomorphic. The tree forms are clearly derived, and so are candidates for adaptations. Among the tree forms, there are a number of morphological changes present, and though they belong to different clades, these characters allow them to live in higher elevations. As far as this goes, adaptation is a plausible hypothesis. We have a derived form, within several lineages; and it is correlated with significant environmental differences. It is in fact not clear whether an adaptive explanation is right. Diploidy is the ancestral condition among the *Montanoa*, but the trees and only the trees are polyploids. Polyploids are typically larger than diploids within the group, and polyploids often arise from diploids in the group. The distinction between trees and shrubs may in fact be a simple by-product of changes in ploidy level. It may, that is, be a spandrel (for more discussion of the case, see Brooks & McLennan 1991, pp. 159 ff.). If so, then even though this is a derived trait, it is still not an adaptation.

The principal point is that phylogenetic information provides a useful and important independent constraint for framing and evaluating evolutionary explanations. It can be used to rule out some traits as adaptations. It can also lend varying degrees of support to the hypothesis that a trait is an adaptation. The comparative method offers no simple litmus test for adaptation. It may support the claim that a trait is an adaptation, or it may undercut it; but there are a number of more specific morals that can be drawn. First of all, unless a trait is derived within a lineage, it cannot be an adaptation within that taxonomic group or at that taxonomic level. Secondly, the claim that a trait is an adaptation is strongest if there is more than one independent evolutionary origin associated with comparable selective regimes. If a character evolves within a specific selective regime more often than would be expected by chance, then that supports the claim that it is an adaptation within that group. Toepads, for example, have evolved in at least three lineages of lizards. This can be subjected in some cases to a statistical test, evaluating whether an evolutionary transition takes place within a common environment more often that would be expected by chance. Larson and Losos (1996) estimate that the probability of three origins for toepads is roughly 0.028, which supports the claim that it is indeed an adaptation to an arboreal habitat. There is a third moral. Since historical variables concerning rates of evolution, ancestral environment, and branching of lineages, are more available for some taxa than others, there is no guarantee that a phylogenetic study will yield a definitive answer; nonetheless, adaptation does indicate a number of results that would not generally be expected under alternative evolutionary scenarios.

6. PHYLOGENY AND ADAPTATION: HUMAN EVOLUTION

There is a remarkably good fossil record relevant to the advent of modern humans, and there is a great deal known about our primate kin. This might seem a reason to be optimistic about evolutionary psychology. We do, after all, have independent information about the relevant phylogenies based on the fossil record and on contemporary species. There are, moreover, some reasons for viewing human intelligence and human language as adaptations that are largely consonant with comparative methodology. They are unique to the human lineage and thus are derived characters within the hominid clade. However, the grounds for optimism are illusory. On closer inspection, they evaporate. Our psychological capacities are surely the product of evolution. They may be adaptations. The explanations offered, though, are chimera. The sort of information that is available does not support the adaptive explanations common within evolutionary psychology. A look at the state of the art in hominid evolution should suffice to turn us toward agnosticism with respect to the claims of evolutionary psychology, though not, of course, to human evolution. What is known of hominid phylogeny and what is known of the evolution of these traits is not enough for the program of evolutionary psychology.

Within the primates, apes form a well-defined group. There are, unfortunately, only five living genera and of those only two have more than one living species. The fossil record for the apes is disappointing, with few fossil remains. The fossil remains we do have shed less light than we would like on crucial evolutionary transitions. The great apes probably originated in the middle Miocene (11-16 mya). The most famous specimen is probably *Proconsul*, though there are a number of genera representative of ancestral hominoids (see Ward 1999 for a current discussion). Some apes are more closely related to humans than to other apes, so from a cladistic perspective we need to be included among the apes.[12] It is clear that the Orangutan (*Pongo*) and Gibbons (*Hylobates*) are relatively distant relations of the gorillas (*Gorilla*), common chimpanzees (*Pan troglodytes*), pigmy chimpanzees (*Pan paniscus*), and humans (*Homo sapiens*). The four latter species form an evolutionary clade, though there is less unanimity concerning the pattern of relationship among them; that is, it is controversial whether, within the hominoid primates, the African apes are more closely related to one another than any is to their human sister group. The problem is that the branching events which distinguished gorillas, chimps and humans were nearly simultaneous; as a result, there is a great deal of similarity among all of them from genetic, morphological, and behavioral standpoints. Morphological evidence tends to favor a grouping of chimpanzees with gorillas, some of which are plausibly construed as shared derived characters (Kluge, 1983). Molecular phylogenies, on the other hand, favor a grouping of chimpanzees with humans (Sibley & Ahlquist, 1984). Insofar as there is a consensus, it is that the chimps and hominids form a clade, with gorillas branching off first (see Andrews & Martin, 1987; Figure 2).

Within the hominids, there are a number of fossil forms of australopithecines and several species within the genus *Homo*. All are characterized by a bipedal gait. Like other apes, the hominids are sexually dimorphic, probably with long gestation periods and extended maternal care. Australopithecines have relatively large teeth with small brains. All the species of *Homo* are broadly distinguished by an enhanced cranial capacity in comparison with the australopithecines. The relationships among these various hominid forms are uncertain and subject to considerable divergence of opinion (cf. Chamberlain & Wood, 1987; Strait, Grine, & Moniz, 1997; and Fleagle, 1999; Wood & Collard, 1999). The several species of *Homo* do at least form a clade distinct from the australopithecines, though some phylogenies do not even render *Homo* as monophyletic. Some treat *A. afarensis* as a sister group, though more commonly it is *A. africanus* or one of the "robust" australopithecines that is the sister group. Within the australopithecines, relations among *A. bosei, A. robustus, and A. aethiopicus*, and, more recently, *A. garhi*, are especially controversial, though *bosei, aethiopicus* and *robustus* appear to be more closely related to one another than to *africanus* or *afarensis*.[13] There are at least four forms of *Homo* commonly recognized, and often more than four (see table 2). It is generally conceded that *H. habilis* is distinct

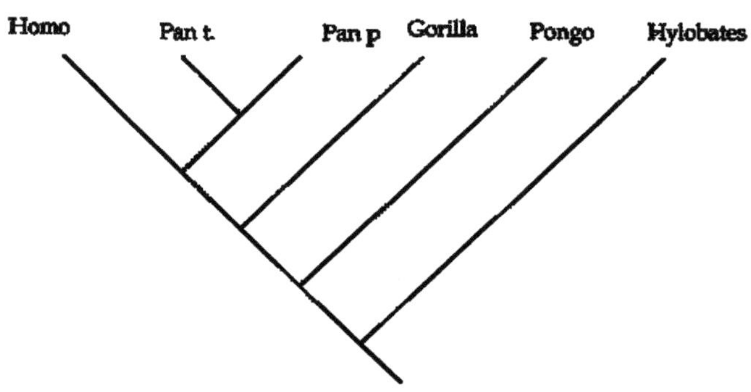

Figure 2. A phylogeny for living apes. Though the proper phylogeny is controversial, this represents the most commonly held phylogeny among living apes. It is uncontroversial that humans, chimps and gorillas are more closely related to one another than to either Orangs or Gibbons. On the basis of genetic similarity, humans and chimpanzees would form a clade, with gorillas branching off earlier. This is the phylogeny represented above. There are alternative phylogenies, according to which chimpanzees and gorillas form a clade, with the human line branching off earlier.

Species	Height	Cranial Capacity Dates	Known
H. habilis	1-1.5 m.	500-800 ml.	2-1.6 mya
H. erectus	1.3-1.5 m.	750-1250 ml.	1.8-0.3 mya
H. neanderthalis	1.5-1.7 m.	1200-1750 ml.	130,000-60,000 years
H. sapiens	1.6-1.9 m.	1200-1700 ml.	130,000 years

Table 2. Features of four species within the genus *Homo*. The most marked characteristic of the clade is a larger cranial capacity; many other traits, including a bipedal gait are shared with australopithecines.

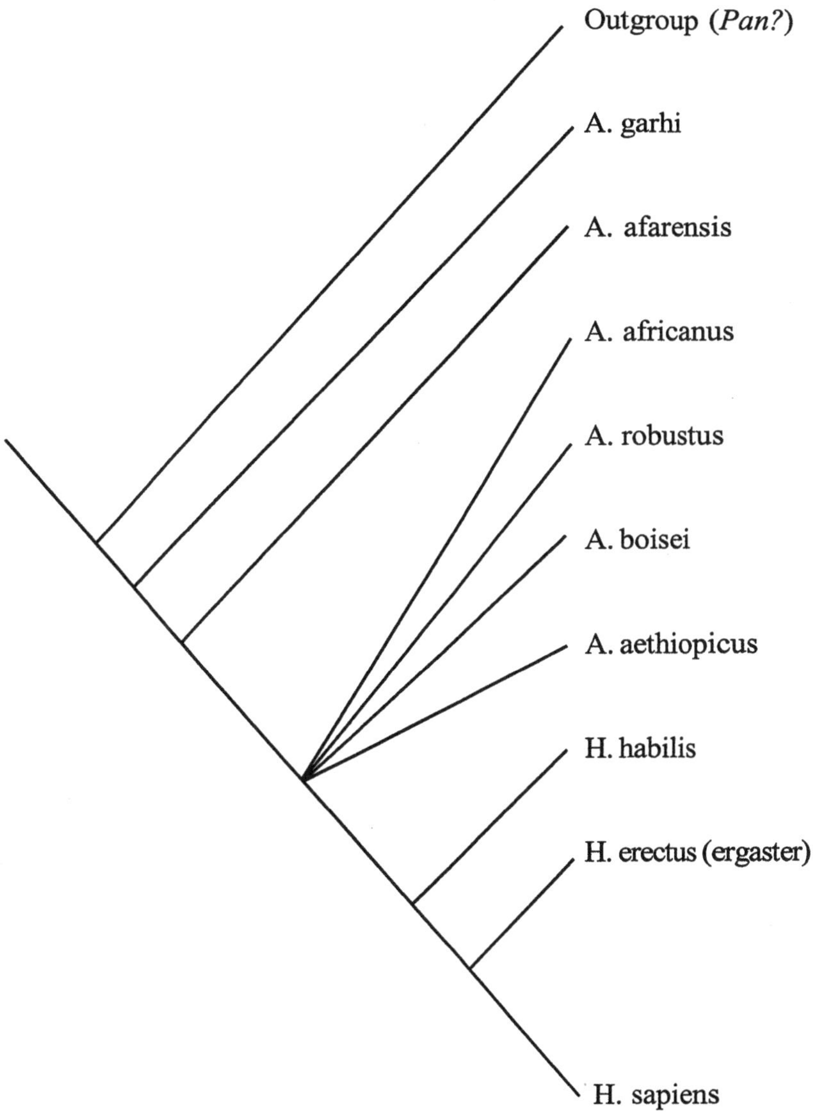

Figure 3. A plausible (but partial) hominid phylogeny. A number of phylogenies have been proposed between the various hominid forms; the resolution of some relationships is not clear. Even the number and rank of differences is contested. It is generally agreed that *H. erectus (ergaster)* is closely related to *H. sapiens*, that *habilis* is more primitive than either, and that *A. robustus*, *A. boisei* and *A. aethiopicus* are closely related. Some assign these latter three to a distinct genus (*Paranthropus*), though there is controversy surrounding whether they even form a clade. I have depicted this as an unresolved node. There is no clear consensus on whether *A. africanus* or *A. afarensis* is more closely related to the robust forms of the australopithecines. What is given above is thus a plausible phylogeny, though emphatically only one of many defended in recent discussions.

from the remaining groups within the genus. The relationships among *H. erectus*, *H. rudolfensis*, *H. sapiens*, and *H. neanderthalis* are less clear; in some accounts the Neandertals are a subspecies of human and in others they are a distinct species. Insofar as any distinguishing features are behavioral, the fossil record is not decisive. As a result, hominid phylogenies are very much matters of controversy. Figure 3 presents a plausible hominid phylogeny, though by no means the only one defended in recent literature.

The general trends within the genus *Homo* do suggest that intelligence has increased within the clade. Brain size has certainly increased, and the pattern of its evolution supports the conclusion that natural selection was important. Though the brains of australopithecines fall within the range to be expected within the great apes, by roughly 1.5 million years ago the brains of *Homo erectus* averaged more than twice as large. Gross brain size by itself is unreliable as a measure of intelligence, and must at least be scaled to body size if we are to make real sense of these trends.[14] Allometric scaling (cf. Gould, 1966), which relies on a measure of brain size relative to body size rather than absolute size, further supports the conclusion that the relative cranial capacity of *Homo sapiens* is a derived character and that there was a trend toward greater cranial capacity within the lineage. Within mammals generally, brain size can be reliably predicted from body size as a power function. As a rough rule, brain size scales as a power function of body size within taxonomic groups: For mammals, the scaling factor is approximately 0.75, and for primates the estimates vary somewhat but are more or less consistent with this value. Though there is again some uncertainty concerning such comparisons, it is generally agreed that at any stage of development primates have a larger brain size for a given body size than would be expected from the mammalian index alone, by roughly a factor of two. Within primates, the growth curve for human infants follows the primate norm until birth; in most primates, the relative growth rate of the body then increases after birth while in humans the brain grows at an elevated rate postnatally. As a result, adult humans also have a brain size that is markedly higher than would be expected given the primate norms (see Deacon, 1992a and 1992b). Primates are therefore exceptional within mammals. Humans are unusual even for primates. Such comparisons are suggestive, and are consistent with selection affecting brain size, but they must be regarded with caution. The comparisons among primates are less problematic than are comparisons across broader taxonomic groups. Structurally, primate brains are very similar. They have the same gross cortical and subcortical structures, with a similar architecture. However, the similarity is not perfect. Aside from its larger size, the human brain is more asymmetrical, and olfactory centers are reduced. Endocranial values alone do not reflect such structural differences. None of the differences in overall size, in any case, do anything to explain the marked cognitive differences between humans and other primates, including, most obviously, our specialized linguistic abilities. If we take seriously the idea, favored by many evolutionary psychologists, that our cognitive organization is modular, then overall differences would not directly reflect the evolution of human cognitive capacities. What would be needed, at a minimum, would be estimates for specific neural structures. Furthermore, organizational and functional differences are surely more important than sheer bulk in assessing human intelligence and our communicative facility, and on this the fossil record is largely silent. It is not clear how even the allometrically corrected measures would explain such differences in ability. Realistically, encephalisation quotients are crude measures of intelligence at best. They reflect overall size, but gloss over the kinds of specializations that are likely to explain major differences among species and that are likely to be important to human cognitive capacities. Nonetheless, this enhanced cranial capacity is a derived feature within the genus. That is at least consistent with selection for enhanced intelligence.

Language provides a similar case. Once again, there is much evidence supporting the view that human language is an adaptation. The complexity of our vocalizations and the rule-governed character of language are unlike any other systems of communication. The facility of first language acquisition suggests, moreover, that there is significant specialization for language acquisition (cf. Pinker, 1994). Once again, the neurophysiological changes are the clearest. Humans have a larger prefrontal cortex than would be expected. Cortical regions adjacent to the Sylvian fissure are directly involved in language use. Studies of patients with trauma, as well as stimulation studies involving these regions, show that they are specialized for language use. Studies involving electrical stimulation and brain imaging support the general picture, though they have shown that language functions rely on a diverse range of brain systems, especially in the left hemisphere. There are functional and morphological asymmetries between hemispheres in other species as well, but the extent of the asymmetry in humans is certainly more marked than in other extant primates. Moreover, other primate calls are controlled in the midbrain and the forebrain rather than in the motor cortex. As we turn to comparisons with other hominids, the picture is still less clear, largely because we must rely on indirect evidence. Fossil remains do allow us to discern some of the gross features of hominid brains since the surface features of the brain leave some imprints on the skull. Endocasts of fossil skulls display these features, though, again, changes in size and other changes resulting from upright posture make the interpretation of these features difficult. The Sylvian fissure does show some asymmetry in archaic humans, including both *H. habilis* and *H. erectus*. That would support the view that there is an enlarged left hemisphere, and an asymmetry similar to that common in humans. Some researchers conclude on the basis of such evidence that these other species of hominid had specializations associated with language, but though such asymmetries are consistent with the conclusion, they are at best equivocal evidence for language use in these species. Yet others, most notably Deacon (1997), claim that the evolution of human language and cognition are ineluctably intertwined. The most reasonable conclusion, in my view, is that we simply do not know whether language use is a derived character specifically for *Homo sapiens*, or whether it is a characteristic of other hominids as well. It is, though, a derived trait within the hominid line, and an adaptation characteristic of *Homo* to the exclusion of the australopithecines.

Thus, it is plausible to hold that language use and human intelligence are derived traits, but this provides only marginal comfort to evolutionary psychology. Our knowledge of the cladistic structure of the relevant species is imperfect, and subject to considerable controversy. There is good reason to hold that both human intelligence and human language are apomorphic; that is, they are both derived from and differ from the ancestral conditions. This is at least consistent with the claim that they are adaptations.[15] One thing that is strikingly absent is knowledge of the relevant homologous structures or behaviors in ancestral groups. In assessing evolutionary adaptation, this provides a critical feature in understanding the variation, and thus the relative advantage that structure or behavior provides. It is not clear, furthermore, whether language use is a derived trait within *Homo sapiens*, or a more general feature of the hominids. It does appear that through the lineage there was a tendency toward greater cranial capacity. This enhanced cranial capacity is not limited to *Homo sapiens*. The increase within the genus therefore provides little information concerning the evolution of human intelligence and is even less relevant to the sorts of changes necessary for language. This makes a comparative assessment particularly problematic. The ecological settings for the several species, insofar as we can reconstruct them from paleoecological data, are varied. During the Pliocene and early Pleistocene (roughly 4 to 1 mya.), australopithecines were widely distributed in southern and eastern Africa. Their habitats included wooded regions, open savanna and grassland. *Homo habilis* overlapped temporally with australopithecines, but is evidently associated more with wooded habitats. With *Homo erectus*, we find more use of open habitats, perhaps indicating increased reliance on hunting. By roughly one million years ago, toward the middle of the Pleistocene, the range of hominids had spread out into Europe and Asia. It is anybody's guess what physical environment is associated with the trend toward increased cranial capacity. The ecological information we do have is varied, indicating a wide range of environments, and little information is available concerning what ecological associations would be involved in the evolution of increased intelligence or the use of language. We are left considering traits that may be ancestral in *Homo sapiens*, or may be specialized, with an uncertain ecological connection. It is hardly an ideal case for a comparative analysis, and provides little substance supporting any particular adaptive explanation.

When we turn to language, the very feature that makes it plausible to treat it as an adaptation—its uniqueness relative to other forms of communication—makes it an especially poor subject for comparative analysis. Terence Deacon (1997) observes that the problem with evolutionary accounts for the origin of language is the lack of continuity with other forms of communication. The same observation would be appropriate when we turn to human intelligence, especially if it is not a general-purpose mechanism.[16]

Language acquisition provides a useful model for seeing the depth of the problem. Linguists have recognized that the problem with treating language acquisition as a matter of learning—as some sort of social conditioning or generalization—is that the sort of evidence available to the child could not result in what we see. The human child develops language of remarkable complexity, with an ease and facility that is also stunning. Linguists often emphasize the strides children make in learning language at an age at which other cognitive skills are not especially well developed, suggesting that language acquisition is mediated by distinctive mechanisms. Human language acquisition is so unlike what we think of as learning in other contexts that many developmental linguists are reluctant to count the acquisition of language as *learning* at all. The uniqueness of what happens in language acquisition may be overstated somewhat, but not in comparison to the kinds of mechanisms simple models of learning can offer. We lack the kinds of incremental stages that would explain its acquisition in these terms. Accounts of the evolutionary origins of language are in an analogous position, for much the same reason. The very lack of homology between human language and other forms of communication—even with the forms of communication present in other primates—leaves us unable to correlate it in any interesting way with any environmental variables, or to compare it with other forms of communication.[17] Social organization is surely involved somehow in the evolutionary origins of language, if language is the product of natural selection. Our primate cousins also participate in complex social arrangements, but there is no real analogue of human language in the rest of the extant primates. We do not know what forms of communication were present in ancestral hominid species, or even whether they had linguistic capacities, for the simple reason that the sort of information available in the fossil record is indecisive. The uncertainty concerning the relevant taxonomic level for the emergence of language only exacerbates the problem.

The mere fact that social coordination is facilitated by the use of human language provides no comfort; for though coordination and social interaction favor some form of communication, there is no basis for thinking that they would favor the sorts of complex features that are integral to human language. If there is any generalization concerning primate social organization, it is that it is highly variable. Gibbons are largely monogamous, orangutans are largely solitary, gorillas are polygamous, and chimpanzees exhibit a fluid and variable social structure. This makes the extrapolation of ancestral social organization highly problematic. There is virtually nothing that would help in understanding the evolution of language. If we were seriously asked how social organization could favor, for example, the particular lexical and phrasal categories of human language, or how the rules underlying meaning and structure would contribute to enhanced fitness, it is hard to imagine how to address the question. In any case, all the hominids, as well as primates, are social creatures; and little is known concerning the specific social organization of ancestral hominids. As Deacon says,

> A full evolutionary account cannot stop with a formal description of what is missing or a scenario of how selection might have favored the evolution of innate grammatical knowledge. It must

> also provide a functional account of why its particular organization was favored, how incremental and partial versions were also functional, and how structures present in nonhuman brains were modified to provide this ability (1997, p. 38).

No doubt tool use, hunting, social coordination, and communication were significant features in the shaping of human capacities. Beyond such vague observations we are offered little or nothing concrete to aid in explaining the specific features of human cognition or language use. If the goal is one of explaining the specific organizational patterns of human cognition, such an explanation is lacking in evolutionary psychology. Likewise, Pinker, Tooby, and Cosmides emphasize the modular character of cognition; but no explanation of its modular structure is forthcoming. Pinker, Bloom and Dennett are fond of emphasizing the need for communication, and extol the virtues of social coordination; but this suggests no explanation whatsoever of the unique features characteristic of human language use, or of the remarkable cognitive capacities that are responsible for it.

The record of hominid evolution assembled by anthropologists is remarkable for its quality. It leaves no serious doubt that humans are the result of evolution. Though that record sheds a tremendous amount of light on our ancestry, it does not favor the sort of speculations concerning the evolution of human intelligence and human language offered within evolutionary psychology. In the ideal cases for the comparative method there is a sufficiently diverse taxonomic group in which we can see the same traits evolving in several lineages under similar environmental contingencies. Convergent evolution, resulting in a polyphyletic group displaying analogous features, is suggestive of natural selection, though even convergence can be the result of alternative mechanisms (see Leroi, Rose & Lauder, 1994). Under the best of circumstances, comparative studies may in fact not be able to disentangle the various features affecting evolution, including selection, drift, mutation and migration. Even with extensive data and many related taxa, it is difficult to disentangle one evolutionary explanation from another. Hominids and primates simply do not provide the ideal group for comparative analysis. We are a small group with few extant species, and the fossil record does not offer the sort of evidence we need. Moreover, much of what we are most interested in is unique to the hominid line. Much as we might care about our origins, much as we might want to understand the origins of sociality, intelligence, or language, the case we provide to evolutionary biology is not an easy one. The record is simply not sufficient for the questions posed. This may be as good as it gets.

7. CONCLUSION: EXPLANATORY AGNOSTICISM

The program advanced by advocates of evolutionary psychology is an ambitious one. The goal is not just to defend the idea that humans evolved, but also to offer evolutionary explanations of specifically human features in terms of natural selection. These explanations are intended to reflect the evolutionary origins of the features they study. Human language and cognition are certainly complex capacities. As with other capacities, we may reasonably inquire into their evolutionary origins. Evolutionary psychologists offer us explanations of psychological mechanisms in terms of the specific environmental demands that shaped them in our evolutionary past. An understanding of these capacities, and of their evolutionary functions, will in turn illuminate the psychological mechanisms.

Understood as an evolutionary explanation, what we are offered is inadequate. The complexities of human language and cognition do suggest they are adaptations, and it is these complex capacities we wish to explain. The explanations we are offered, though, are often so general that they cannot approach the problem we begin with. Pinker asks, "Why did brains evolve to start with?" and offers a simple and straightforward response: "The answer lies in the value of information, which brains have been designed to process" (1997, p. 175). Dennett similarly touts the virtues of anticipation over trial and error. It is, of course, uncontroversial that brains do serve to process information, and it is uncontroversial that we do sometimes anticipate the consequences of our actions. It is a long way from such simple observations to an evolutionary explanation of human intelligence, much less to the features more specific to human intelligence. It is as if we began by noticing that the knee is a complex mechanism, and accordingly sought its evolutionary explanation. It would be no explanation at all to be told that the knee is as it is because it is used for locomotion, or that there are virtues in being mobile rather than sedentary. It is, of course, true that the knee is involved in locomotion. It is uncontroversial that mobility is important in human evolution. These generalities leave us uninformed about that peculiar and complex structure we began with. Pinker's and Dennett's proposals are similarly uninformative even if they are true. They tell us nothing specific concerning human intelligence. Pinker explains that though there is a common "plan" to the organization of mammalian nervous systems, the human brain is unique in a number of ways. He says:

> The human brain ... tells an evolutionary story. Even a quick side-by-side comparison shows that the primate brain must have been considerably re-engineered to end up as a human brain. Our brains are about three times too big for a generic monkey or ape of our body size. The inflation is accomplished by prolonging fetal brain growth for a year after birth. If our bodies grew proportionally during that period, we would be ten feet tall and weigh half a ton. (1997, p. 183).

It is true that our brains are strikingly different from those of our primate ancestors and cousins. It would be important to have an evolutionary explanation of such differences. There undoubtedly is an evolutionary story to be told, at least in describing the evolutionary changes within the hominid line. That story does not, by itself, tell us how or why these evolutionary changes came about. There have also been structural changes in the size of the human brain by comparison with our primate cousins or with the mammalian norm. There are, more importantly, changes in its organization. This is at best indirect evidence concerning the functional differences. What explanation is offered for these more specific differences in size or fetal growth patterns? None. What explanation is offered for such features uniquely characteristic of the human brain such as why, for example, the olfactory lobes are reduced? Again, none. What explanation is offered for the features characteristic of human language? None, again. What Pinker and Dennett do "explain" moves from the general to the specific; for example, from human language as a system of communication to human language with all its foibles. Systems of communication no doubt facilitate communication. What does that tell us about the evolution of human language? Nothing at all. It tells us no more about the evolution of human intelligence.

Again, considered as an evolutionary hypothesis, the question should be what the right explanation of human language or cognition is. The evolutionary psychologist's goal is to provide an evolutionary explanation of human traits as complex adaptations—as the products of natural selection—and not merely as evolved features of the human lineage. It is important to recognize that there is more than one possible evolutionary explanation even for complex characters, and not all would make the results adaptations. It is also important to recognize that even if natural selection plays a significant role, there will still be importantly different explanations possible. When we are told that human language and cognition are adaptations, we should ask what the alternative to this hypothesis might be. Pinker suggests at one point that the opposition to an adaptationist position is the view that human ingenuity is "a by-product of blood vessels in the skull that radiate heat" (p. 187). If this is the opposition, Pinker could have written a much shorter book. Buss (1999, pp. 35-36) tells us that there are three theories concerning the origin of complex mechanisms: evolution by natural selection, creationism, and cosmic seeds (which he acknowledges is not really an alternative to selection). He sees that given such alternatives, there really is no contest. Natural selection is the only prospect for explaining human nature. Dennett's declared opponents in defending the view that our language is an adaptation are Gould and Noam Chomsky, or in any case his rather extreme version of what he takes their views to be.[18] Dennett sides with Pinker and Bloom (1992), claiming that "although Gould has heralded Chomsky's theory of universal grammar as a bulwark against an adaptationist explanation of language, and Chomsky in return endorses Gould's anti-adaptationism as an authoritative excuse for rejecting the obvious obligation to pursue an evolutionary explanation of the innate establishment of universal grammar, these two authorities are supporting each other over an abyss" (1995, p. 391). It is an important and interesting question whether our linguistic capacities are adaptations. It is an equally important and interesting project to develop an evolutionary explanation of these capacities, whether or not they are adaptations. One might reserve skepticism over whether they are adaptations, and one might be skeptical about the specific adaptationist explanations proposed, without doubting for a moment that these are complex evolved traits.

The issue is not the importance of evolution or of natural selection, but the right explanation for evolutionary changes we see and for our cognitive capacities. Advocates argue as if the issue is whether there is evolution, or argue as if the question is whether evolution can result from natural selection. Both claims are uncontroversial. They are also distinct. The conflation of the two is evident. Alternatives to natural selection within evolutionary theory are generally not mentioned, and not considered. Dennett's rendering of Darwin's "dangerous idea" blinds him to the difference. He says that Darwin's fundamental idea is that "Life on Earth has been generated over billions of years in a single branching tree—the Tree of Life—by one algorithmic process or another" (1995, p. 51). Darwin, to the contrary, saw natural selection as the natural outcome of competitive struggle, driven by superfecundity. Evolution and speciation were the results, but Darwin's fundamental idea was the centrality of natural selection in the understanding of adaptation. Natural selection is one thing. Evolution is another. To treat Darwin's "fundamental idea" as Dennett does is to obscure the difference between evolution and evolution by natural selection.[19] The issue in understanding human evolution is then identified mistakenly as whether there *is* evolution, or whether human evolution *can* result from natural selection. With affirmative answers to the uncontroversial questions, advocates then feel confident in inferring that evolutionary psychology is secure: Our psychological profile is properly explained as an adaptation to our Pleistocene heritage. The inference is an error. Neither evolution nor the importance of natural selection is the critical issue. It is simply not enough to argue that a trait is evolved, and then conclude that it is an adaptation, much less that it is an adaptation to some particular environmental feature. For the argument that human psychology is the consequence of natural selection to be taken seriously, we cannot afford to gloss over the difference between evolution and its mechanisms. We likewise cannot afford to gloss over the differences between natural selection and its mechanisms. We cannot afford inattention to the details required of respectable evolutionary explanations. We cannot afford to ignore the importance of history, settling for evolutionary theories without a historical context. What we require is a more developed and more articulated understanding of evolutionary biology if we are to shed light on the adaptive significance of human cognition and language.

Once we recognize the contributions that evolutionary biology offers to evolutionary questions, the fact is that the resources of systematics and comparative biology offer little comfort to evolutionary psychology. Neither do the resources of population genetics. In both cases, the available biological and paleoanthropological record leaves the relevant ecological, social, phylogenetic and populational variables sufficiently indeterminate that we do not know enough about the factors that shaped human evolution to evaluate the proposals offered by advocates of evolutionary psychology. We do know that language and intelligence are derived forms within the hominids, but it is not clear from the record whether they are features of *Homo sapiens*, or of a broader taxonomic class.[20] If we treat these as specifically human adaptations, for all we know we are focussed on the wrong taxonomic level. We do know that social organization and the environment were relevant to the evolution of human language and intelligence, but we know little concerning the actual social structure or even the relevant physical environment. It is true that language facilitates communication, but that is hardly sufficient to develop an evolutionary explanation. It is true that human intelligence facilitates the management of information, but that does not give any substance to an evolutionary explanation of our intellectual capacities. Lacking knowledge of the forms of communication used by our ancestors, lacking knowledge of the sorts of social organization present, lacking knowledge of the sort of ecological problems these ancestors confronted, lacking even knowledge of which ancestors are the proper focus of investigation, we lack the ingredients for seriously explaining our psychological capacities in evolutionary terms. We are left either with empty generalities or unconstrained speculation. In considering evolutionary psychology, we should ask whether its explanations are grounded in observation, or whether it makes contact with the relevant historical and biological record. If the answer to these questions is no, we should follow David Hume's (1777) advice, "Commit it then to the flames; for it can contain nothing but sophistry and illusion."

NOTES

This work has been discussed at the Boston Colloquium for Philosophy of Science, The Southern Society for Philosophy and Psychology, Florida State University and Washington University. Those opportunities and the discussion in those cases improved the work in a number of ways. The work in this paper also has benefited from discussions with Peggy DesAutels, Paul Davies, Peter French, and Thomas Kane. I am grateful for their input, though that does not imply that they share my views.

[1] It is not clear to me how comfortably this conception of the function of reason fits with the modular architecture so common among evolutionary psychologists. The roots of the modular conception lie with the "organology" of Gall, and are opposed, at least historically, to the more Cartesian picture of the function of rationality, a picture explicitly repudiated by Cosmides and Tooby as the "standard social science model." This would be an interesting thread to follow, but not one I will follow here. See de Jong and van der Steen (1998), section 4, for some discussion of this view.

[2] In fact, I expect that the problem of the origin of life is far more tractable than is the origin of human language. This is in part because we can independently assess the environmental conditions.

[3] Dennett unfortunately offers more in the way of discussions of cathedrals

and cranes than of organisms. Dennett's treatment of architectural spandrels has been decisively criticized by Fred Gifford (1997).

[4] Evolutionary ecology is replete with equilibrium models. Under the leadership of Hutchinson and MacArthur ecologists explored a variety of problems, including questions of the number of species that can be supported in an ecosystem. The problem became one of understanding how competition leads sympatric species to structure the niche in a way that enhances stability. What is missing in much of this literature is an interest in history and phylogeny as a consequence of their focus on equilibrium states. Sharon Kingsland refers to this as the "eclipse of history" (1985, ch. 8; cf. Ricklefs, 1987).

[5] It is often thought, mistakenly, that this would preclude traits maintained by stabilizing selection. That it does not is easily seen. If a trait is present due to stabilizing selection, then it must have been maintained in a lineage by selection. That too is a comparative issue, depending on whether, in phylogenetic ancestors, there was variation or whether, in phylogenetic ancestors, that variation would have been maintained.

[6] Such phylogenetic trees should be based on a large number of characters, and if they were they would very robust. Including the characters at the focus of an analysis is generally harmless, since the topology of a tree will not depend critically on any one character.

[7] Essentially the same points could be made using continuous variables, though the analysis is more complicated.

[8] Choice of an outgroup is difficult and important. It cannot be among the species being considered, but needs to be closely related if it is to offer useful information. If it is possible to use a close relative, then that provides the tool for deciding which characters were present in the ancestors to the clade, and which provide useful phylogenetic information. It is also directly relevant to questions of polarity of change.

[9] Once we accept that some characters may evolve more than once, there are a number of procedures for selecting the best tree (see Felsenstein, 1982 for an extensive discussion). Allowing derived characters to evolve more than once, the most parsimonious tree might be the one with the fewest character state derivations. If we hold that complex characters cannot evolve more than once but can be lost repeatedly, the most parsimonious tree would be that with the fewest losses of derived characters. We might hold simply that the most parsimonious tree is the one with the fewest character state transitions. The various criteria result in significantly different choices, but such disputes do not centrally affect the points I will be making here.

[10] A trait may be ancestral and still require an evolutionary explanation if we know, for example, that variations arose regularly. Since we would expect that in the absence of selection, common variations would sometimes be established in a set of lineages, if we know there is variation for a trait, then we might properly explain the ubiquity of a trait in terms of selection. A phylogenetic analysis can also reveal information about the extent of variation.

[11] The topology of the tree for the family Rhinoceratidae is similar to our toy example. If species D represents the one- horned form and the remaining species are all two-horned, then what we can see is that the one-horned form is a derived form; what we would then seek is an explanation of the one-horned variety as evolved from an ancestral two- horned form.
[12] This does not mean that humans evolved from any of the modern apes. Humans and apes evolved from a common ancestor. There are some fossil remains from the Miocene that are plausible candidates for a common ancestor, and it is the record in the late Miocene and early Pliocene that is especially lacking. This is evidently the critical period for the divergence between apes and hominids.
[13] These robust australopithecines share a number of cranial features, probably related to diet. However, these craniodental features are evidently not independent developmentally and so cladistic analyses which treat them as independent may overweight them. For interesting discussions from a cladistic perspective, see Skelton & McHenry, 1992; Lieberman, 1995; Lieberman, Wood, & Pilbeam, 1996; Skelton & McHenry, 1997; and McCollum, 1999.
[14] Even if endocranial capacity were a reliable measure of intelligence, the measures of endocranial capacity are problematic. In recent work, computerized scans of an *A. africanus* skull were used to generate a three-dimensional computer model. The resulting value for endocranial volume was substantially lower than previous more conventional estimates (518 cc as opposed to values greater than 600 cc). This opens the prospect that endocranial values have been systematically overestimated. See Conroy et al. 1998 for discussion.
[15] Adaptations are limited to derived, or apomorphic, characters, though of course not all apomorphic characters are adaptations. A plesiomorphic character may, of course, turn out to be derived within a larger taxonomic group.
[16] If Deacon is right that human intelligence and human language co-evolved, this is exactly what we should expect. This raises a number of interesting issues, discussed at length by Deacon in *The Symbolic Species: The Co-evolution of Language and the Brain* (1997).
[17] By contrast, there is no reason to think that the human brain includes structures wholly lacking in other primates, though there undoubtedly are modifications in architecture specific to humans.
[18] His interpretation is a distortion of both Chomsky and Gould. So far as I can tell from either of their writings, or from the passages Dennett cites, their position is more moderate and more defensible than Dennett recognizes. Of course, the ideal opponent, rhetorically, is one whose position is extreme and untenable. Strategic advantage is no defense for rhetorical excess.
[19] Dennett's "algorithms" have guaranteed results, executed consistently. This means that evolution (or perhaps natural selection) would have deterministic consequences. Evolutionary explanations, though, are fundamentally probabilistic. I have not emphasized the probabilistic character of evolutionary theory here, but it is absolutely central in understanding any of

the discussions concerning the significance of natural selection (see Richardson & Burian, 1992). It would be possible to introduce probabilistic algorithms, of course, but that would require that Dennett abandon some of his criticisms of Gould's commitment to contingency (cf. Dennett 1995, ch. 10).

[20] In other cases, as when we turn to parental care, family structure, or the moral sentiments, we can be certain that a concentration on *Homo sapiens* is much too narrow; in one form or another, these characterize the hominids more generally and often our primate cousins.

REFERENCES

Alberch, P. 1982. "Developmental Constraints in Evolutionary Processes", in Bonner, J.T., ed., *Evolution and Development, Dahlem Conference Report No. 20*, New York: Springer Verlag, pp. 313-32.

Amundson, R. 1994. "Two Concepts of Constraint: Adaptation and the Challenge from Developmental Biology", *Philosophy of Science*, 61, 556-78.

Amundson, R. 1996. "Historical Development of the Concept of Adaptation", in Rose, M.R. & Lauder, G.V., eds., *Adaptation*. New York: Academic Press, pp. 11-54.

Amundson, R. & Lauder, G.V. 1994. "Function Without Purpose: The Uses of Causal Role Function in Evolutionary Biology", *Biology and Philosophy*, 9, 443-69.

Andrews, P. & Martin, L. 1987. "Cladistic Relationships of Extant and Fossil Hominoids", *Journal of Human Evolution*, 16, 101-118.

Barkow, J.H., Cosmides, L., & Tooby, J., eds. 1992. *The Adapted Mind: Evolutionary Psychology and the Generation of Culture*. New York and Oxford: Oxford University Press.

Baum, D.A. & Larson, A. 1991. "Adaptation Reviewed: A Phylogenetic Methodology for Studying Character Macroevolution", *Systematic Zoology*, 40, 1-18.

Beatty, J. 1980. "Optimal-Design Models and the Strategy of Model Building in Evolutionary Biology", *Philosophy of Science*, 47, 532-561.

Bodmer, W.F. & Cavelli-Sforza, L.L. 1976. *Genetics, Evolution and Man*. San Francisco: W.H. Freeman.

Brandon, R. 1978. "Adaptation and Evolutionary Theory", *Studies in History and Philosophy of Biology*, 9, 181-206.

Brandon, R. 1990. *Adaptation and Environment*. Princeton: Princeton University Press.

Brooks, D.R. & McLennan, D.A. 1991. *Phylogeny, Ecology and Behavior: A Research Program in Comparative Biology*. Chicago: University of Chicago Press.

Burian, R. 1983. "Adaptation", in Grene, M., ed., *Dimensions of Darwinism*. Cambridge: Cambridge University Press, pp. 287-314.

Chamberlain, A.T. & Wood, B.A. 1987. "Early Hominid Phylogeny", *Journal of Human Evolution,* 16, 119-133.

Coddington, J.A. 1988. "Cladistic Tests of Adaptationist Hypotheses", *Cladistics,* 4, 5-22.

Conroy, G.C. et al. 1998. "Endocranial Capacity in an Early Hominid from Sterkfontein, South Africa", *Science,* 280, 1730-1.

Cosmides, L. & Tooby, J. 1987. "From Evolution to Behavior: Evolutionary Psychology as the Missing Link", in Dupré, J., ed., *The Latest on the Best: Essays on Evolution and Optimality.* Cambridge: MIT Press, pp. 277-306.

Cosmides, L. & Tooby, J. 1992. "Cognitive Adaptations for Social Exchange", in Barkow, J.H., Cosmides, L., & Tooby, J., eds., *The Adapted Mind: Evolutionary Psychology and the Generation of Culture.* New York and Oxford: Oxford University Press, pp. 163-228.

Cracraft, J. 1981. "The Use of Functional and Adaptive Criteria in Phylogenetic Systematics", *American Zoologist,* 21, 21-26.

Darwin, C. 1859. *On the Origin of Species.* A Facsimile of the First Edition, edited with an introduction by Ernst Mayr. Cambridge: Harvard University Press.

Dawkins, R. 1982. *The Extended Phenotype.* Oxford: W.H. Freeman & Co.

de Jong, M.L. & van der Steen, W.J. 1998. "Biological Thinking in Evolutionary Psychology: Rock Bottom or Quicksand?", *Philosophical Psychology,* 11, 183-206.

Deacon, T.W. 1992a. "The Human Brain", in Jones, S., Martin, R., & Pilbeam, D., eds., *The Cambridge Encyclopedia of Human Evolution.* Cambridge: Cambridge University Press, pp. 115-123.

Deacon, T.W. 1992b. "Biological Aspects of Language", in Jones, S., Martin, R., & Pilbeam, D., eds., *The Cambridge Encyclopedia of Human Evolution.* Cambridge: Cambridge University Press, pp. 128-33.

Deacon, T.W. 1997. *The Symbolic Species: The Co-evolution of Language and the Brain.* New York: W. W. Norton.

Dennett, D.C. 1995. *Darwin's Dangerous Idea: Evolution and the Meanings of Life.* New York: Simon and Schuster.

Emerson, S.B. & Hastings, P.A. 1998. "Morphological Correlations in Evolution: Consequences for Phylogenetic Analysis", *The Quarterly Review of Biology,* 73, 141-62.

Endler, J.A. 1986. *Natural Selection in the Wild.* Princeton: Princeton University Press.

Endler, J.A. & McLellan, T. 1988. "The Processes of Evolution: Towards a Newer Synthesis, *Annual Review of Ecology and Systematics,* 19, 395-421.

Farris, 1983. "The Logical Basis of Phylogenetic Analysis", in Platnik, N. & Funk, V., eds., *Advances in Cladistics.* NY: Columbia University Press, 2, pp. 7-36.

Felsenstein, J. 1982. "Numerical Methods for Inferring Evolutionary Trees", *Quarterly Review of Biology,* 57, 397-404.

Fisher, R.A. 1930. *The Genetical Theory of Natural Selection*. Oxford: Clarendon Press.
Fleagle, J.C. 1999. *Primate Adaptation and Evolution*, second edition. London: Academic Press.
Frumhoff, P.C. & Reeve, H.K. 1994. "Using Phylogenies to Test Hypotheses of Adaptation: A Critique of Some Current Proposals", *Evolution,* 48, 172-80.
Gifford, F. 1997. "Why Not Spandrel Rather than Adaptation?", presented to the American Philosophical Association.
Gillespie, J.H. 1991. *The Causes of Molecular Evolution*. New York: Oxford University Press.
Gould, S.J. 1966. "Allometry and Size in Ontogeny and Phylogeny", *Biological Review,* 41, 587-640.
Gould, S.J. 1977. *Ontogeny and Phylogeny*. Cambridge: Harvard University Press.
Harvey, P.H. & Pagel, M.D. 1991. *The Comparative Method in Evolutionary Biology*. Oxford: Oxford University Press.
Hennig, W. 1966. *Phylogenetic Systematics*. Urbana: University of Illinois Press.
Hume, D. 1777. *Enquiries Concerning Human Understanding and Concerning the Principles of Morals*, edited by L.A. Selby-Bigge and revised by P.H. Nidditch. Oxford: Clarendon Press, 1975.
Jones, S., Martin, R., & Pilbeam, D., eds., 1992. *The Cambridge Encyclopedia of Human Evolution*. Cambridge: Cambridge University Press.
Kingsland, S. 1985. *Modeling Nature*. Chicago: University of Chicago Press.
Kluge, 1989. "A Concern for Evidence and a Phylogenetic Hypothesis of Relationships among *Epicrates* (Bovidae, Serpentes)", *Systematic Zoology,* 38, 7-25.
Krebs, J.R., Erichsen, J.T., Webber, M.I., & Charnov, E.L. 1977. "Optimal Prey Selection in the Great Tit (*Parus major*)", *Animal Behavior,* 25, 30-38.
Kreitman, M. 1983. "Nucleotide Polymorphism at the Alcohol Dehydrogenase Locus of *Drosophila melanogaster*", *Nature,* 304, 412-7.
Kreitman, M. & Hudson, R.R. 1991. "Inferring the Histories of the *Adh* and *Adh-dup* loci in *Drosophila melanogaster* from Patterns of Polymorphism and Divergence", *Genetics,* 127, 565-82.
Larson, A. & Losos, J.B. 1996. "Phylogenetic Systematics of Adaptation", in Rose, M.R. & Lauder, G.V., eds., *Adaptation*. New York: Academic Press, pp. 187-220.
Lauder, G.V. 1981. "Form and Function: Structural Analysis in Evolutionary Morphology", *Paleobiology,* 7, 430-442.
Lauder, G.V. 1996. "The Argument from Design", in Rose, M.R. & Lauder, G.V., eds., *Adaptation*. New York: Academic Press, pp. 55-91.
Leroi, A.M., Rose, M.R., & Lauder, G.V. 1994. "What Does the Comparative Method Reveal About Adaptation?", *American Naturalist,* 143, 381-402.
Levins, R. 1963. "Theory of Fitness in a Heterogeneous Environment. II.

Developmental Flexibility and Niche Selection", *American Naturalist,* 97, 75-90.
Lewontin, R.C. 1969. "The Bases of Conflict in Biological Explanation", *Journal of the History of Biology,* 2, 35-45.
Lewontin, R.C. 1974. *The Genetic Basis of Evolutionary Change.* New York: Columbia University Press.
Lewontin, R.C. 1978. "Adaptation", *Scientific American,* 239, 212-230.
Lewontin, R.C. 1990. "The Evolution of Cognition", in Osherson, D.N. & Smith, E.E., eds., *An Invitation to Cognitive Science: Thinking.* Cambridge: MIT Press, first edition, pp. 229-46.
Lewontin, R.C. &. Gould, S.J. 1979. "The Spandrels of San Marco and the Panglossian Paradigm", *Proceedings of the Royal Society of London,* B, 205, 581-598.
Lieberman, D. 1995. "Testing Hypotheses About Recent Human Evolution from Skulls: Integrating Morphology, Function, Development and Phylogeny", *Current Anthropology,* 36, 159-97.
Lieberman, D., Wood, B., & Pilbeam, D.R. 1996. "Homoplasy and early *Homo*: An Analysis of the Evolutionary Relationships of *H. habilis sensu stricto* and *H. rudolfensis*", *Journal of Human Evolution,* 30, 97-120.
Losos, J. G. et al. 1998. "Contingency and Determinism in Replicated Adaptive Radiations of Island Lizards", *Science,* 279, 2115-18.
MacArthur, R.H. 1957. "On the Relative Abundance of Bird Species", *Proceedings of the National Academy of Sciences,* 43, 293-5.
MacArthur, R.H. 1960. "On the Relative Abundance of Species", *American Naturalist,* 94, 25-36.
Maddison, W.P., Donaghue, M.J., & Maddison, D.R. 1984. "Outgroup Analysis and Parsimony", *Systematic Zoology,* 33, 83-103.
Maynard Smith, J., Burian, R., Kauffman, S., Alberch, P., Campbell, J., Goodwin, B., Lande, R., Raup, D., & Wolpert, L. 1985. "Developmental Constraints and Evolution", *The Quarterly Review of Biology,* 60, 265-287.
McCollum, M.A. 1999. "The Robust Australopithecine Face: A Morphogenetic Perspective", *Science,* 284, 301-5.
Nozick, R. 1993. *The Nature of Rationality.* Princeton: Princeton University Press.
Pinker, S. 1994. *The Language Instinct: How the Mind Creates Language.* New York: William Morrow & Co.
Pinker, S. 1997. *How the Mind Works.* New York: Norton.
Pinker, S. & Bloom, P. 1990. "Natural Language and Natural Selection", *Behavioral and Brain Sciences,* 13, 707-784.
Pinker, S. & Bloom, P. 1992. "Natural Language and Natural Selection", in Barkow, J.H., Cosmides, L., & Tooby, J., eds., *The Adapted Mind: Evolutionary Psychology and the Generation of Culture.* New York and Oxford: Oxford University Press, pp. 451-493.
Reeve, H.K. & Sherman, P.W. 1993. "Adaptation and the Goals of Evolutionary Research", *The Quarterly Review of Biology,* 68, 1-32.

Richardson, R.C. 1996. "The Prospects for an Evolutionary Psychology: Human Language and Human Reasoning", *Minds and Machines,* 6, 541-57.

Richardson, R.C. & Burian, R.M. 1992. "A Defense of Propensity Interpretations of Fitness", in Forbes, M. & Hull, D., eds., *PSA 1992*, vol. 1. East Lansing: Philosophy of Science Association, pp. 349-62.

Ricklefs, R.E. 1987. "Community Diversity: Relative Roles of Local and Regional Processes", *Science,* 235, 167-71.

Rose, M.R. & Lauder, G.V., eds., 1996. *Adaptation*. New York: Academic Press.

Schank, J.C. & Wimsatt, W.C. 1988. "Generative Entrenchment and Evolution", in Fine, A. & Machamer, P.K., eds., *PSA-1986, volume II.* East Lansing: The Philosophy of Science Association, pp. 33-60.

Skelton, R.R. & McHenry, H.M. 1992. "Evolutionary Relationships Among Early Hominids", *Journal of Human Evolution,* 23, 309-349.

Skelton, R.R. & McHenry, H.M. 1997. "Trait List Bias and a Reappraisal of Early Hominid Phylogeny", *Journal of Human Evolution,* 34, 109-113.

Strait, D.S., Grine, F.E., & Moniz, M.A. 1997. "A Reappraisal of Early Hominid Phylogeny", *Journal of Human Evolution,* 2, 17-82.

Tooby, J. & Cosmides, L. 1992."The Psychological Foundations of Culture", in Barkow, J.H., Cosmides, L., & Tooby, J., eds. 1992. *The Adapted Mind: Evolutionary Psychology and the Generation of Culture.* New York and Oxford: Oxford University Press, pp. 19-136.

Tooby, J. & De Vore, I. 1987. "The Reconstruction of Hominid Evolution Through Strategic Modeling", in Kinzey, W.G., ed., *The Evolution of Human Behavior: Primate Models.* Albany: SUNY Press.

Vogel, S. 1988. *Life's Devices: The Physical World of Animals and Plants.* Princeton: Princeton University Press.

Ward, S. 1999. "*Equatorius*: A New Homonoid Genus from the Middle Miocene of Kenya", *Science,* 285, 1382-86.

Wimsatt, W.C. & Schank, J.C. 1988. "Two Constraints on the Evolution of Complex Adaptations and the Means for their Avoidance", in Nitecki, M., ed., *Evolutionary Progress.* University of Chicago Press, pp. 231-273.

Wood, B. & Collard, M. 1999. "The Human Genus", *Science,* 284, 65-71.

CHAPTER 10

EVALUATING EVOLUTIONARY PSYCHOLOGY: CHECK YOUR LENS

HARMON R HOLCOMB III
University of Kentucky

Abstract. My purpose here is, first, to motivate practitioners of evolutionary psychology to take seriously the phylogenetic critique of adaptationism, and second, to motivate less informed critics to acknowledge why they may misinterpret the field.

TAKING THE PHYLOGENETIC CRITIQUE SERIOUSLY

Practitioners think that they have already won the argument over phylogeny and have ready responses that deflate the problem. To clarify our reasoning, I urge readers to think carefully over five basic dimensions of the critique. First, there is the meaning of the criticism, what it does and does not charge. Second, there are the arguments behind the criticism, whether the specific examples they use do make their case. Third, there are the distinctions inherent in the criticism, for instance, between substantive and methodological adaptationism and between methodological adaptationism and methodological selectionism. Fourth, there are the criticism's implications, such as: should we respond to it by ignoring it as practitioners do, by throwing evolutionary psychology to the flames as Richardson urges, or by developing new research methods that combine adaptationist and phylogenetic methods as Griffiths advocates? Fifth, there are the criticism's presuppositions: Does it presuppose a view of scientific method implying appropriate expectations about what evolutionary psychology can achieve or does it expect too much given the constraints on what evidence we can develop in the best possible evolutionary research?

First, consider the criticism's meaning. On one side, critics like Griffiths and Richardson insist that adaptationist research strategies

H.R. Holcomb III (ed.), Conceptual Challenges in Evolutionary Psychology, 375–392.

give us no good reasons to believe evolutionary conclusions in the absence of comparative methods that deliver specific phylogenetic information. On the other side, evolutionary researchers like Pinker and the contributors to this volume tend to set aside problems over adaptationism. They say that critics of adaptationism make an unreasonable demand that evolutionary psychologists provide a precise timeline of our evolutionary history of when and how selection occurred, which cannot be done by special design arguments.

Here's the rub: this response begs the question. The critical thesis is that special design arguments can't give us good reason to believe evolutionary psychological conclusions because phylogenetic information pertinent to the assumptions about selection made in the special design arguments is also needed. The response is that special design arguments cannot supply that phylogenetic information, and that such information is not needed when we can identify selection pressures on human traits without knowing when they first appeared in our ancestry. Further, even if organisms do not exhibit special design, as supposed, due to historical contingencies, that possibility is not evidence that they do not exhibit the supposed special design. However, this reply is forceful against the critical thesis that "the claims about what special designs exist are false", whereas the evolutionary agnosticism of Richardson is not the assertion that the special design claims are false but that we have no compelling reason to believe they are true. So, the response is not an effective counterargument against the original criticisms: that special design arguments alone cannot give reason to believe evolutionary conclusions as Richardson says, or that they have only heuristic value as Griffiths puts the point, or that evolutionary psychology is incapable of giving us rigorous functional explanations without evolutionary historical methods as I have said (Holcomb, 2000, p.606; Gangestad, 2000, p.628). The methodological problem is: can we do evolutionary statics (special design arguments) without evolutionary dynamics (phylogenetic arguments) to successfully identify exactly what was selected for?

We have seen that the empirical, scientific studies in this volume advocate or embody a host of different methods suitable for their subject matter, research situation, and questions. In addition, there are foundational methodological issues that arise in assessing the research value of these studies that are treated explicitly in Part One of this volume. I shall limit my discussion here mainly to two questions of interpretation. First, if the phylogenetic critique is on mark, can a "selectionist" method finesse it by minimizing commitments about adaptation, and therefore about what was selected for, thereby minimizing the need for phylogenetic information? Second, what reasons have led evolutionary psychologists to generally ignore the criticism that they must do phylogeny? Then, I shall raise basic questions about interpretation itself, calling attention to the fact that background beliefs of evaluators act as a lens through which we see the field and lead either to accurate interpretations or misinterpretations.

SELECTIONISM AS A BROADER METHOD THAN ADAPTATIONISM

Pinker (1997) argues that natural selection explains life, but restricts the sorts of explanations he is willing to countenance to explanations of "improbable adaptive complexity", e.g., of how the human mind works. This thesis concerns the subject matter of evolutionary psychology: what it can explain. Having argued that although it studies a narrower set of attributes of the mind than the innatist interpretation admits (not all traits have an evolutionary past), I maintain that it can explain a broader array of attributes than Pinker thinks. Evolution is broader than adaptive complexity and even selection.

Ocean larvae supply a good example of how adaptive complexity versus simplicity is in the mind of the beholder. Marine biologists have described larvae of lobsters, crabs urchins, and starfish as living or dying as a matter of chance. Born near land, they are swept adrift at birth, carried by tides and ocean currents for hundreds of miles, helpless in a turbulent ocean. Recent findings show them to be more than the animal equivalent of seeds. They can swim up or down to catch currents that are flowing in the right direction, so that they swim out to sea for months or years and return to a good place to settle and change into adults. They can hunt. They can nimbly deter predators with hooked spines that pop up in a "porcupine effect", flex to make their bodies look bigger, dodge or hide, pass through the guts of predators when eaten, or curl up and sink out of harm's way. These organisms are said to be "exquisitely adapted" for solving the problems they encountered in their evolutionary past in a vast and turbulent ocean. So, even some of nature's simplest organisms seem complex and adapted, and improbably so, given what we know from ordinary life and existing theories other than evolution by natural selection.

If marine biologists were not chided for speaking of the exquisite adaptation of such simple creatures, how inconsistent it would be for us to chide evolutionary psychologists to speak of the exquisite adaptation of human beings! We are far more complex than these sea creatures, and the fact that we are still present on the earth is reason enough to think that underneath both our simple and our complex traits there exists a huge repertoire of adaptations. Anorexia, a seemingly simple trait involving the mere expression of low self-esteem by eating too little, is revealed as an extremely complex disorder that is improbable given its associated self-destructive dimension. In any case, for critics to deny the substantive assumption that the mind involves a host of specialized adaptations would be to endorse a pre-Darwinian view that humans are so unique that we are beyond nature. All species biologists have studied carry many adaptations, and, roughly, the more complex the organism the more complex the adaptation. Critics of adaptationism would have to explain how it is possible that humans are the only species that are an exception to the rule.

Thus, a sympathetic alternative to Pinker's view that may be called the "selectionist method" or "selectionism" is defined in terms of two subtle but important differences. The first concerns the relativity of simple versus complex: whereas "adaptationism" invokes natural selection only in cases of complex design, according to "selectionism" the ability of natural selection to explain evolutionary changes in trait frequencies is independent of whether the traits studied are at first sight simple or complex. The second concerns the need to identify adaptations: whereas "adaptationism" holds that traits must be studied in ways that must identify an "adaptation" (either the trait itself or another trait related to it), "selectionism" dismisses this as a requirement on the grounds that evolutionary theory is structured so that we can explain organic design and diversity in terms of the ways they reflect the operation of natural selection without incurring a commitment to say whether what has been affected by natural selection is a genuine adaptation or not. The main arguments for the use of selectionism in evolutionary psychology are: first, that critics like Griffiths and Richardson are right about their critique of adaptationism; second, that evolutionary ecologists and ethologists who study nonhuman animals actually practice selectionism rather than adaptationism; and third, that by adopting selectionism evolutionary psychology will be just as legitimate and scientific as evolutionary biology generally if evolutionary psychologists also do phylogeny when possible and admit limitations when doing phylogeny is not possible.

A response to their critique is not a serious response, I submit, if it does not modify basic methodology. An example of a serious response is to try to construct a methodology which: develops methods that are as valid as possible when used without also using phylogenetic methods; develops phylogenetic methods as far as possible; uses adaptationist and phylogenetic methods together whenever possible and fruitful; clarifies the topics and conditions where phylogenetic methods will probably not be very fruitful (e.g., cases where the differences between the traits studied do not vary with phylogenetic differences); clarifies what phylogenetic commitments a study does and does not make; identifies exactly which parts of a study's results are placed into jeopardy by the lack of pertinent phylogenetic information; and does justice to the relative validity of results (e.g., a suggestive conclusion in distinction to a compelling conclusion) in terms of differences in the reliability of the total evidence that converges from various fields. The basic idea is to minimize the target of the phylogenetic critic so that it does not apply or else it makes the least significant difference to the validity of the studies guided by these injunctions.

Allow me to clarify this contrast between adaptationism and selectionism. Pinker's adaptationism focuses upon traits exhibiting adaptive complexity. Such traits, according to Pinker, almost certainly are adaptations—traits the functions of which have been produced or at least maintained by natural selection. An alternative—what may be called "selectionism"—drops the requirement of adaptive complexity. Even relatively simple traits, and even traits the adaptive value of which is difficult to discern, may have evolved as a consequence of natural selection. Selectionism is more substantive than Pinker's adaptationism, extending to traits that Pinker leaves aside as legitimate items to be explained. Like Pinker's approach, selectionism underwrites a form of methodological adaptationism. While Pinker applies Mayr's maxim to traits exhibiting adaptive complexity, selectionism applies it further to traits that are relatively simple in their operations. Even relatively simple traits can have important adaptive functions. Hence, we can adopt a more liberal methodology. This does not mean, however, that selectionist hypotheses are mere "Just-so" stories. Claims about the selective history of a capacity are constrained by informed hypotheses and experimental data about the current, proximate mechanisms that underwrite the capacity and phylogenetic information.

The crucial point is this: adaptation implies selection but selection does not imply adaptation. Hence, selectionism is broader than adaptationism in a related respect: natural selection combined with other forces of evolutionary change explains why the frequency of a trait increases, decreases, or remains the same regardless of whether the trait qualifies as an adaptation or not. Pinker invokes natural selection only to explain adaptation, with the caveat that the adaptations laid down while the human mind evolved in ancestral conditions may not lead to fitness-enhancing (adaptive) behavior in modern conditions of life. The logical point that Pinker's commitment to explaining adaptation is dispensable rests on a difference in quantifiers. For each trait that counts as an adaptation, natural selection has operated on it so intensely that it shows design for reproductive success. But natural selection operating on a trait need not make it into an adaptation. For each trait that is a product of evolution, it is potentially subject to natural selection, genetic drift, migration, immigration, meiotic drive and other forces of evolutionary change, but whether a given evolved trait subject to all these forces is made into an adaptation or not is a fact of history. Natural selection is just as explanatory when it constructs adaptations as when it does not. This point is rooted in Darwin's theory.

Darwin's theory has the structure that natural selection (an outcome of heritable variation in fitness) explains changes in the frequencies of traits in a population. In modern terms, natural selection explains evolution, that is, changes in gene frequencies, where phenotypic traits are linked to genes through complex hereditary and developmental processes. Darwin also generated his theory of natural selection to explain the diversity of life. It explains the divergence of characters: when the limiting factors on a population change, natural selection may or may not keep pace, and the traits of subsequent generations diverge from ancestral traits, the population members may become more adapted, less adapted, or remain equally adapted. Natural selection, therefore, explains the three main logically possible trends of life, roughly, as follows. It explains adaptational stasis: When limiting factors remain constant and adaptations have already been built up, if new variations do not occur or new variations occur but are maladaptive or neutral relative to existing adaptations, or if new adaptive variations occur but are not heritable, then the adaptive design of the organisms in the population will remain relatively constant. It also explains the extinction of less-improved forms of life: when natural selection fails to keep pace, new forms do not diverge in relevant ways (namely, ones that match changes in the limiting factors), and the result over many generations is a gradual extinction of less-improved forms, culminating in extinction of existing species. It also explains the origin of species: when natural selection keeps pace, new forms diverge from old ones in ways that match changes in the limiting factors, and the result over many generations is a gradual evolution of divergent forms, so that new varieties may diverge so radically from old ones that new varieties become new species. (Even though Darwin's species concept was too "typological", an analogous statement could be made for species defined as reproductively isolated populations.) Thus, selectionist methodology draws on the Darwinian heritage to go beyond adaptationism, so that selection is an explanatory factor not only in adaptation, but also in extinction and in evolution. This applies equally well to humans as nonhumans.

Selectionist methodology can be applied in a variety of ways, in which evolutionary adaptation may be assumed or else may attempted to be shown, but commitments about adaptation can be minimized in order to get on with a project. A good example is Daly's and Wilson's (1988) seminal evolutionary study of the psychology of murder. They asked, "Who kills whom, and under what conditions?" They considered parents killing children, children killing parents, siblings killing each other, spouses or mates killing each other, unrelated men or women killing each other, and kin killing nonkin. Nowhere in that book do they assume or try to prove that murder is an adaptation. Instead, they cautiously only commit themselves to the view that murder is an extreme expression of various evolved strategies for reproductive success, where the strategies are differentiated in virtue of genetic relations between the killer and victim. In some sense they do commit themselves to our having adaptations specific to these evolved strategies, but they do not speculate on what, precisely, in human nature constitute these underlying strategic adaptations.

A similar example is found in my study of anorexia (this volume). I asked, "Why are young females so at risk for acquiring anorexia, and in what conditions do they acquire it? The reproductive suppression model was defined and developed in the chapter so as to specify a host of kinds of conditions that lead to anorexia, some becoming symptomatic of anorexia as the eating disorder increases in severity. Nowhere did I assume or try to prove that anorexia is an adaptation. Instead, I was committed only to the view that anorexia is an extreme expression of an evolved reproductive strategy for human female control of her timing of reproduction, and possibly of dominant females manipulating subordinate females to suppress and delay reproduction against the fitness interests of subordinates. Just as evolutionary psychologists acknowledge that Daly and Wilson made real progress using their selectionist methodology that is not adaptationist methodology, so I attempted to show how selectionist methodology can make progress using the example of eating disorders.

The adaptive value, past and present, of human characteristics and whether they qualify as adaptations or not, are open questions for evolutionary research directed by Mealey's proposed methodology (this volume). Whichever is the case, natural selection, and particularly limits on the effectiveness of natural selection in fashioning perfect designs, is a primary explanatory factor in human phenomena that have strong effects on reproduction, and therefore have evolutionary significance. Evolutionary researchers focus on such phenomena, and anorexia counts as having evolutionary significance for several reasons. It is found throughout mammals and affects their reproductive success. In humans it has large effects on reproductive success by destroying health, leading sometimes to death. It postpones mating success in many cases through the mammalian mechanism of suppressed ovulation, and does so in humans as well. Crawford's (1989) method fails to incorporate the fact, of which he is certainly aware, that even traits neutral with respect to fitness can evolve. Evolution does not imply evolution by natural selection. Therefore, evolutionary significance need not imply (a) adaptive significance, or (b) a past history of natural selection, or (c) traceability to an adaptation. Each of (a), (b), and (c) are the key ideas in Crawford's method; each may be sufficient for evolutionary significance, but none is necessary for it. On theoretical grounds, "evolutionary significance" is even broader than "influenced by natural selection".

Most evolutionary psychologists say that they are studying adaptation, not phylogeny, but that critics misread "phylogeny" when they see "evolution". This is a large reason why evolutionary psychologists have not paid much attention to the criticism that their explanations have no scientific value unless backed up by careful phylogenetic reconstruction. Both Griffiths and Richardson make that type of criticism. Two features stand out in their ways of making that criticism that may be reasons why practicing evolutionary researchers may ignore it. First, it is focused on adaptationism, not selectionism. Second, it is critical without being constructive in the sense that scientists in the criticized field regard it as constructive. Let me explain.

Adaptation is quantitative because the concept of a selection pressure is a quantitative trait. Most traits studied lie on a continuum of responses to one or more selection pressures such that the response endpoints are maximally fitness enhancing (yield highest possible relative reproductive success) or maximally fitness-detracting (yield lowest possible relative reproductive success). In order to demonstrate that something is adaptive, we would have to identify all of its effects, then measure all the fitness costs and benefits of each of these effects, and do so in relation to all the actual alternatives. As a rule, this is just not done in evolutionary ecology or evolutionary behavioral biology of nonhumans, because it is too hard to accomplish. Instead, we usually study one or two factors of fitness at a time in an empirical study. To show that something is an adaptation, one would have to demonstrate improvements in adaptiveness by measuring all these values for each past generation back to the adaptation's incipient form. As a rule, this cannot be done because humans cannot make observations on the relevant time scale, except in controlled laboratory experiments morally acceptable for application to humans. Further, we must calculate the outcome of the interaction of selection with genetic drift, migration, immigration, meiotic drive, and any other forces of genetic change. That calculation must take into account the many constraints on natural selection's efficacy and the circumstantial constraint set within which a trait can be said to be "optimal" or "sub-optimal." In simple terms, it is a category mistake to say, "Adaptation explains traits"; adaptation is not an explanatory factor. Instead, natural selection in concert with other evolutionary forces explains both adaptation and lack of adaptation, and so it is not necessary to identify adaptations in order to explain our traits using evolutionary theory.

Practicing researchers know about these principled sources of incomplete evidence, and find special methods to develop feasible studies despite the lack of perfect evidence. We often begin our study on the assumption that natural selection has been acting on the trait in question, where one or more selective factors shaped it, not with a strict adaptive/maladaptive or adaptation/maladaptation dichotomy. Thus, "selectionist methodology" is more apt a term than "adaptationist methodology" to describe the practice of evolutionary psychology based on analogous practice in evolutionary biology.

FRUITFUL PHYLOGENETIC METHODS: NECESSARY OR UTOPIAN?

This practical choice to do evolutionary psychology using "feasible' though not "ideal" methods is relevant to the second issue, namely, why evolutionary psychologists have not embraced the critics' requirement that they do phylogenetic reconstruction. We have already said that, given Darwin's theory, one can distinguish three different phenomena to be explained: adaptational stasis, extinction, and evolution. Once we do so, and study adaptation (or responses to selective pressures, more generally) rather than extinction or evolution, we have opted out of doing phylogenetic reconstruction as a methodological gambit. Not all gambits are equal, however; it could be a bad gambit. The problem here, which Griffiths and Richardson push us to confront, is that the adaptationist explanation for a given trait can appear highly plausible—better than any available alternative explanation—and nevertheless be false.

In thinking about the phylogenetic critique we need to think about the epistemological point in contention. Responses vary with different presuppositions made about scientific method and ensuing expectations for what counts as achievement in evolutionary psychological science. Responses vary with different implications drawn about whether and how we should change research methods. Let me give some examples.

Here is one response. Lack of pertinent phylogenetic information is a legitimate source of principled doubt. Yet scientific debate over deep explanations of why we do what we do is not a matter that is subject to proof or disproof. Although principled doubt makes it reasonable to disbelieve these deep explanations, it still remains reasonable to believe them. Neither side is being irrational. The phylogenetic critique is reduced to this result, which applies to any science of human behavior and not just to evolutionary psychology. Here is another way people respond. Many people hanker after something called "scientific method" that upholds science as attaining a relatively secure, though uncertain, form of knowledge in virtue of its distinctive method. Their skepticism over evolutionary explanation may be reduced to a discovery that evolutionary psychology violates the expectation of such relative security, a result that applies to the other sciences if only we were to push them hard enough. Because we don't reject traditional psychology on these grounds, we shouldn't reject evolutionary psychology on these grounds. Here is a third way. Suppose evolutionary psychologists had done their work using only phylogenetic methods, and no adaptationist methods. Then there would arise skeptics like Griffiths who would say that phylogenetic methods only provide heuristics rather than evidence because they leave out adaptationist methods and skeptics like Richardson who would say that phylogenetic methods provide no compelling reason to believe their conclusions in the absence of unattainable adaptationist information. So, their critiques arise from privileging phylogenetic over adaptationist methods, which is indefensible. Consider two types of information relevant to evolutionary conclusions. One type can always be counted as necessary (or its value otherwise be privileged) and the other type can be counted as unnecessary (or its value otherwise be demeaned). This thought experiment implies that it is unfair to treat phylogenetic methods as necessary; of course, evolutionary psychologists are eager to use any phylogenetic methods that are fruitful, but the critics haven't shown how to use them fruitfully in cases under study.

The method of inference to the best explanation in science takes into account a basic fact underlying all three responses. Scientific method is no royal road to truth; any explanation may be false, not just hypothetically, but in the sense that in our actual circumstances we know of research methods that could provide falsifying information. But the problem with adaptive explanation can be rephrased. If an explanation is better than its alternatives, in terms of the qualities that make for worthwhile explanations, then it is rationally preferable to those alternatives. Many studies in evolutionary psychology try to show this, e.g., the evolutionary explanation is better than alternative traditional explanations, given the amount of supporting evidence it enjoys, e.g., Cummins, Gangestad, myself, and Jankowiak and Woodman in this volume. Yet a preferable explanation may not have enough empirical support to be worthy of belief. This is the standard that, I think, Griffiths and Richardson, are using (although they do not talk explicitly in terms of inference to the best explanation). An explanation is rationally believable only if it is preferable to all alternative explanations and its evidential basis is strong enough to warrant a conclusion about its worthiness to become a belief. Without phylogenetic reconstruction, there is not the right kind of evidence for evolutionary explanations to be rationally believable; that is our charge.

Interpreted in these terms, the charge is very hard to rebut, I think. But evolutionary psychologists do have methodological strategies for finessing this charge. Here are some. First, the charge asks that evolutionary explanations of human mind and behavior be held to a standard of "beyond a reasonable doubt" when they are only being proposed to be backed by enough evidence to make a judgment worthy of belief. This response applies the trial analogy to scientific testing, which inspired the idea of inference to the best explanation in the first place, and appeals to our common sense. Second, because the charge applies equally well to evolutionary ecology and behavioral ecology of nonhumans as to evolutionary psychology of humans, and yet the critics do not infer that those biological studies are not worthy of belief, by parity of reasoning, they should not infer that the psychological studies are unworthy of belief. This response utilizes a principle of fairness to undercut the implication that evolutionary psychology is uniquely illegitimate. Third, phylogenetic reconstruction is standard practice in evolutionary biology when studying the body, but behavior leaves no fossils, and so the nature of any reconstruction of behavioral or mental phylogeny will be speculative. This response argues that phylogenetic reconstruction will not close the evidential gap that the critics require to be closed. Fourth, lack of phylogenetic reconstruction does supply an important source of error, but reconstructing the history of humans back to our hominid and primate ancestors has always been the domain of evolutionary anthropology. This response accepts the need for doing whatever phylogenetic reconstruction we can do, but modestly accepts that it will be done in a better way by evolutionary anthropologists, who base their methods on more of evolutionary theory than do evolutionary psychologists.

Fifth, and perhaps most importantly, in studying human mind and behavior with important reproductive outcomes that have a physiological basis, there is no guiding methodological assumption less likely to be wrong than "the phenomenon is influenced, directly or indirectly, by natural selection". In the example of explaining anorexia, what better evolutionary alternative is there than to hypothesize that, given the ability of mammalian females to control the timing of their reproduction, human females may be waiting to reproduce until conditions get better? In other words, when phylogenetic information is unavailable and not forthcoming arguments from special design constitute the approach least likely to commit one to false conclusions: remember that the criticism is not that adaptationist conclusions are false, just than they lack necessary evidence to warrant truth. This last response calls for the critic to come up with a better alternative than what they criticize.

It is up to the reader to decide whether the critique of evolutionary psychology from phylogeny is on the mark or not. I invite you to decide which is stronger: the critiques or the defenses? Do the detailed criticisms of Griffiths and Richardson escape all previous rebuttals to the criticisms of adaptationism, or must evolutionary psychologists combine adaptationist with phylogenetic methods whenever possible, and when not possible, accept that our research is severely limited?

EVALUATING EVOLUTIONARY PSYCHOLOGY

It is handy to have philosophers around, as we are trained to get to the heart of an issue and to push problems often presented in superficial terms on to deeper levels of thought. People typically think of philosophy as a separate field from science, but a main trend in philosophy of science in the past decades has been to work much closer to actual science. Griffiths and Richardson work closely with the evolutionary psychological literature to raise a perennial problem in philosophy, namely, the possibility of genuine knowledge of the ultimate nature of ourselves and our world, which takes the form: can an effective methodology for achieving evolutionary psychological knowledge be worked out, or do the things we do not and cannot know about past human evolution from evidence that we can gather today prevent the possibility of that knowledge?

Of course, many volumes in evolutionary thought about humans simply present some of the best new science and let outsiders say what they want about it, which normally is to criticize it. There are two kinds of critics, one with constructive criticism and the other with destructive criticism. One will explain why you can't do what you want to do. The others will get at the heart of what you're trying to do and then figure out another way you can do it. It is the second kind of thinking we'll need if we are to live with the problem of phylogenetic construction and still conduct evolutionary research on human mind and behavior. I worry that evolutionary psychology may suffer a fate similar to sociobiology: a stalemate emerges between two independent camps of "practitioner-advocates" and "outsider-critics". Such criticisms can be very destructive of the field, as they suggest that we stop doing evolutionary research on humans, make it harder to get funding for new evolutionary research, and reinforce myths about human nature and individual differences. A criticism of the type, "You can't do what you are trying to do" gives scientists nowhere to go in the future, whereas a criticism of the type, "Here's how you can better achieve your aims" or "Here are realizable aims that should replace your present aims" is much more constructive. So, to advance the field it is best to bring the criticisms back inside the field. I propose an alternative to this "independent camp" mode of adversarial discourse, namely, an ideal of "mutual dependence community" of scientists and nonscientists, of science and meta-science, of doing science and critically discussing science. This ideal has twin benefits. On one hand, the field's practitioners can exert some control over the critical evaluations in order to add solid criticisms to their pedagogical tool-kits, learn from their mistakes in doing science, and free other scientists and nonscientists from faulty criticisms, helping them give better, more constructive criticisms in the future that will advance the field. On the other hand, readers of evolutionary psychology can exert some control over the direction of the science itself, learn from their mistakes in criticizing the science, and free the general academic community and public from obsolete preconceptions about the import of evolution for human mind and behavior.

But there is an overwhelming obstacle to our aim of advancing the discipline. Between statements within science and evaluations of those scientific statements lies interpretation: what is being attributed to the science. Those attributions—how we read it and what we read into it—involve the often hidden context of the statements themselves, their presuppositions, and their implications. The interpretation of evolutionary psychology is a fundamental problem that is philosophical and inescapable. This is not to deny that there is a mundane sense in which evolutionary psychology does not need in interpreting. The studies in this volume stand on their own merits. In the previous introductions to each Part, my intent is not to provide an interpretation that trumps the way the authors understand their own work; it is to facilitate critical discussion because people generally do not know how to go about interpreting and evaluating evolutionary works on humans.

Having been immersed in assessing the polemical "sociobiology debate" that raged throughout academia for two decades (Holcomb, 1993), I have tracked the progress of evolutionary psychology as a successor field to human sociobiology. My concern is that the reception of evolutionary psychology turns on readers' obsolete preconceptions drawn from sociobiology or pre-sociobiology. Current science is read through a lens of everyday views prominent in our culture that treats evolution as irrelevant to human mind and behavior and if relevant then just plain false. When that lens is uncritically assumed in understanding the import of these studies, the result is a confused misinterpretation of current science and complacency in having no positive alternative view of the evolution of mind and behavior. By systematically misreading the meaning, presuppositions, and implications of the scientific studies, outsiders combine incompatible elements (thought that severely limits the import of evolution for our sense of ourselves and thought that pushes that import as far as possible). This sort of misreading inevitably leads to misinterpretation, trumpeted as "criticisms of the science," but merely reduces to absurdity the reader's combination of nonevolutionary and evolutionary thought. An example of this sad rhetorical situation, evident from simple but informed inspection of the documents, is found in a recent review (Schmitt, 2001) of an anthology entitled, *Alas, Poor Darwin: Arguments Against Evolutionary Psychology* (Rose & Rose, 2000). In a field where practitioners find that most objections are based on false assumptions or misunderstandings, such objections should be viewed as requests for clarification.

Why do such misinterpretations abound when people read Darwinian studies of human mind and behavior? People simply rely on their preconceptions. Somehow we were taught in our culture that any evolutionary study of adaptation is committed to genetically based limits to human nature that interfere with our sociopolitical and moral goals. So, people who have not subjected this culturally transmitted denigration of science to their own critical thought just force all evolutionary psychology into the box of that preconception. To evolutionists, doing so commits "the moralistic fallacy", the fallacy of assuming that "What ought to be is" or "What ought to be can be". The prominent example here is, "Sex differences in mating strategies and tactics ought not to exist; therefore, they do not exist; hence, anyone finding such differences must be using poor research methods or be politically motivated or prejudiced in their research". Readers of our anthology on Part Two dealing with mating will no doubt make such damaging interpretations of it. But if we move from the "opposing camp" model of science and its evaluation to a "mutual dependence" model, we are forced to confront the problem that our interpretation might be misleading or wrong, and that the moralistic fallacy is a formula for generating bad interpretations, so that we will try to catch ourselves falling into the moralistic fallacy in order to prevent misinterpretations that lead to needless miscommunication and controversy. How do we know that the statements in evolutionary psychology "are really untrue" without engaging the science itself? We don't and we can't. Evaluators of science have as much need to fully engage science as scientists have to fully engage evaluations of science.

The intense debate and needless controversy that arises from misinterpretations provides a rationale for the nature of my introductions. This unusual research situation makes it appropriate for an editor to provide Introductions to each of Parts One, Two, and Three that engage the controversy through clarifying analysis and evaluation, as I do, in order to help to improve the quality of critical discussion of these papers. It is each reader's responsibility to do some soul searching in order to identify their own preconceptions and ensuing effects on which interpretation in the range of possible ones to adopt, and then critically reflect on their own lens and decide whether it should be replaced. In this sense, one cannot even rationally read evolutionary psychology without doing a bit of philosophy.

There are at least five main sources of systematic error in interpreting evolutionary psychology, sources that should prompt us to critically reflect on our own preconceptions and ensuing interpretations and evaluations, treating them as problematic. It is easy difficult to recognize the multi-disciplinary character of the field and to discern how various disciplinary starting points affects the approaches in particular studies. False preconceptions that provide the driving force for criticisms of the field, even though they have already been undermined or abandoned during prior history of evolutionary research into human behavior. Not having read the voluminous literature in the history of Darwinism or the hundreds of articles and books that the contributors to a volume such as this have been informed by, readers go on first impressions, leading to criticisms that combine obsolete notions with current studies in a way that criticizes caricatures of what these studies say. Then, there are debatable presumptions about the goals, methods, and quality of results of the evolutionary approach, together with their presuppositions and implications, which are serious issues for philosophical analysis. At a deeper level, there are controversial but hidden expectations about how solid the theoretical reasoning, empirical findings, and ensuing theoretical explanations of the empirical findings must be in order for evolutionary psychology to qualify as legitimate, good science. That is, beliefs about the scientific method regulate expectations that determine how we assess the scientific status of the field. Finally, neglect of the research context of individual studies that leads the authors to take for granted things readers not immersed in that specialized context do not know, leading to false attributions.

Given a synergistic effect between these five sources of criticisms based on misunderstandings, overrepresentation of "innatist-adaptationist" interpretations, underrepresentation of other work and other interpretations, and exaggerated oversimplifying media coverage, we have a perfect recipe for an unfair and inaccurate overall image of the field. Misinformation synergy makes it difficult to reach consensus on what counts as a proper or improper interpretation. Yet evolutionary psychology can no more be critically assessed, advanced, or even understood without a presupposed interpretation than we can discuss the Holy Bible without interpreting it. So, I have spoken in my introductions directly to some of the sources of wrong-headed interpretations-plus-ensuing-evaluations of these chapters to prompt critical discussion, clearing away the "bad criticisms" so that the reader will be in a position to make "good criticisms". When we criticize the field, we should ask: Are we criticizing our own misinterpretation? We can raise consciousness about the foundational problem, "How should evolutionary psychology be interpreted?" by flagging likely responses that see the field through a lens assumed by the reader which itself is problematic and often has already been undermined by previous work in the field.

People generally become interested in these interpretive problems—what are these studies saying or not saying—only after they have engaged in debate, where issues rest on whether a criticism is based on an accurate understanding or a misunderstanding. Differing interpretations of the field are something we have to live with. If we turn a blind eye to them, we will end up with opposing pro and con views because we attribute different underlying research commitments to individual studies and to the field. The standards of criticism have just been raised—they should be as high as the standards we expect the studies to fulfill. A good criticism has to justify what it attributes to the studies. Interpretation precedes evaluation. Practice precedes interpretation. Future practice is guided by past evaluation. At stake is the very human issue of making up our minds about how we shall act. Now that evolutionists are taking human development into account, the old idea of an evolution-imposed fixed human nature is replaced by the idea of evolved predispositions that may or may not arise during individual development and be expressed according to natural, familial, social, and cultural factors of human nurture. The dichotomy between nature and nurture is obsolete: Human nurture is part of human nature. Traditional connotations of "evolutionary" that ties it to "innate, not learned", "nature, not nurture", and so on, should be treated as so much worn out baggage that should be tossed in the garbage heap. The old question about human nature arises in a new form that takes into account evolutionary studies of human nurture, now that ideas about human nature should be expressed in terms of human nurture: Is human nurture malleable enough to become what we hope for when we think of "using evolutionary ideas to make a better world" or does evolution impose limits on human nurture that prevent us from changing to fit our ideals?

REFERENCES

Crawford, C. 1989. "The Theory of Evolution: Of What Value to Psychology?", *Journal of Comparative Psychology*, 103, 4-22.

Daly, M. & Wilson, M. 1988. *Homicide*. Hawthorne, NY: Aldine-de-Gruyter.

Pinker, S. 1997. *The Language Instinct*. New York: Morrow.

Rose, H. & Rose, S. 2000. *Alas, Poor Darwin: Arguments Against Evolutionary Psychology*. NY: Harmony Books.

Schmitt, D. 2001. "Review of *Alas, Poor Darwin: Arguments Against Evolutionary Psychology*", Human Ethology Bulletin, 16, No.1, 11-13.

INDEX OF NAMES

INDEX OF SUBJECTS

STUDIES IN COGNITIVE SYSTEMS

. J.H. Fetzer (ed.): *Aspects of Artificial Intelligence.* 1988
ISBN 1-55608-037-9; Pb 1-55608-038-7

. J. Kulas, J.H. Fetzer and T.L. Rankin (eds.): *Philosophy, Language, and Artificial Intelligence.* Resources for Processing Natural Language. 1988 ISBN 1-55608-073-5

. D.J. Cole, J.H. Fetzer and T.L. Rankin (eds.): *Philosophy, Mind and Cognitive Inquiry.* Resources for Understanding Mental Processes. 1990 ISBN 0-7923-0427-6

. J.H. Fetzer: *Artificial Intelligence: Its Scope and Limits.* 1990
ISBN 0-7923-0505-1; Pb 0-7923-0548-5

. H.E. Kyburg, Jr., R.P. Loui and G.N. Carlson (eds.): *Knowledge Representation and Defeasible Reasoning.* 1990 ISBN 0-7923-0677-5

. J.H. Fetzer (ed.): *pistemology and Cognition.* 1991 ISBN 0-7923-0892-1

. E.C. Way: *Knowledge Representation and Metaphor.* 1991 ISBN 0-7923-1005-5

. J. Dinsmore: *Partitioned Representations.* A Study in Mental Representation, Language Understanding and Linguistic Structure. 1991 ISBN 0-7923-1348-8

. T. Horgan and J. Tienson (eds.): *Connectionism and the Philosophy of Mind.* 1991
ISBN 0-7923-1482-4

. J.A. Michon and A. Akyürek (eds.): *Soar: A Cognitive Architecture in Perspective.* 1992
ISBN 0-7923-1660-6

. S.C. Coval and P.G. Campbell: *Agency in Action.* The Practical Rational Agency Machine. 1992 ISBN 0-7923-1661-4

. S. Bringsjord: *What Robots Can and Can't Be.* 1992 ISBN 0-7923-1662-2

. B. Indurkhya: *Metaphor and Cognition.* An Interactionist Approach. 1992
ISBN 0-7923-1687-8

. T.R. Colburn, J.H. Fetzer and T.L. Rankin (eds.): *Program Verification.* Fundamental Issues in Computer Science. 1993 ISBN 0-7923-1965-6

. M. Kamppinen (ed.): *Consciousness, Cognitive Schemata, and Relativism.* Multidisciplinary Explorations in Cognitive Science. 1993 ISBN 0-7923-2275-4

. T.L. Smith: *Behavior and its Causes.* Philosophical Foundations of Operant Psychology. 1994
ISBN 0-7923-2815-9

. T. Dartnall (ed.): *Artificial Intelligence and Creativity.* An Interdisciplinary Approach. 1994
ISBN 0-7923-3061-7

. P. Naur: *Knowing and the Mystique of Logic and Rules.* 1995 ISBN 0-7923-3680-1

. P. Novak: *Mental Symbols.* A Defence of the Classical Theory of Mind. 1997
ISBN 0-7923-4370-0

. G.R. Mulhauser: *Mind Out of Matter.* Topics in the Physical Foundations of Consciousness and Cognition. 1998 ISBN 0-7923-5103-7

. K.L. Butler: *Internal Affairs.* Making Room for Psychosemantic Internalism. 1998
ISBN 0-7923-5261-0

. B.A. Thyer (ed.): *The Philosophical Legacy of Behaviorism.* 1999 ISBN 0-7923-5736-1

D. Livingstone Smith. *Freud's Philosophy of the Unconscious.* 1999 ISBN 0-7923-5882-1

M. Perlman: *Conceptual Flux.* Mental Representation, Misrepresentation, and Concept Change. 2000 ISBN 0-7923-6215-2

J.H. Fetzer: *Computers and Cognition: Why Minds are Not Machines.* 2001
ISBN 0-7923-6615-8

STUDIES IN COGNITIVE SYSTEMS

26. H. Cruse, J. Dean and H. Ritter (eds.): *Prerational Intelligence: Adaptive Behavior and Intelligent Systems Without Symbols and Logic, Volume 1.* 2000 ISBN 0-7923-6665-4
    H. Ritter, H. Cruse and J. Dean (eds.): *Prerational Intelligence: Adaptive Behavior and Intelligent Systems Without Symbols and Logic, Volume 2.* 2000 ISBN 0-7923-6670-0
    J. Dean, H. Ritter and H. Cruse (eds.): *Prerational Intelligence: Interdisciplinary Perspectives on the Behavior of Natural and Artificial Systems.* 2000 ISBN 0-7923-6669-7
    ISBN Indivisible Set 0-7923-6666-2
27. H.R. Holcomb III (ed.): *Conceptual Challenges in Evolutionary Psychology.* Innovative Research Strategies. 2001 ISBN 1-4020-0133-9

KLUWER ACADEMIC PUBLISHERS – DORDRECHT / BOSTON / LONDON

MIX
Papier aus verantwortungsvollen Quellen
Paper from responsible sources
FSC® C105338

If you have any concerns about our products,
you can contact us on
ProductSafety@springernature.com

In case Publisher is established outside the EU,
the EU authorized representative is:
Springer Nature Customer Service Center GmbH
Europaplatz 3, 69115 Heidelberg, Germany

Printed by Libri Plureos GmbH
in Hamburg, Germany